The rural areas of Britain and the developed world are currently undergoing massive changes. Over-production of food is forcing governments to reconsider agricultural subsidies; after years of population decline rural areas are now showing a remarkable population turnaround; there has been a tremendous increase in demand for recreational facilities in rural areas; and there is continuing and increased environmental concern.

Rural Politics focuses on the key issues affecting rural areas today and examines them in the light of the current agricultural crisis. Issues such as water pollution, forestry, the greening of agricultural policy, as well as mainstream agricultural policy, are of increasing importance in the development of the countryside. Examining the history of agricultural policies and environmental concerns, the book looks in particular at the political parameters to these issues and how concern for the countryside is essentially part of a wider set of political processes. The author discusses rural problems in the context of the political history of the modern urban-industrial state, and by employing a critical political science approach shows how the content and impact of policy cannot be understood solely by studying legislative provision.

Rural Politics is an important study of the evolution and content of policies affecting the countryside, both in terms of the major land uses and economic development.

Michael Winter is Professor of Rural Economy and Society at the Countryside and Community Research Centre, Cheltenham and Gloucester College.

RURAL POLITICS

Policies for Agriculture, Forestry and the Environment

MICHAEL WINTER

ROUTLEDGE

LONDON AND NEW YORK

First published 1996
by Routledge
11 New Fetter Lane, London EC4P 4EE

Simultaneously published in the USA and Canada
by Routledge
29 West 35th Street, New York, NY 10001

Routledge is an International Thomson Publishing company

Typeset in Garamond by
Solidus (Bristol) Limited
Printed and bound in Great Britain by
Biddles Ltd, Guildford and King's Lynn

British Library Cataloguing in Publication Data
A catalogue record for this book is available from the British Library

Library of Congress Cataloguing in Publication Data
Winter, Michael.
Rural politics: policies for agriculture, forestry, and the
environment / Michael Winter.
p.   cm.
Includes bibliographical references and index.
1. Rural development–Government policy–Great Britain.
2. Agriculture and state–Great Britain. 3. Agriculture and state–
Environmental aspects–Great Britain. 4. Forest policy–Great
Britain. 5. Environmental policy–Great Britain. 6. Great Britain–
Rural conditions. I. Title.
HN400.C6W56   1996
307.1'412'0941–dc20   95-38688

ISBN 0-415-08175-0 (hbk)
ISBN 0-415-08176-9 (pbk)

Dedicated to the memory of
David Winter

CONTENTS

FIGURES

TABLES

ACKNOWLEDGEMENTS

My research on rural politics owes much to collaboration with Philip Lowe and Graham Cox. Philip Lowe's confident empirical approach to the study of politics did much to revive my own interests in empirical social science, correcting the influence of the somewhat abstract trends in sociological thinking prevalent in the early 1980s. At the same time, Graham Cox's abiding interest in theoretical issues helped me to avoid any retreat into empiricism. My gratitude to both is therefore deep. For several years the three of us worked together on many of the issues covered in this book and some of the conclusions drawn here inevitably owe much to that partnership.

Philip Lowe and Wyn Grant have read and commented on the entire manuscript and I am enormously grateful to them; and to Sadie Ward for her comments on the chapters on agricultural history. Data for Figures 6.3 and 6.4 were kindly supplied by Berkeley Hill and for 9.1 by Ian Hodge. I am indebted to Nigel Curry for his support and encouragement during the final stages of preparation and to Elizabeth Orme for advice and guidance on various aspects of agricultural and environmental regulations. I alone remain responsible for the contents.

I am grateful to Nicky Greenhill for typing various sections and to both Nicky and Jan MacLaren at Cirencester for much 'admin' support during the years in which this book was written. Having moved to Cheltenham during the course of finalising the volume, I then received word-processing and computer help from Mary Mitchell, Edward Collier and Daphne Comfort and research assistance from Amanda Stone and Hilary Winter. At Routledge Tristan Palmer and Matthew Smith have been helpful and patient editors.

At different times during its preparation I received financial support from a variety of sources. Lloyds Bank provided a grant in the early stages of preparation which freed me from other duties at the Royal Agricultural College to pursue some of the research on which this book is based. Special thanks are due to their agricultural manager Pat Oakley for his support and patience. Parts of the book draw on research projects supported over many years by the Economic and Social Research Council; and other work on agricultural policy was supported by a grant from the Esmeé Fairbairn Foundation Trust.

Various papers not readily available in libraries were provided by Mrs E.

Day, widow of the late Mr Wallace Day. I must also acknowledge the help of library staff at Cheltenham and Gloucester College, Exeter University, and the Bodleian in Oxford and especially for the unstinting and cheerful help of Rachel Rowe and her library staff at the Royal Agricultural College, Cirencester.

I gratefully acknowledge permission to reprint the following material: Figure 1.3 by permission of Basil Blackwell Publishers; Table 11.3 by permission of Christopher Helm Publishers; Figure 9.2 by permission of HMSO and MAFF. Crown copyright is reproduced with the permission of the Controller of HMSO; Figure 3.1 by permission of HMSO and the Treasury. Crown copyright is reproduced with the permission of the Controller of HMSO; Tables 5.2, 6.1 and 6.3 by permission of Methuen & Co; Figures 10.2, 10.4 and 10.5 by permission of the National Rivers Authority; Figure 1.2 by permission of Professor Philip Norton; Table 1.1 by permission of Oxford University Press; Table 2.1 and Figure 1.6 by permission of Prentice Hall; Table 5.1 and Figure 7.2 by permission of Routledge for Allen & Unwin; Figure 1.4 by permission of Sage Publications; Table 8.4 by permission of the University of Wales Press. Table 6.5 is reprinted from *Land Use Policy* (A. Jones 1991) with the kind permission of Butterworth-Heinemann journals, Elsevier Science Ltd, The Boulevard, Langford Lane, Kidlington, Oxford OX5 1GB, UK.

Most of the book was written at home where Hilary provided moral support and, with Emily and Benedict, welcome distraction. Finally, I have dedicated this book to the memory of my father, David Winter who died in 1989. His first-hand experiences of the vagaries of farming and then of the world of agricultural education gave him an ambivalent, even sceptical, outlook on the rural world, which acted as a powerful corrective to my own youthful eulogising of the countryside. Moreover, his concern for social justice and his Liberal politics first excited my own twin interests of sociology and politics.

Exbourne, Devon, May 1995

INTRODUCTION

PREAMBLE

In writing *Rural Politics* I have had two main audiences in mind: those involved in the world of policy and politics who might wish to know more about specific aspects of rural policy and institutions, especially regarding the genesis and historical antecedents for current policies; and students concerned with rural issues who need to know more about the policies which affect, and have affected, rural land use and life in the countryside generally. I have in mind, particularly, undergraduates taking courses in Agricultural Economics, Rural Land Management, Countryside Planning, Geography and Environmental Sciences. Most of the chapters could be expanded to form books in themselves! My aim has been to condense large volumes of information to produce an accurate picture of the development of key aspects of rural policy. Inevitably this has involved a good deal of selectivity and the text is fully referenced so as to provide guidance to appropriate further reading.

I have defined 'rural' in an entirely pragmatic and geographical sense to refer to areas dominated by extensive land uses and primarily, in this instance, by agriculture and forestry. It is as well to point out at this stage that there is no such thing as a 'rural policy' in the UK. There are no departments of rural affairs, nor does 'rural' command a legal and statutory definition, except with regard to settlement size.

Although I deplore some of the consequences of separating agriculture and environment, I decided to organise the book along these lines. There is no conceptual justification for such a decision, only the pragmatic one that the worlds of policy making and policy implementation are organised in this way. The structure of the book merely reflects that empirical reality.

THE HISTORICAL APPROACH

My approach is unashamedly historical. I have resisted the temptation to catalogue the vast array of rural policy prognoses that have poured from the pens of politicians, academics, journalists and pressure groups in recent years. All too often students' views of rural policy are based almost entirely on the most recent proposals for reform as though debate on these issues only began during their undergraduate careers! My aim is to show how contemporary debates are shaped by forces deeply rooted in political, cultural, social and economic history. Consequently there is a good deal of history in this book, for Britain is an old country, and living in an old country requires a disciplined attention to the past.

So much has been written in recent years on the preoccupation of Britain with its past – what Patrick Wright (1985: 256) terms the 'elegant but also grievous culture of national "decline"' – that attention to the past might seem neither original nor necessary. However, the fragmentation of academic undertaking is such that texts covering aspects of contemporary rural policies are often devoid of historical context or, worse, re-invent history to suit a particular interpretation of the present. Indeed the very continuity of the British *ancien régime* (Gamble 1994) can, paradoxically, detract attention from aspects of the past, the past becoming heritage rather than history. This identification of heritage with rurality means that rural studies are particularly prone to trace their own appropriate and particular histories in, for example, the history of 'the countryside' rather than in the development of a managed or corporatist *urban* polity. Wright (1991: 94) talks of a rural world 'where history is venerated as tradition and culture is based on ancestry and descent'. But the history thus venerated is both particular and uneven. Rural history as part of the history of an urban managerial and imperial state is often neglected.

Whether countries which have experienced abrupt policy ruptures are more likely to play host to informed and critical historical reflexivity than those characterised by the incrementalism of continuity is open to debate. But certainly, sectoral policies in Britain have been presented, on occasions, with scant regard for their antecedents, as though continuity removes the need for critical historical examination. The Second World War, in particular, is used as a convenient watershed. The various committees set up during the war to examine how best the country might be planned and managed in peacetime resulted in a battery of legislation enacted by the Labour administration in office from 1945 to 1951. These covered, *inter alia*, town and country planning, agriculture, conservation and recreation; and set the policy agenda firmly for the next three to four decades. In many respects, the legacy remains in the 1990s. Neither before nor since has there been such a wide-ranging set of rural policy innovations over such a short period of time.

Thus 1945 provides an easy starting point for many an account of rural policy, but these legislative cornerstones marked the culmination of processes dating back to the previous century. To understand post-1945 legislation as the starting point is profoundly to miss the point. For example, the contrasting manner in which recreation/access, landscape and ecology are treated in the National Parks and Access to the Countryside Act 1949 seems merely quirky if its roots in the class antagonisms of inter-war Britain are left unexposed. Similarly the protectionism of the Agriculture Act 1947 can so easily be seen as an opportunistic triumph for the farming lobby in the light of wartime and post-war food shortages if the rise of a managed economy, the dramatic demise of landlordism and the debate on empire (all political events preceding the 1939–1945 war) are ignored.

Of course, with too much licence, such a line of reasoning can be pursued relentlessly to pre-history, with diminishing returns. Rules have to be made. In my case, with only occasional lapses, I take the 1830s and 1840s as the point of departure. The repeal of the Corn Laws in 1846 marks a convenient point to launch a discussion of rural policy, ushering in, as it did, an uninterrupted era of *laissez-faire* economics for agriculture which lasted until 1916. The Great Reform Act of 1832 provides another convenient starting point, symbolising the beginnings of a parliamentary state based on universal suffrage.

KEY THEMES

An important influence on this book is *The State and the Farmer* by Peter Self and Herbert Storing. This subtle and sophisticated analysis of just fifteen years of agricultural politics and policy in Britain (1945–1961) has been matched for no other period. It remains a mine of information on many aspects of agricultural politics which I have turned to again and again to check hunches and confirm facts. Above all, although they do not use the word themselves, Self and Storing provided the foundations for the *corporatist* understanding of twentieth-century agricultural politics which has dominated much of my research.

Corporatism is one of the key themes of this book. In recent years, the corporatist position has come under sustained attack and has almost disappeared from view within contemporary political science. It has been replaced by the policy networks approach, originating in an explicit attack on corporatist analysis, and an 'overly desiccated academic exchange' (Judge 1993: 121) on the distinctions between policy communities, policy networks and issue networks. Whatever language is deployed (and some might argue that my use of terms such as corporatist and policy communities are more interchangeable than they should be), it is important not to lose sight of the wider significance of this area of work. Stated simply, policy is a process and within that process there are many actors, including the representative groups of affected interests. The manner in which these actors interact is at the heart of corporatist and policy networks approaches. It provides a key sub-theme to this volume. It is through examining the policy process in historical context that we can begin to answer some important questions of rural policy, such as how and why an urban-industrial-commercial country like Britain came to have a highly protected agriculture; or how various peculiarities of the British environmental movement originated.

Finally, notwithstanding the historical and political science foci, the book does contain much material related to contemporary policy. Key policy developments in agriculture, rural environmental and forestry policy areas are described and analysed and cover the period up to approximately summer

1994, by which time most of this book was written. In a final edit carried out in the spring of 1995 I have attempted, where possible, to add information on key subsequent policy developments. But my coverage of post-1994 events is inevitably rather more limited than the coverage of earlier policy. As this book goes to press, the rural policy world awaits with curiosity – I would hardly call it avid interest – the publication of the government's 1995 Rural White Paper. At the same time the Environment Bill continues its passage through Parliament. The policy process continues.

Part I

APPROACHES TO POLICY

1

THE POLICY PROCESS

INTRODUCTION

Few would deny that the actions of governments have had important consequences for all aspects of the rural scene in recent decades. Yet the approaches to policy adopted by many students of rural affairs have paid scant regard to the concepts and methods developed within political science. This is because, in disciplinary terms, rural studies has little in the way of a coherent identity, and is best seen as a collection of various approaches to rural matters. Rural studies embraces subjects as diverse as ecology and soil science on the one hand and sociology and social anthropology on the other. Students of rural affairs tend, therefore, to come from a diversity of disciplines, where the ways in which policy is understood and analysed vary dramatically. This is not the place to provide a detailed critique of how policy is conceptualised by geographers, economists, environmental scientists, and so forth. Suffice it to say that some of the studies of policy in these disciplines might have benefited from rather more attention to the insights available from political science.

Perhaps one of the reasons that some geographers, economists and others have chosen to develop their own approaches to policy is that political science itself can appear rather diverse and unapproachable. It is, after all, a discipline that has emerged relatively recently from its origins in history, philosophy and law (Jones and Moran 1991). With such roots, it is not surprising that its literature contains much which appears far removed from the practicalities of policy and its influences on the ground. Political theory, in particular, is firmly rooted in problems of moral philosophy. Philosophical questions about the nature of the state and democracy, and notions of justice and freedom continue to dominate, quite rightly, the concerns of many in the discipline. Those political scientists who are concerned with investigating the mechanics of policy making have adopted theoretical approaches, often deriving from sociological theory, which may seem abstract and unapproachable to those of other disciplines.

By contrast, those whose interests are perhaps more legal and constitutional, have tended to work within a 'public administration' genre which while seriously considering governmental processes does so in a descriptive, practical and non-theoretical manner and might be accused of being 'neither rigorous nor cumulative' (Rhodes 1991). This book has relatively little to say about political or constitutional theory. And its comments on the machinery of government, on the one hand, and the impact of policy on the other, are important primarily in so much as they cast light on our understanding of policy and politics and, in particular, the policy process. It follows that this is not a handbook for those who wish to know exactly what laws apply to particular activities in the countryside, nor for those who require a full description of the formal machinery of government. Such books exist but they say little about why particular laws came into being, how they are interpreted and understood, or what their consequences are.

The focus is on the policy process – how it is that certain policies come to be adopted and pursued by those seeking to influence or to formulate policy; and the consequences these political decisions have for the policies that come to be implemented and ultimately to affect rural land use and the lives of those concerned with the countryside. Some of what follows in this chapter may seem heavy-going, even obscure, to readers without a social science training. I ask them to persevere for, hopefully, the relevance and applicability of the key ideas will become apparent later. Certainly some of the concepts explored here are taken up and used in subsequent chapters when the real world of rural policy is discussed. The purpose of the political science explored in this chapter is not to provide proof of academic credentials nor to indulge in any form of intellectual one-upmanship. Rather it is because it provides the best set of tools available to do a particular job, namely to develop an appreciation of what agricultural and rural environmental policies are and why and how they have developed along certain lines.

WHAT IS POLICY?

Defining policy is not easy. In a memorable phrase, a former civil servant once commented that 'policy is rather like the elephant – you recognise it when you see it but cannot easily define it' (Cunningham 1963: 229). Cunningham also makes the crucial point that policy analysis requires something more than merely attending to the detailed content of legislation:

> Sometimes policy ... gets written down in an Act of Parliament, or in statutory instruments made under an Act. Sometimes it gets itself recorded in a memorandum or circular. But quite often it emerges from departmental practice in dealing with some particular type of business, or is determined by the way in which a Minister or a public authority settles an individual case.

(Cunningham 1963: 229)

From this emphasis upon the range of means by which policy can be promulgated it is a relatively small step to the view that policy is a process, that is, something that is dynamic and changing rather than a single action, decision or piece of legislation. It is this sense of process which characterises the following key definitions:

A policy may usefully be considered as a course of action or inaction rather than specific decisions or actions.

(Heclo 1972: 85)

A policy consists of a web of decisions and actions that allocate values.

(Easton 1953: 130)

Policy is a set of interrelated decisions concerning the selection of goals and the means of achieving them within a specified situation.

(W.I. Jenkins 1978: 15)

Thus policy is best seen as a web or network of decisions and actions that take place over a period of time. This *policy process* approach is different in key respects to the understanding of policy that emerges from studying either 'law' or 'administration', although there is inevitably considerable overlap with the concerns of both, particularly perhaps public administration. *Law* provides the details of each piece of legislation and the case-law which has determined how separate pieces of legislation are combined and interpreted. *Administration* shows, amongst other things, how laws and their interpretation impinge on the ground. Thus, in many instances an understanding of the technicalities of administration is a crucial element in land management decisions. However these approaches tell us little about the origins of policy; the interests and bargaining positions of the different individuals, nations, civil servants and lobbyists; or how the manner in which a policy is implemented and managed on the ground may subsequently influence amendments to legislation or new legal provision in the future.

Ham and Hill (1993) suggest that a dynamic understanding of policy leads to five main implications about the nature of policy, all of which emphasise the case that to study policy is to study something rather different to either law or administration. Their five points can be paraphrased as follows:

1   A web of decisions may take place over a long period of time, thus extending far beyond any formal initial policy-making process.
2   A policy usually involves a series of decisions rather than a single decision.
3   A policy may change over time.
4   Policy may involve nondecisions as well as decisions, especially if the context for policy shifts over time with no corresponding fresh decision taking.
5   Actions rather than, or as well as, formal decisions are important in defining policy. This may be particularly true for understanding the content of policy in the context of actions taken by those responsible for implementing policies rather than formulating them.

Thus, this book considers policy from a *policy process* perspective. It examines the manner in which policy decisions for rural areas are arrived at, and how they change over time. Where relevant it looks at *nondecisions* – that is, situations where a *lack* of action by policy makers has a profound influence on some aspect of rural land use – as well as decisions (Bachrach and Baratz 1963), and asks questions about policy implementation and what actually happens 'on the ground'. It is an approach which endeavours to make as few assumptions as possible about what might be expected to happen from an examination of legislative provision alone. To do this an inter-disciplinary approach – drawing on political science, sociology, environmental studies, history and geography – is adopted and, indeed, this can be seen as an essential element in the nature of policy analysis: 'The purpose of policy analysis is to draw on ideas from a range of disciplines in order to interpret the causes and consequences of government action' (Ham and Hill 1993: 11).

MODELS OF THE POLICY PROCESS

We turn now to an examination of various ideas that have been put forward by political scientists to explain and describe the policy process. Figure 1.1 portrays the policy process as a system which has as its input both political demands and resources and as its output different kinds of policy decisions and consequences; the decisions in turn have an impact upon society and consequently influence future inputs (Burch 1979). Thus the process is both circular and continuous. In between the input and output, and within what can loosely be defined as 'government', are three main stages – policy initiation, policy formulation and policy implementation (B. Jones 1991; sub-divided into eight stages by Hogwood and Gunn 1984). How policy makers within government actually operate these stages of the policy process is itself a complex question. For example, the extent to which a series of rational and logical decision-making steps are taken has been disputed by incrementalists, such as Charles Lindblom (1959, 1965, 1979) who argues that policy makers muddle through in response to the pressures brought to bear upon them; and that any resemblance to a rational and sequential approach is likely to be a consequence of *post hoc* rationalisation. Figure 1.1 illustrates the cyclical and dynamic nature of the policy process. On the input side are the political demands of society expressed through parties, pressure groups, election results and so forth. These political demands, together with the resources available to government, combine to determine policy programmes and output. These, in turn, influence society thus affecting future political demands and, through the economy, the resources available to future governments. However the model says little about precisely how policy inputs are fed into government and, once inside, how they are processed. To understand this we need to consider the different elements or interests within the political system. So far the policy process has been described in a somewhat abstract manner with scant reference to crucial questions about which groups or individual people are in a position to influence policy initiation, the design of policy instruments, and so forth. In considering this aspect of the literature we have identified five key models of the policy process: *formal structural, pluralist, elite, marxist* and

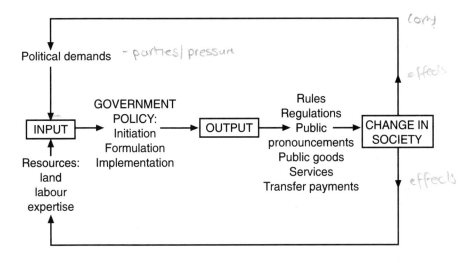

Figure 1.1 The policy process

Sources: after Burch 1979; B. Jones 1991

corporatist. These models should not be seen as necessarily mutually exclusive.

Formal structural model

Essentially the formal structural model embodies the official or constitutional version of the policy process as presented by the Central Office of Information and, sometimes, by civil servants and politicians in public utterances. This approach to policy making 'maintains that Parliament represents and interprets the public will through its representatives, who formulate executive policies which are faithfully implemented by civil servants' (B. Jones 1991: 505–506). Thus, the two key actors in the policy process are Parliament and the Civil Service, and the key concept is parliamentary sovereignty. The approach is perhaps best summed up by considering briefly a constitutional representation of British government as illustrated in Figure 1.2. The diagram is useful in so far as it lists the diverse sources for the framework of laws and customs which comprise a central element of policy. However it says little about how laws and conventions come into being apart from through parliamentary initiative. In short, it is not a model of policy making, nor to be fair, is it intended to be. Figure 1.3 represents an attempt to present the constitution in such a manner as to give a stronger indication of the policy-making functions and, crucially, the centrality of the House of Commons. The schema is based on a version of the constitution derived from the middle years (1832–1867) of the last century, an era described as the 'Golden Age' of Parliament when the House of Commons displayed the 'characteristics of an active, or policy-making,

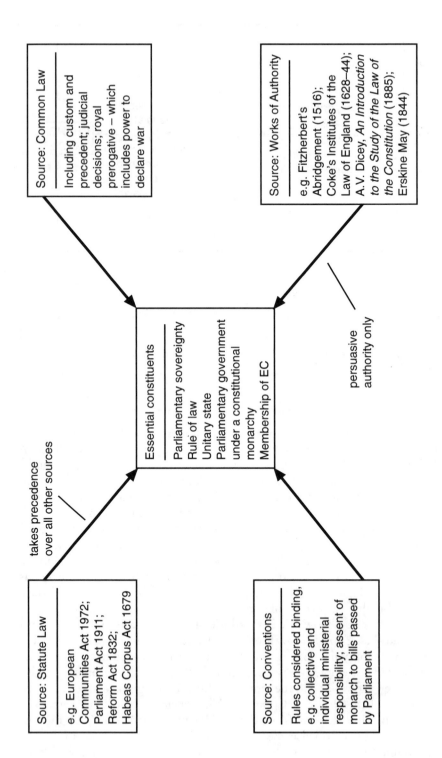

Source: Common Law

Including custom and precedent; judicial decisions; royal prerogative – which includes power to declare war

Source: Works of Authority

e.g. Fitzherbert's Abridgement (1516); Coke's Institutes of the Law of England (1628–44); A.V. Dicey, *An Introduction to the Study of the Law of the Constitution* (1885); Erskine May (1844)

Essential constituents

Parliamentary sovereignty
Rule of law
Unitary state
Parliamentary government under a constitutional monarchy
Membership of EC

persuasive authority only

takes precedence over all other sources

Source: Statute Law

e.g. European Communities Act 1972; Parliament Act 1911; Reform Act 1832; Habeas Corpus Act 1679

Source: Conventions

Rules considered binding, e.g. collective and individual ministerial responsibility; assent of monarch to bills passed by Parliament

Figure 1.2 The British constitution: sources and constituents

Source: B. Jones 1991: 278

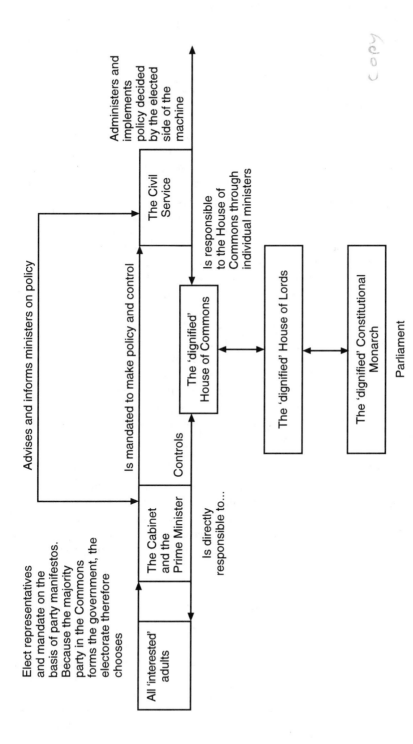

Figure 1.3 The liberal-democratic constitution of Cabinet, or prime ministerial government within a parliamentary democracy

Source: Dearlove and Saunders 1984: 30

legislature' (Norton 1991b: 316). During this period, the House of Commons removed governments and ministers without dissolution of Parliament, as party organisation was weak and party voting rare. In order to stay in power ministers conceded information and legislative initiatives to individual members of parliament. More than during any other period of British political history, the House of Commons created the conditions under which the government operated.

After the Reform Act 1867 had widened the franchise, extensive party organisation was needed to reach new voters; consequently the focus switched from the role of individual MPs to the election of party-based governments. The age of cabinet government had arrived and, by the twentieth century, even the most cursory glance at the nature of parliamentary procedures, the complexities of legislative provision and the bureaucratic machinery required to formulate and implement policy, should convince an impartial observer of the implausibility of the formal structural or constitutional model as a helpful description of policy making.

Parliament, for example, is thoroughly organised on party lines with the government of the day usually comprising members, but not all members, of a particular party. Time constraints alone mean that much of the groundwork for policy making is done by political parties, particularly by the party that is in government, or by civil servants in Whitehall. Thus, rather more plausible variants of the formal structural model have emphasised the important role of *political parties*, as channels of policy making, and of the *Cabinet* and the *Civil Service* in formulating policy.

We will return to some of these matters in the next chapter, which considers the various checks and balances within the core executive of government. Suffice it to say at this stage that the formal structural model (and its variants) concentrates attention on these actors in the central policy arena. But the model says very little about how issues become policy issues and what influences the policy agendas of politicians and civil servants. With neither a tradition of referenda nor particularly frequent elections, the British policy agenda must clearly be influenced by other external factors. Essentially the remaining models are attempts to understand what these factors might be, through looking beyond Parliament and the executive to consider the role of pressure groups and/or private interests in determining the policy agenda.

Pluralist model

The pluralist model is probably the single most influential account to have been developed by political scientists seeking to explain the complex nature of the modern policy process. It is based on empirical investigation and is critical of 'the constitutional fictions about representative government accepted by nineteenth century liberals or twentieth century legal theorists' (Dunleavy and O'Leary 1987: 24). Indeed some British writers have gone so far as to characterise policy making as 'non-parliamentary' (Jordan and

Richardson 1987: 28), a view criticised by Judge (1993). Pluralists assign a particular importance to the dispersed and non-cumulative character of the distribution of power in society, and within that framework to the role of interests and of pressure groups. This view is not entirely a recent one: 'The development of social interest – and that is democracy – depends not only on adult suffrage and the supremacy of the elected legislature, but on all the intermediate organisations which link the individual to the whole' (Hobhouse 1911: 233; quoted in Harrison 1980: 66).

Notwithstanding Hobhouse's early recognition of the existence of pressure groups within the political process (see also Bentley 1949), it was not until after 1945 that their importance was elevated to provide the basis for a fully developed theory of politics. In Britain, MacKenzie (1955) and McKenzie (1958) both claimed that pressure groups had become a more important channel of political communication than political parties. But it was in American political science that pluralism really took hold, particularly through the work of Robert Dahl (1961; but see in addition Latham 1952 and Truman 1951). The following comments illustrate some of the key elements in the pluralist model:

> The laws issuing from the government are shaped by the manifold forces brought to bear upon the legislators. Ideally, Congress merely reflects these forces, combining them – or 'resolving' them, as the physicists say – into a single social decision. As the strength and direction of private interests alters, there is a corresponding alteration in the composition and activity of the great interest groups – labor, big business, agriculture. Slowly the great weathervane of government swings about to meet the shifting winds of opinion.
>
> (Wolff et al. 1965: 11–12)

> Pluralist theory ... argues that power in western industrialised societies is widely distributed among different groups. No group is without power to influence decision-making, and equally no group is dominant. Any group can ensure that its political preferences and wishes are adopted if it is sufficiently determined.
>
> (Ham and Hill 1993: 28)

> [Pluralism is] a theory of bargaining between autonomous, often competing, groups and a fragmented state in which the emphasis is on the flow of influence from the groups to the state with an inbuilt set of checks and balances which supposedly prevent any one group becoming too powerful.
>
> (Grant 1985a: 21)

> The pluralist case ... rests on the argument that the essential thing is competition and participation among organised *groups*, not among individuals.
>
> (Presthus 1964: 19)

Taking the example of agricultural policy, pluralists would examine the various groups, such as the National Farmers' Union (NFU), the Country Landowners' Association (CLA), and food and environmental groups, in order to chart the policy process in this sector. Particular emphasis would be given to the lobbying strengths of these organisations, their level of resources, and the ideas and views they contribute to the policy-making process. The response of government itself is clearly an important indicative factor in such an approach. Some pluralists see government as neutral (e.g. Latham 1952), essentially acting as a referee in a contest between groups, but the majority, following the arguments of Dahl (1961), consider government agencies to act, in effect, as pressure groups themselves alongside non-governmental groups. It should be remembered, however, that this position derives largely from American research:

> One consequence is that pluralist theory often seems to reflect a more open, fragmented political system than applies in the case of Britain. In particular, government is often presented as highly fragmented. Such a picture has considerable validity in the US with its autonomous executive agencies, but less so in Britain.
>
> (Grant 1989: 26–27)

Even in the American literature, it has never been asserted that pressure groups have equal power and access (Grant 1989; Smith 1990a). However fragmented the political structure, some government agencies are more powerful than others and some groups enjoy closer relations with those agencies than do others. In Britain this is quite clearly the case and has led to the idea of *policy communities* and *networks* as means of explaining the particular features of pluralism within Britain (Richardson and Jordan 1979; Jordan and Richardson 1987). Although originally associated with the adaptation of pluralist ideas to suit British conditions, policy communities have come to be associated with both pluralist and corporatist positions in Britain, and further discussion of them is postponed until later in this chapter after the corporatist position has been described.

Elite model

As the name implies the elite model emphasises the importance of a relatively small ruling group within society, 'a unified and all-pervasive ruling class' (Scott 1982: 179). Thus for Mosca (1939), all societies are characterised by a class that rules and one that is ruled:

> The first class, always the less numerous, performs all political functions, monopolises power and enjoys the advantages that power brings, whereas the second, the more numerous class, is directed and controlled by the first, in a manner that is now more or less legal, now more or less arbitrary and violent.
>
> (Mosca 1939: 50; quoted in Ham and Hill 1993: 31)

And for Bottomore,

> The political elite ... will include members of the government and of the high administration, military leaders, and, in some cases, politically influential families of an aristocracy or royal house and leaders of powerful economic enterprises.
>
> (Bottomore 1966: 14)

In the United States elite theory was developed partly in reaction to the pluralist perception that power was widely distributed within society. Those advocating the elite model, such as C. Wright Mills (1956), argued that, in reality, American society was controlled by those occupying the higher circles of power (in government, business and military positions) who could accept or reject forces for change which might emerge from pressure groups from below. However, Mills has been criticised for exaggerating the degree of harmony within the elite:

> Struggles and clashes between different factions are the rule rather than the exception in the higher echelons of the economic order; nothing is more out of accord with reality than to present a 'conspiracy' picture of an unbroken cooperative consensus (as critics of western society, like Mills, have tended to do of capitalist societies, ...). Moreover, in terms of level of solidarity between elite groups, rather than within the economic elite itself, there can be little doubt that in the United States there is a considerably greater degree of fragmentation, if not necessarily of overt conflict, between elite sectors than in most other societies.
>
> (Giddens 1973: 171)

While Mills does not deny the existence of competing groups within the elite – he highlights, for example, the tendency of competing factions to coalesce around the main political parties – he does suggest an ultimate unity of purpose amongst the elite. In Britain, Bottomore (1966) has drawn a distinction between the *political elite*, comprising those exercising power, and the *political class* which encompasses all those involved in the policy process, including the elites of organisations not wielding power as such, particularly opposition political parties. Thus most exponents of the elite model suggest that in democratic systems there are competing elites, often based around political parties. Pressure groups exist of course, as in the pluralist model, but their influence is dependent upon the extent to which their interests are consistent with the interests of the elite in power. In other words, however determined and resourceful a group might be its success will be severely limited if its interests are opposed to those of the elite. There is none of the relative equality of opportunity assumed in the pluralist model. There is, however, some opportunity for pressure groups to exert their influence due to the existence of competing elites and the lack of cohesion which this can entail, especially in times of social and political change. The marxist model allows for no such possibility.

Marxist model

The crucial distinction between the elite model and the marxist model is that the latter not only identifies a ruling elite but also equates it with a class defined by its economic position within society. Thus in the *Communist Manifesto* first published in 1848, Marx and Engels (1952) observed that 'the executive of the modern state is but a committee for managing the common affairs of the whole bourgeoisie'. Much twentieth-century marxist scholarship has been dedicated to examining this proposition, especially in the light of the greatly increased complexity and size of the state since Marx's day. For some political scientists, the basic proposition holds true:

> In the Marxist scheme, the 'ruling class' of capitalist society is that class which owns and controls the means of production and which is able, by virtue of the economic power thus conferred upon it, to use the state as its instrument for the domination of society.
>
> (Miliband 1973: 23)

However Miliband's account is more sophisticated and subtle than might appear at first sight. He is at pains to point out that the government is not synonymous with the business elite, which would plainly be an absurd claim for an advanced democratic and bureaucratic society:

> the dominant economic interests can normally count on the active good-will and support of those in whose hands state power lies.... But these interests cannot, all the same, rely on government and their advisers to act in perfect congruity with their purposes.... governments may wish to pursue certain policies which *they* deem altogether beneficial to capitalist enterprise but which powerful economic interests may, for their part, find profoundly objectionable; or these governments may be subjected to strong pressures from other classes which they cannot altogether ignore.... In other words, the initial good-will and general support which capitalist interests may expect to find inside the state system does not remove the need for them to exert their own pressure for the achievement of their immediate and specific goals.... these interests bring to the task resources far greater, in a variety of ways, than those of any other interest in capitalist society.
>
> (Miliband 1973: 130)

The achievement of these goals is accomplished in many ways through the exertion of direct and indirect pressure on government and through the legitimation of certain ideas and actions through control of the media. According to Miliband, the bourgeoisie uses, wherever possible, its special power to promote its own interest through personal contacts with civil servants and ministers. As the bourgeoisie and members of the state executive (politicians and civil servants) have similar social backgrounds, this is not a difficult task. Moreover, it is argued that even socialist leaders are likely to be rapidly assimilated. Finally in Miliband's scheme, the freedom of independent action by politicians or civil servants is severely limited by the economic power of capital. Miliband's position has been described as an *instrumentalist* one in that he sees the state as an instrument of the bourgeoisie; a view shared by other marxists such as O'Connor (1973) and Gough (1979).

It should be noted, however, that this instrumentalist version of marxism has been subjected to lengthy criticism by others within or influenced by the *structuralist* marxist tradition whose view of the state is quite different – indeed Miliband himself shifts his arguments to some extent in later work (Miliband 1977). There is a vast and complex literature associated with these arguments but key references include the following: Hirst (1977), Jessop (1982), Offe (1972, 1975) and Poulantzas (1973, 1975). In their attempt to understand how the interests of subordinate classes are reflected, at least to some extent, within the state, these writers may come close to seeing the state as an independent force standing above and outside society, and to that extent may even share some ground with pluralists. Unlike pluralism, though, this kind of marxism retains a firm commitment to class analysis as the basis from which to understand politics.

Corporatist model

In different ways, corporatism represents a response to elite, marxist and, in particular, to pluralist theories. Moreover, it represents the most recent and, until recently, the most vibrant of the research traditions briefly reviewed in this chapter (key overviews of the subject include: Cawson 1986; Grant 1985b; Lehmbruch and Schmitter 1982; Schmitter and Lehmbruch 1979; Williamson 1985, 1989). Sometimes termed *liberal, bargained, modern* or *societal* to distinguish it from earlier 'neo-fascist' ideas of the corporate state, corporatism has become one of the most persuasive theories attempting to explain the complex relationships that surround modern government. Some of the roots of modern corporatist theory can be found in Beer's accounts of British political parties and pressure groups (Beer 1956, 1965). However, a single article by Schmitter (1974) probably did more than any other published work to develop modern corporatist theory. He certainly provided the most often quoted definition of corporatism:

> Corporatism can be defined as a system of interest representation in which the constituent units are organised into a limited number of singular, compulsory, noncompetitive, hierarchically ordered and functionally differentiated categories, recognised or licensed (if not created) by the state and granted a deliberate representational monopoly within their respective categories in exchange for observing certain controls on their selection of leaders and articulation of demands and supports.
>
> (Schmitter 1974: 93–94)

The pivotal role accorded to pressure groups is indicative of the corporatist response to pluralist theory. In acknowledging the importance of pressure groups in policy formation, corporatist thinkers are anxious to avoid the pitfalls of pluralism, in particular its inadequate treatment of government. For corporatists, government agencies are in no way neutral or impartial recipients of policy pressures but are active in sponsoring and recognising particular pressure groups:

> This model perceives an alliance between Ministers, civil servants and the leaders of pressure groups in which the latter are given a central role in the policy making process in exchange for exerting pressure upon their members to conform with government decisions. In this view therefore interest groups become an extension – or even a quasi form – of government.
>
> (B. Jones 1991: 506)

Conversely, the role of the pressure groups is also important for, thus recognised by the state, they are expected to behave in a certain manner. Crucial to corporatism,

> is the direct link with regulation, whereby representative groups assume some responsibility for the self-regulation and disciplining of their own constituency in return for the privileges afforded by their relatively close relationship with government.
>
> (Cox, Lowe and Winter 1986a: 475–476)

---

1 Organised interests representing functional interests show a tendency towards a position of monopoly.
2 Certain functional interests are granted privileged access to the state's authoritative decision-making processes and in other ways supported by the state, but such 'licensing' is granted on the basis of adherence to certain norms.
3 Membership of such associations may cease to be wholly voluntary, while the associations' privileged monopoly position deprives the members of alternative effective channels.
4 In addition to performing a representative function, interest associations also perform a regulatory function over their members on behalf of the state.
5 Interest associations and state agencies show increasing bureaucratic tendencies so sectors in society tend to be regulated through hierarchical structures.
6 Functional interest associations and state agencies enter into a closed process of bargaining over public policy where consciously or not the associations do not fully pursue their immediate interests but act in a 'system-regarding' manner.

Figure 1.4 A formal model of corporatism

Source: Williamson 1989: 68

Some of the key characteristics of the corporatist model are illustrated in Figure 1.4. Its applicability to the world of agricultural policy will be considered further in chapter 5. With the advent of Thatcherism after 1979 corporatism came under increasing strain as New Right thinkers attacked the supplanting of Parliament by interest groups, particularly the TUC and the CBI.

Competing or complementary models?

Although there are important differences between the models discussed, they are not necessarily stark alternatives. Acceptance of the corporatist framework, for example, does not preclude the possibility of retaining elements of marxist theory (see, for example, Panitch 1980, 1981) or of elite theory. There may well be strong elements of a ruling elite within a government which incorporates interest groups in order to achieve certain goals. So too the interests of a capitalist or business class may be particularly important in determining the outcome and shape of corporatist policy making.

Pluralism and corporatism place a common emphasis on pressure groups which may be seen as complementary, even if the practitioners of political science find much to argue about. Indeed, 'most corporatist writers agree that in practice corporatist and pluralist arrangements are often found *side by side* or in *some mixed combination*' (Williamson 1989: 64). Thus in some sectors, relationships between government and groups will be pluralist and in others corporatist. There is a tendency, it has been argued, for corporatist arrangements to develop over issues concerned with production and for pluralist politics to prevail over consumption issues (Cawson and Saunders 1983). For example, we might conclude that the relations between the NFU and the Ministry of Agriculture, Fisheries and Food (MAFF) constitute a corporatist political arrangement in the arena of agricultural commodity policy, whilst MAFF receives pluralist pressure from a wide range of other groups over environmental matters or food policy (consumption issues).

Clearly, the main distinction between corporatism and pluralism lies in the manner in which pressure groups are involved in the policy process, and this is illustrated in diagrammatic form in Figure 1.5. Cawson's representation

Liberal ⟷ Consultation ⟷ Pluralism Corporatism		
formal role in policy formation and implementation	Role in policy formation but not implementation	'parliamentary role': influence, but no formal role in formation or implementation

Figure 1.5 The corporatist–pluralist continuum

Source: Cawson 1986: 40

draws on the work of Martin (1983) who highlights the extent to which organised groups are incorporated within the policy arena as the key element defining whether a particular sector of policy making is corporatist or pluralist.

PRESSURE GROUPS AND POLICY NETWORKS

The importance attached to the role of pressure groups leads us, in the final section of this chapter, to consider in more detail the role of pressure groups and how they combine with the state and its agencies to form policy communities and issue networks, which may exhibit pluralist or corporatist characteristics. Figure 1.6 provides a typology of pressure groups developed by Wyn Grant from his experience over many years of studying pressure groups across a wide range of sectors.

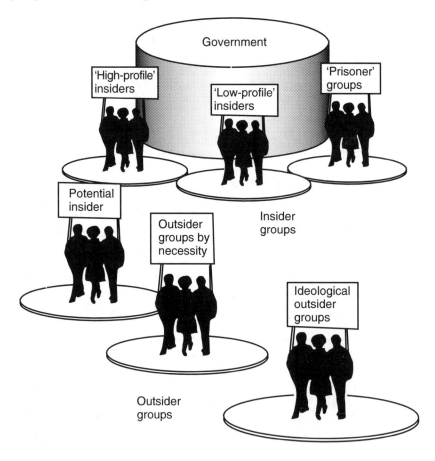

Figure 1.6 A typology of pressure groups

Source: Grant 1989: 15

Insider groups are those which 'are regarded as legitimate by government and are consulted on a regular basis' (Grant 1989: 14). It is important to note, however, that insider status does not necessarily carry with it 'substantive influence over policy outcomes' (Saward 1990: 588). Outsider groups are not directly involved in the policy process. It is worth briefly discussing the applicability of this typology, using some of the main examples that will be known to those familiar with the contemporary rural policy scene.

High-profile insiders

The National Farmers' Union is a good example of a high-profile insider group. It achieved its consultative status after a period of time, in the inter-war years, when it was an outsider group. In the war and in the post-war years, as a result of both changed circumstances and the consequences of its own lobbying, the Union acquired insider status. However, its public profile has continued to be high, even though much of its important work is conducted in private with government officials. This is because of the emphasis the Union places upon the importance of favourable public opinion as a means of reinforcing its position within government and, also, as a way of providing evidence of its activity to rank and file members.

Low-profile insiders

These groups do not court publicity, choosing instead to exert influence through establishing appropriate high-level contacts. Low-profile insiders are often informal groups of employers or specialists, and therefore have no significant mass memberships to consider. An example within the agricultural world is the society, RURAL, which was established in the early 1980s by a number of leading figures in the agricultural world who, whilst sharing some of the concerns of the emergent environmentalist critique of agriculture, were also anxious to present a 'balanced' picture. The organisation's early attempts to attract members and become a conventional pressure group were soon replaced by an emphasis upon closed 'invitation-only' seminars with senior officials within MAFF and the Department of the Environment.

Prisoner groups

Similar in many ways to low-profile insiders, prisoner groups are those which,

> find it particularly difficult to break away from an insider relationship
> with government, either because they are dependent on government for
> assistance of various kinds (e.g. secondment of staff or office accom-
> modation), or because they represent part of the public sector (e.g. local
> authority associations or chairmen of nationalised industries).
> (Grant 1989: 16)

A rural policy example of a prisoner group of this type would be the chairmen of the five UK Milk Marketing Boards, prior to their abolition in 1994. (The specific corporatist characteristics of milk marketing are discussed in greater detail in chapter 5.)

Potential insiders

Groups wishing to have insider status are numerous as it is commonly perceived, correctly or not, that this would result in the furtherance of a group's cause. However, only those with the potential to do so, in terms of political skills and resources, fall into this category. The Farmers' Union of Wales was in this position until the government granted it formal recognition in the annual review process in 1978. The Tenant Farmers' Association might now justly claim to occupy that position. Some long-standing and well-organised environmental pressure groups, such as the RSPB and the CPRE, are potential insider groups. Indeed, on certain issues they might already be said to occupy insider status. Such groups might be seen as 'thresholder' groups, intermediate between insider and outsider status and exhibiting 'strategic ambiguity and oscillation between insider and outsider strategies' (May and Nugent 1982: 7, quoted in Grant 1989: 18).

Outsider groups by necessity

These groups may share the same ambitions as potential insiders but lack adequate political skills and knowledge:

> The language of the British civil service is a language of veiled understatement and it is characteristic of politically unsophisticated outsider groups that their demands are presented in strident terms. Their lack of understanding of the political system leads them to make demands which are constitutionally impossible.
>
> (Grant 1989: 18)

Some of the more radical environmental groups, such as Friends of the Earth, certainly fitted into this category in the 1970s. But by the 1980s Friends of the Earth had amassed considerable political expertise. The Smallfarmers' Association provides a good example of an agricultural outsider group by necessity, enervated not so much by its stridency as by political naïvety and inadequate resources.

Ideological outsider groups

These are groups which, for ideological reasons, do not accept the possibility that change may be brought about through participation in the policy process. A good example is the Animal Liberation Front. Groups committed to direct action in opposition to road developments provide another instance.

*   *   *

One problem with the approach adopted here is that it might be taken as based on an assumption that insider status is essential to success in influencing the policy process. Whilst this may often be the case, especially with regard to the achievement of modest incremental policy adaptations, on some issues direct action may be equally or more effective as demonstrated with the widespread opposition to the poll tax in the 1980s and, more recently, with campaigns against the export of live animals from Britain to continental Europe. Some of the more effective and sophisticated environmental groups, such as the CPRE and the RSPB, choose to operate, at least in part, as outsider groups in order to influence political agendas and to challenge existing policy premises. Nor should it be assumed that 'insider status', although a prerequisite for corporatism, necessarily implies a corporatist arrangement. It is possible for a group to be well regarded by government, leading to regular consultations and exchanges of information, even an involvement in the formulation of policy, without that group being in any way involved in the implementation of policy.

Policy networks

In recent years the corporatism/pluralism approach has, in some measure, been superseded by a new focus on policy networks. In part, this reflects conceptual developments in political science, in particular disdain for what were perceived to be the rather ambitious theoretical claims associated with corporatism. But also of great significance were the consequences of the assault on corporatism that characterised the British governments of the 1980s:

> Under Mrs Thatcher's governments parliamentarism successfully challenged corporatism. In the words of Lord Young, former Trade and Industry Secretary: 'We have rejected the TUC, we have rejected the CBI. . . . We gave up the corporate state' (*Financial Times*, 9 November 1988).
>
> (Judge 1993: 120)

However these important developments did not lead to the abandonment of all forms of functional interest representation:

> The practice of government in the 1980s continued to accommodate organised groups in the process of policy making and the response of British political scientists was to invoke a more appropriate empirical-descriptive theory – this time focusing upon policy communities and policy networks.
>
> (Judge 1993: 120)

A number of writers have argued for two main types of network: *policy communities* and *issue networks* (e.g. Marsh and Rhodes 1992b; M.J. Smith 1993). *Policy communities* are networks around specific policy sectors comprising ministers, key civil servants in the relevant government department(s), and the leaders and officers of key interest groups. They are characterised by 'stability of relationships, continuity of a highly restrictive membership, vertical interdependence based on shared service delivery responsibilities and insulation from other networks and invariably from the general public.... They are highly integrated' (Rhodes 1988: 78). *Issue networks* comprise similar groups of policy actors with a shared interest in a particular policy sector or issue. They are less integrated than policy communities:

> The distinctive features of this kind of network are its large number of participants and their limited degree of interdependence. Stability and continuity are at a premium, and the structure tends to be atomistic. Commonly, there is no single focal point at the centre with which other actors need to bargain for resources.
>
> (Rhodes 1988: 78)

It has been claimed that decision making on agriculture and food production has become more complex with new pressure groups obtaining access in recent years, resulting in the replacement of a policy community by an issue network (Jordan and Schubert 1992).

Policy communities and issue networks can be contrasted in many ways. First, participants in a policy community tend to share a deep and direct interest in a policy area. For example, the policy sector may be of interest to the civil servants because it is a key area of government economic policy with implications for economic performance; to the politicians because electoral prospects are so linked with economic factors; and to a key interest group because the livelihood of its members is bound up with sectoral economic performance. Production issues are more likely to give rise to this shared sense of purpose and community. By contrast, issue networks may form around consumption issues, on which government economic performance does not depend. For the pressure groups concerned about a particular issue, it may be one of many concerns.

Second, participants of a policy community tend to believe that more resources can usefully be applied to their policy area. In other words a policy community comes into existence when there is an agreed set of priorities and particularly an agreed need for public expenditure. A government department harnesses a pressure group to assist its case for lobbying the Treasury and other cabinet ministers. The pressure group adopts the department as the most likely channel for advocating its case. An issue network, on the other hand, is more likely to exhibit characteristics of conflict as groups with radically different perspectives on an issue struggle for supremacy. Consequently there are no such shared beliefs.

Third, participants in a policy community tend to share an 'appreciation' of the issues. There is a shared or common 'culture' within a community. Not only is there shared agreement on the need to devote resources, but also a shared understanding of the problems and priorities. This shared culture of

Table 1.1 Types of policy networks: characteristics of policy communities and issue networks

Dimension	Policy community	Issue networks
Membership		
Number of participants	Very limited number; some groups consciously excluded	Large
Type of interest	Economic and/or professional interests dominate	Encompasses range of affected interests
Integration		
Frequency of interaction	Frequent, high-quality, interaction of all groups on all matters related to policy issues	Contacts fluctuate in frequency and intensity
Continuity	Membership, values and outcomes persistent over time	Access fluctuates significantly
Consensus	All participants share basic values and accept the legitimacy of the outcome	A measure of agreements exists, but conflict is ever present
Resources		
Distribution of resources (within network)	All participants have resources; basic relationship is an exchange relationship	Some participants may have resources, but they are limited, and basic relationship is consultative
Distribution of resources (within participating organisations)	Hierarchical; leaders can deliver members	Varied and variable distribution and capacity to regulate members
Power	There is a balance of power among members. Although one group may dominate, it must be a positive-sum game if community is to persist	Unequal powers, reflecting unequal resources and unequal access. It is a zero-sum game

Source: Marsh and Rhodes 1992b: 251

a policy community is vital to its internal success, its cohesion and its closure to other interests. An issue network often exhibits no such stability and shared culture.

Finally, a policy community tends to be relatively stable with continuity of membership. In order for a community to exist there has to be some stability and continuity. In other words there have to be boundaries which are recognised by the members of the community. Theoretically there are few policy issues which do not in some measure impinge upon the majority of people; but it is impossible to envisage a coherent policy community comprising the entire electorate; nor can we envisage a coherent policy community where those concerned are constantly changing as might be the case in an issue network.

The main contrasting characteristics of the two types of policy network are shown in Table 1.1.

CONCLUSIONS

Some readers might be forgiven for thinking that policy communities and issue networks are, in essence, the equivalents, respectively, of corporatist and pluralist models. Although the similarities are obvious this would be a mistaken view. Policy communities and policy networks are essentially descriptive, empirical categories which make far fewer demands of the evidence than do the more rigorous, but abstract, models developed by corporatist and pluralist theorists. A policy community is a prerequisite for corporatism but it is not the equivalent, for the corporatist model requires other characteristics to come into play (see again Figure 1.4, p.20). Some might argue that the real world of policy making is so messy that only the low-level generalisations of policy communities and issue networks will do. Others prefer to play for the bigger stakes of somewhat grander theory, even if the evidence sometimes has rather to be forced to fit the model. M.J. Smith (1993: 73), a firm advocate of the policy network approach because of its empirical strengths, concedes that 'policy networks need a wider macro-theory to explain the sorts of relationships that develop' and suggests that the different traditional theories focus on different aspects of networks. For both marxists and elitists the networks would be closed policy communities dominated by capital and key interests/state actors respectively (M.J. Smith 1993). Pluralism, by contrast, implies a more open issue network or, if policy communities exist, the risk that it will break down as agendas change and new pressures and groups emerge, and as state actors themselves change or develop new perspectives.

In the course of this book we use the concept of policy networks to discuss the nature of agricultural and environmental policy making and politics. On occasions we go further and suggest that the models of corporatism and pluralism might be usefully deployed. But before moving to an examination of aspects of policy making in agriculture and the environment, we need to complete our account of the policy process, for so far we have seriously underestimated a vital ingredient in the policy process – the role of the state itself. The formal structural model gave some indication of the importance of core state actors but these now need to be considered in detail, not as an alternative model but as an empirical exercise, for the success or otherwise of private interests and pressure groups will depend upon the nature of the governmental process itself and, indeed, upon the interests of those involved in that process on a daily basis.

2

THE INSTITUTIONAL
FRAMEWORK

INTRODUCTION

The first chapter deliberately eschewed an institutional approach to policy, arguing that pressure groups and non-parliamentary interests are so important that it is misleading to consider the policy-making process in formal institutional terms. Clearly, however, the *parliamentary* or *legislative process* is important as indeed is the other side of the formal coin – the *executive process*, that is, the work of the Civil Service and government departments. Indeed, these are the processes that pressure groups seek to influence. And if pressure groups are to some degree incorporated in these processes it is partly because such incorporation is explicitly or implicitly sanctioned or permitted by Parliament and/or the executive. Thus, for example, the 1947 Agriculture Act sanctioned the involvement of the NFU in the Annual Review, even if this corporatist arrangement developed very much along extra-parliamentary lines in its day-to-day and year-to-year working.

Having outlined some of the conceptual background to the complex processes by which policy is formed, this chapter turns to an overview of some of the key institutions and actors in these processes. It provides a brief introduction to the main formal institutions within the UK and EC which make policy, and attempts to identify the broad institutional parameters of key policy networks such as agriculture and the environment. In particular it seeks to examine some complex questions concerning the locus of power within the core institutions of the British state. Some readers might feel that this is to turn again to the formal structural model of the policy process, dismissed rather cursorily in the first chapter. This is not the case, for a detailed examination of how central government functions does not preclude the possibility that the policy agenda might be heavily influenced by other factors. Thus the considerations of this chapter should be seen alongside the models of the policy process put forward in the previous chapter.

THE CORE EXECUTIVE

First we turn to a consideration of the central arena of British government, known as the *core executive* or, by others, as the *central executive territory* (Madgwick 1991). The core executive is not a synonym for all aspects of central government. Rather it refers to the innermost centre of central government consisting of a

> complex web of institutions, networks and practices surrounding the PM, Cabinet, cabinet committees and their official counterparts, less formalized ministerial 'clubs' or meetings, bilateral negotiations, and interdepartmental committees. It also includes some major coordinating departments – chiefly, the Cabinet Office, the Treasury, the Foreign Office, the law officers, and the security and intelligence services.
>
> (Dunleavy and Rhodes 1990: 3)

Madgwick (1991) suggests that the 'active inhabitants' of the core executive number only 300 people. A diagrammatic representation of the core executive is given in Figure 2.1. The Prime Minister's Office includes the Private Office, the Press Office, the Political Office and the Policy Unit. Of these the Private Office is 'the single most important section' (Donoughue 1987: 17), providing day-by-day advice to the PM on the routines of government, including the all-important task of deciding which papers deserve the PM's immediate attention and commenting on them. The Political Office advises on 'party and electoral matters and usually plays a crucial role in speech writing' (Hennessy 1990: 387). The Policy Unit has, on the surface, very similar functions to the Private Office. The crucial difference is that whereas the Private Office tends to be staffed by career civil servants, the Policy Unit comprises many experts chosen by the PM for their particular policy strengths. Many of the better known members of the Unit

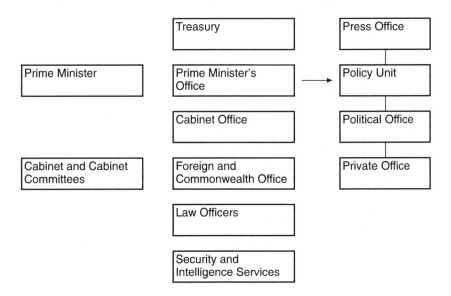

Figure 2.1 The core executive of British government

have been recruited from business or academia. Examples include Bernard Donoughue who worked for both Wilson's and Callaghan's Policy Unit, and Professor Bryan Griffiths for Thatcher. The Unit receives from the Private Office 'all papers covering domestic policy except those dealing with appointments and security matters' (Hennessy 1990: 659).

The Cabinet comprises some 20–25 senior government ministers under the chairmanship of the Prime Minister. The full Cabinet meets approximately once a week for a couple of hours, with much of its remaining work taking place in cabinet committees. The PM has a tight control over the committees and important policy decisions may be taken in small committees and then reported to Cabinet.

The Cabinet Office (CO) is staffed by 100 civil servants, although only the more senior officials number among the 300 identified by Madgwick as making up the core executive. The CO can be seen as 'the servant of the Cabinet and of the Government collectively, its purpose being to promote and assist the discussion and resolution of issues that transcend departmental boundaries' (Armstrong 1986; quoted in Hennessy 1990: 388). However, formally the CO is the responsibility of the PM and Minister of State for the Civil Service, and some have argued that it is much closer to a Prime Minister's office than the name and some of the official claims imply. The Office was introduced by Lloyd-George during his premiership in the First World War and was retained in 1918 as a permanent feature 'for the purpose of collecting and putting into shape the agenda, of providing the information and material necessary for [the Cabinet's] deliberations, and of drawing up the results for communication to the departments concerned' (Haldane Committee quoted in Hanson and Walles 1981: 119). Harold Wilson strengthened the CO by giving it the responsibility for providing most of the chairmen for the high-level interdepartmental committees, thereby providing a means for Downing Street to keep an eye on Whitehall.

The Cabinet Office comprises six secretariats: economic, overseas and defence, European, home affairs, science and technology, and joint intelligence (Hennessy 1990). These, it will be immediately noted, amalgamate and cut across the interests of the key departments within government itself. For example, Ministry of Agriculture issues would be largely discussed within the Economics Secretariat alongside issues from the Departments of Trade and Industry, Energy, and Employment, although clearly the European Secretariat is likely to have some interests here too, and, in times of national emergency, the Overseas and Defence Secretariat may assume responsibility for food security issues. Environmental issues are, in the main, the preserve of the Home Affairs Secretariat. The work of the secretariats is planned each week at a meeting 'chaired by the Cabinet Secretary and attended by the heads of his six secretariats, the PM's Principal Private Secretary and the Head of the Policy Unit' (Hennessy 1990: 391).

Of the remaining chief players in the core executive – the Treasury, the Foreign Office, the law officers, and the security and intelligence services – the Treasury is clearly the one that most influences rural policy as a consequence of its impact on the level of government expenditure and its pre-eminence amongst other government departments: 'the Treasury has a window into every ministry and departmental activity across Whitehall. The Chancellor of the Exchequer, if he is assiduous in his reading, is probably even better informed than the Prime Minister about his colleagues' business' (Hennessy 1990: 394). It is a complex department with key sections covering all of the main areas of public expenditure as well as numerous other financial matters as diverse as the rating of government property and aid and export finance.

Essentially there are three schools of thought as to the relative strengths of the chief actors in the core executive, and which of them exerts the most influence over the central governmental process: prime ministerial government, cabinet government and bureaucratic government.

Prime ministerial government

The Prime Minister's power is considerable, and can be wielded in several distinctive arenas. A Prime Minister can expect to wield power over his or her own political party, through a high degree of control over the policy-making machinery of the party (more the case for Conservative than for Labour PMs). Modern governments comprise no less than 110 cabinet and non-cabinet junior ministers. The vast majority of these are members of the House of Commons, and thus the PM has considerable powers of preferment within the party. Even those members at odds with the PM are expected to have regard to party loyalty, especially at election times. That a Prime Minister's power over the party is not absolute was shown in dramatic manner with the precipitate departure of Mrs Thatcher from office in 1990, at the hands of her parliamentary party rather than the electorate.

Linked to power over party is power over Parliament and, in particular, a considerable influence over the control of parliamentary business. The PM also has influence over the electorate through his or her control and management of the dissemination of public information and publicity to the media, which can be used to personal political advantage. Considerable power over the Civil Service is wielded to the extent that the PM decides which civil servants will hold key positions. Power over the Cabinet lies in rights of appointment and dismissal and also the control of agendas and meetings, even the power to take decisions without votes (which are rare in cabinet meetings), and the power to appoint cabinet committees. The PM has great control over whether or not issues come to cabinet and in what form: 'for all prime ministers the business of government is pushed forward for the most part outside Cabinet – by minute, bilateral conversation and by telephone' (Madgwick 1991: 84). As well as powers of appointment and dismissal, the PM can reorganise departmental responsibilities, so allowing a particular Minister to accrue power or lose it. There are also powers of patronage (including the honours system) and the right to take the decision on when and how to terminate the life of a government.

Of course, this does not add up to absolute power for there are clear limitations of time and energy confronting any individual in such a position. Moreover, the PM does not necessarily receive all information required for a complete grasp of all the relevant policy issues. This is particularly the case with regard to access to interdepartmental or Cabinet Office papers at a preliminary stage of drafting. Thus a modification of the notion of prime ministerial government is the PM clique view, which emphasises the importance of a small coterie of senior ministers, civil servants and advisers. This group is not necessarily drawn exclusively from within the core executive and certainly will not encompass all core executive members. Mrs Thatcher's premiership, for example, was characterised by extensive use of individual advisers on economic and foreign affairs, speech writers, advertisers, PR specialists, a strong reliance on the Political Office at No.10, and centralisation of Whitehall press and information services under her Press Secretary, Bernard Ingham. In an attempt to address what she perceived as inefficiencies within the Civil Service she developed an Efficiency Unit charged with the task of generating ideas for reforming the processes of government. She also did much to foster and encourage external policy advisers and consultants, particularly emergent 'New Right' think-tanks such as the Adam Smith Institute, the Institute of Economic Affairs and the Centre for Policy Studies. Thus to some extent members of Thatcher's clique were drawn from outside the formally defined core executive of government.

Cabinet government

The study of the core executive arose primarily in response to the somewhat discredited notion of cabinet government, which it has been argued is a normative ideal which is no longer consistent with the complex operations of the central state in the UK (Dunleavy and Rhodes 1990). None the less it is important to identify the key notions that make up this ideal. These include cabinet unanimity, solidarity and confidentiality:

> Originally, the rationale of ministerial unity was to provide a defensive shield against the crown and so prevent the victimisation or dismissal of individual ministers. As the locus of responsibility passed from the crown to the Commons, however, the advantages of cabinet unity came to be recognised as a defence for the executive against unnecessary political embarrassment in the developing adversarial context of the House.... At the time when collective responsibility became 'a cardinal feature of British politics' (Birch 1964: 135), in the middle decades of the nineteenth century, it was feasible that cabinet members both should and *could* collectively deliberate upon policy. In these conditions prescription and practice coalesced. In the twentieth century, however, the sheer scope and activism of government reduced the capacity of cabinet to act as *the* collective point of decisions and

increased the centripetal tendencies for decisions to be made elsewhere, most particularly within departments or within 10 Downing Street.

(Judge 1993: 141)

Thus very few believe that the ideal of collegiate cabinet government has worked in practice, certainly not since 1981, when Mrs Thatcher forced through a severely deflationary budget opposed by a majority of her ministers (S. James 1992). However such a view has to be seen in the context not only of Thatcher's obvious strength of character but also of the fact that all her governments were built on considerable parliamentary majorities. Such circumstances clearly make cabinet opposition to unpopular prime ministerial policies difficult to sustain. The processes of dialogue, conflict and compromise, essential to any kind of collegiate decision making cannot operate where power is so unequally distributed amongst the individuals concerned.

This is not to say that cabinet ministers wield no power at all. Thus an adaptation of this model, rather than emphasising the collective side of cabinet affairs, focuses upon the powers of individual ministers within their own spheres of responsibility, especially, as happens so often, when small and specialist cabinet committees are responsible for policy development. Ministers are able to defend their own scope for action and their own departments. The seeming electoral invincibility of the mid-1980s assisted this process. As party discipline became less important so the opportunity for sectional lobbies within the party to influence specific Ministers became greater. At times decisions may be taken outside cabinet, especially in relatively uncontroversial areas upon which the PM has not focused attention.

Bureaucratic government

This view lays stress on the power of the government departments and particularly of the senior civil servants. It is the civil servants in government departments whose role is crucial in advising ministers. They draft legislation, statutory instruments, and government green and white papers; they provide instructions to their regional and other offices on policy implementation; they play a crucial role in both policy formulation and evaluation; they possess considerable technical expertise, much of which is not readily available to back-bench MPs. Of course, in discharging these functions they are answerable to ministers who are themselves answerable in cabinet and to Parliament.

Despite this high degree of influence and the fact that many commentators recognise that the state is increasingly bureaucratic, the bureaucratic model was given relatively less attention in the early and mid-1980s when the Civil Service came under serious attack through the reforms initiated by Mrs Thatcher. However it might be contended now that this model still has much to commend it, especially in policy areas such as agriculture which are not at the centre of UK political debate in the same way as issues such as the economy, employment and the National Health Service.

When the Conservative Party was elected in 1979, Mrs Thatcher initiated a radical break with bipartisan support for the Civil Service by touring ministries 'and dressing-down senior civil servants for alleged failings in their departments' policies' (Dunleavy 1990: 110). A number of ministers publicly chastised the Civil Service for waste, with the clear implication that the Civil Service was, at best, a necessary evil. Departmental efficiency was put under the microscope when Mrs Thatcher invited Sir Derek (now Lord) Rayner, Chairman of Marks and Spencers, to head her newly established Efficiency Unit. A full account of Raynerism in practice is given by Hennessy (1990), and the so-called 'Rayner Raiders' were much feared by government departments in the early 1980s.

In an effort to limit public expenditure the linking of civil service pay to private sector pay was broken in 1982 alongside a 15 per cent random manpower cut over six years (Dunleavy 1990). The Civil Service Department, which had been founded in 1968 with key responsibilities for pay and conditions, was abolished in 1981 and replaced by a small and ineffective Management and Personnel Office (Fry 1986). And the Conservative government waged a campaign to de-privilege the service following the failure of the 1981 pay strike by the Civil Service unions.

The Civil Service, however, was not lacking in response. It received a considerable fillip through the logistical triumph of the Falklands War (Dunleavy 1990). This was based on what Rayner would almost certainly have seen as unnecessary contingency planning. The efficiency with which the war was administered did much to restore confidence in the Civil Service and to weaken the government's case that the Civil Service was inefficient and over-staffed. In a more subtle manner senior civil servants were also adroit at dealing with some of the criticisms levelled against them. For example, the 'Financial Management Initiative' developed by heads of department to improve efficiency in line with government wishes, through better costing, budget control, line management, and so forth, is seen by some as a largely successful attempt to tone down and 'civilise' ministers' assaults on Whitehall. In the words of a civil servant at the time, 'Whitehall absorbed and neutered the attempt to change the culture and make it more managerial' (Ponting 1986: 224).

Consequently, after the initial burst of reform in the early 1980s, the middle years of the decade were relatively uneventful ones. However, in 1988 the Efficiency Unit issued a report concluding that the Financial Management Initiative objective of getting top civil servants to concern themselves with routine management had failed because of short-term political pressures and the sheer size of the Civil Service. Again the Civil Service responded, this time suggesting a new Management Board. But Mrs Thatcher supported the Efficiency Unit, thus heralding a new attempt to force potentially very radical reforms on Whitehall. A key idea in the new approach is that the central Civil Service should consist of a relatively small core to service ministers and manage departments. The rest of the service should become executive agencies contracted to deliver services. They may or may not be Crown employees and it is estimated that 75–90 per cent of civil servants might work in such agencies by the end of the century. This is, in effect, privatisation, for as one commentator has argued:

> even where agencies start out as public sector bodies they could easily metamorphose into firms. A chief executive recruited from the private sector could hire in several private sector deputies, who in turn hire in their own assistants. Within a few years a management buy-out of the agency's operations or franchise could be feasible.
>
> (Dunleavy 1990: 115)

The effects of these changes upon the delivery of policies in agriculture is already being felt. These changes in the Civil Service are hard to assess in terms of relative power, but certainly the scale and pace of the change imply that an elected government has considerable scope to initiate unwelcome changes within the bureaucracy:

> the thrust of the government's policy, the FMI, agencies and the ending of national pay settlements, is to fragment the civil service, so that there may be doubt whether in future there will exist anything that can be called the national government's unified civil service. From the 1870s the British created a unified civil service: from the 1970s the British sought to break it up.
>
> (G.W. Jones 1989: 258)

However this fragmentation runs alongside an 'intensification of central control through accounting systems' (Gray, Jenkins, Flynn and Rutherford 1991: 58).

Of course, it has to be remembered that such changes were greatly facilitated by four successive Conservative victories. This continuity of government meant that some ministers served longer than their own senior departmental civil servants, the direct reverse of what had been the case in the past. It also meant that government had the opportunity to promote to senior positions those civil servants with a greater degree of sympathy for government policies. Consequently the Labour Party accused the government of fostering a departure from the British tradition of a neutral civil service. Not surprisingly this was strongly contested by Conservative ministers, but it would be surprising if there had been no convergence between the views on certain subjects of senior civil servants and government ministers after fifteen years of one party in government.

PARLIAMENT AND PARTIES

In the light of the above discussion some readers may wonder whether the Houses of Parliament and the political parties, which provide so much of the popular focus for political interest, actually have any influence over the formulation of policy. Of course they do, and in recent years there has been something of a return to these traditional foci of concern for political scientists. For example, Jack Brand (1992) has produced a powerfully argued indictment of those who, in their anxiety to give a rightful place to pressure groups and behind-the-scenes policy deliberations, have ignored the role of Parliament and the parties:

> When he introduces the idea of a policy community, Richardson explicitly denies a place to Parliament. It comes into play only when the normal processes have broken down – 'the fuse is blown' – and even then it is often ineffectual. I believe this is a crude and inaccurate account of what Parliament does and is seriously deficient in its account of the policy process.
>
> (Brand 1992: 341)

In some ways this represents a misunderstanding of the policy community concept, for certainly MPs may be recognised members of a particular community in certain circumstances, and even more so of issue networks. None the less Brand is probably right to claim that Parliament's role was seriously under-emphasised in much of the literature in the 1970s and 1980s. The claim by Richardson and Jordan (1979: 121) that 'the significance of Parliament is its very insignificance' certainly seems somewhat overstated. Brand rests his case on a detailed examination of the role of the parliamentary parties in six areas of policy, one of which is agriculture. His argument is that Parliament continues to play an important role in legitimating policy. Cabinet ministers, and by extension civil servants and, indeed, pressure groups, cannot afford to ignore the likely response of Members of Parliament to their legislative intentions. Evidence of their importance, argues Brand persuasively, is to be seen in the significance which pressure groups themselves attach to the influence they have with individual MPs. The powers of patronage exercised by a Prime Minister can be seen as evidence of prime ministerial power but the reverse can also be the case. Back-bench MPs can radically limit such power if their support is withdrawn, as Mrs Thatcher found to her cost in 1990 and John Major in the difficulties he experienced in 1994 and 1995. Brand is careful not to overstate his case. He acknowledges that policy communities are crucial to the policy-making process but claims, in effect, that these communities can only function with the tacit approval of MPs, who can withdraw such approval in certain circumstances:

> There is no question of back-benchers contributing to the ordinary making of policy, and this is what the concept of a policy community refers to. When these same MPs become involved in the debate on policy, however, they have the potential to make the most fundamental decisions. The occasions are rare since members of the same party and close colleagues in Parliament will share the same ideas and the discipline enforced by the adversarial system will predispose back-benchers towards loyalty. Ambition for preferment will have the same

effect. There come occasions when other considerations, such as a threat to survival at the next election or the violation of an important party principle, will take precedence, and when this happens, the back-benchers of the government party are the final arbiters

(Brand 1992: 347)

Brand also highlights the role MPs can play at times of policy shift, when the easy relationship between actors within a policy community is disturbed. For example, the salmonella in eggs issue in 1988 could not be contained within the agricultural policy community. As MAFF and the Department of Health, at ministerial and Civil Service level, appeared to be locked in combat, so back-bench MPs asserted their own views; one of the results being a ministerial resignation (Doig 1989). It is important to recognise, moreover, that party discipline is essential to the model of party government that has operated in Britain in recent decades. Without this, prime ministerial or cabinet government, dependent as they are on the support of Parliament, would be very different indeed. At times factionalism within parties may supersede party unity and loyalty. This was certainly the case after the 1906 General Election when the Opposition Conservatives were split into three distinct camps on the question of tariff reform, with the minority of free traders therefore being closer to the Liberal government on a key policy issue than to others in their own party. If this happens to a party in government the consequences can be dramatic, as experienced by the Labour Party under Ramsay MacDonald in the 1930s. Such pronounced factionalism is perhaps less likely today as a result of the strengthening of party discipline and the increasing importance of parties in the formation of governments in the late twentieth century. None the less, the divisions in the Conservative Party in the 1990s on the question of Europe, which greatly contributed to the downfall of Margaret Thatcher and threatened her successor John Major, are certainly reminiscent of such earlier factionalism.

At a more mundane level, there are several ways in which minor policy changes can be initiated through the involvement of ordinary MPs in the parliamentary process. Another commentator recently to have attempted a positive reassessment of the role of Parliament has given, in rank order, ten main 'normal' functions of Parliament of this nature (Norton 1993: 203; Packenham 1970), as shown in Figure 2.2.

Scrutiny

Numbers 3–7 in Figure 2.2 all depend to some extent on the scrutiny functions of Parliament. Crucial to this are the opportunities offered by the legislative process itself, the basic details of which are shown in Table 2.1.

The first reading of a bill is a formality and even the second reading debate normally takes only a day. Only three times this century has a government lost a bill at second reading, most recently the Shops Bill (Sunday trading) in 1986. By contrast the committee stage of a bill offers at least some

1 Manifest legitimation	formal seal of approval to all legislation; confidence votes
2 Recruiting, socializing and training ministers	
3 Latent legitimation	scrutiny of ministers in the house; work of select committees
4 Acting as a safety valve 5 Achieving a redress of grievance 6 Interest articulation	contributes to support for Parliament and its legitimacy
7 Administrative oversight	miscellaneous means of scrutiny
8 Law making 9 Mobilizing and educating citizens 10 Conflict resolution	

Figure 2.2 The functions of Parliament

Source: after Norton 1993

opportunity for the expression of back-bencher opinion in a manner which can induce the government to accept changes to the legislation. It is much more likely to respond to pressure from its own side than to Opposition demands; it is rare for government to be defeated in committee. It is usual for a standing committee to be appointed afresh for each bill and to comprise between 16 and 50 members. The bill is considered clause by clause, presenting much opportunity for pressure group lobbying of the committee members. The report stage gives the government the opportunity to introduce changes agreed to in committee. It is important to note that bills often contain powers for detailed regulations to be made once the bill is enacted. Such regulations or statutory instruments may or may not require the formal approval of the House. Most bills are government-sponsored bills originating from particular departments or ministries, and most such bills are successfully passed. Between 1987 and 1992, of 213 such bills only 11 failed, compared to 519 failures out of 584 private members' bills (Norton 1993: 55).

The functions of the House of Lords broadly mirror those of the Commons, although its powers are considerably less than those of the Commons, with only limited rights of veto remaining, for example of private bills or statutory instruments (Shell 1992). Party discipline is far less tight in the House of Lords than in the Commons so that revisions during committee stage are not infrequent. The Lords is by no means powerless and is not, as sometimes implied, merely a rubber-stamping chamber.

The work of Parliament is not confined to legislative matters. There are,

Table 2.1 Legislative stages in the House of Commons

Stage	Where taken	Comments
First Reading	On floor of the House	Formal introduction only. No debate
Second Reading	On floor of the House (non-contentious Bills may be referred to a Second Reading Committee)	Debate on the principle of the measure
(Money resolution)	(On floor of the House)	
Committee	Standing committee (constitutional and certain other measures may be taken in Committee of the Whole House)	Considered clause by clause. Amendments can be made
Report	On floor of the House (no Report stage if Bill reported unamended from Committee of the Whole House)	Reported to the House by the Committee. Amendments can be made
Third Reading	On floor of the House	Final approval of the Bill. Debate confined to its content
Lords amendments	On floor of the House	Any amendments made by the House of Lords considered, usually on motion to agree or disagree with them

Source: Norton 1991b: 330

of course, many other ways in which opinions can be expressed. Questions and motions provide opportunities for debate and sharp exchanges on policy matters. Rarely do they directly bring about a fundamental policy shift, but the mood of the House is occasionally such as to contribute to policy amendments. Early-day motions are expressions of concern signed by a small or large number of MPs. They virtually never lead to an early debate as such, but rather are a device for signalling areas of policy concern. Very occasionally they lead to government action, as when, during the passage of the 1988 Education Reform Bill, Michael Heseltine and Nicholas Ridley tabled an EDM demanding the abolition of the Inner London Education Authority. Another avenue open to MPs, often acting through an EDM, is to urge government to set up a royal commission or a committee of inquiry on a particular issue.

Another important aspect of the scrutiny role is the work of the various committees. These can be divided into three main types: party committees, all-party groups and select committees. The first two are often unsuccessful in influencing policy directly; nor do they occupy a formal place within the conduct of parliamentary business (J.B. Jones 1990). Their role is to provide a focus for back-benchers to reveal their particular interests and to develop levels of knowledge and expertise that may help them to be more effective in debate and in select committees, or to attract attention to themselves as potential members of the government. However, on occasions they can provide a considerable and influential policy input. For example, the Conservative parliamentary party's Agricultural Committee exercised a considerable influence on Conservative policy in the post-war period, especially after the Conservatives resumed power in 1951 (Flynn 1989).

Select committees occupy a potentially significant role, in that they are established by the House itself and have a specified function to perform. For most of the period before the 1960s there were only two select committees, Public Accounts and Estimates. More were established in the 1960s but most failed for lack of power, resources and purpose. The Expenditure Committee was seen as quite effective, though, and provided a stimulus for a fresh attempt to introduce a wide-ranging select committee system. Thus, in 1978 the Select Committee on Procedure recommended the appointment of a series of select committees with wide terms of reference and the power and

Table 2.2 Departmental and key policy select committees,* April 1994

House of Commons	House of Lords
Agriculture	Broadcasting
Broadcasting	
Defence	European Communities
Education	+ sub-committees as follows:
Employment	A. Finance, Trade and External Relations
Environment	B. Energy, Industry, Transport
European Legislation	C. Environment and Social Affairs
Foreign Affairs	D. Agriculture and Food
Health	E. Law and Institutions
Home Affairs	
National Heritage	Science and Technology
Northern Ireland Affairs	+ sub-committees as follows:
Public Accounts	A. International Investment in UK Science
Science and Technology	B. Defence Research Agency
Scottish Affairs	
Social Security	Sustainable Development
Statutory Instruments	
Trade and Industry	
Transport	
Treasury and Civil Service	
Welsh Affairs	

* i.e. excluding a number of committees concerned with parliamentary procedures and other administrative matters
Source: Vacher's Parliamentary Companion

resources to appoint specialist advisers. Fourteen new departmental select committees began in 1980 (this had increased to sixteen in line with departmental changes by 1992) and are now an important feature of policy debate. There are other non-departmental committees and in addition some influential House of Lords committees (see Table 2.2). For example, the House of Lords European Communities Committee deals with many agricultural issues and has a high reputation in Whitehall, Westminster and Brussels for its scrutiny work.

As ministers, even junior ones, are not members of select committees, the committees do give an opportunity for ordinary members to have some influence on policy debate and for pluralist pressure groups to be heard (all evidence is published) within a parliamentary forum. In an examination of the first years of the Agriculture Committee, Giddings concluded that its deliberations, for example, on farm animal welfare may have 'moved it up a notch or two in the Minister's priorities' (Giddings 1985: 63). Subsequent analysis of the committees, whilst recognising their important role within a wider public debate and the potential significance for democratic processes, suggests that their direct impact on policy remains slight (Drewry 1989; Hawes 1992; Jogerst 1991; Judge 1990 and 1992).

THE STRUCTURE OF GOVERNMENT

We turn now to a consideration of the departments within government itself, looking beyond the major institutional arrangements of the core executive that we have already considered. This is a complex area of investigation and one that can only be touched upon in this section, which attempts to give a general indication of the division of responsibilities within UK government. A measure of the complexity of modern government can perhaps be gauged by the length of Peter Hennessy's seminal examination of the Whitehall system and government departments, which comprises no less than 850 pages of text (Hennessy 1990). The standard text-book on sub-central government in Britain runs to 450 pages (Rhodes 1988). There have been many attempts to classify the range of national and local government agencies and we adopt here a classification developed by Hennessy (1990) who distinguishes between central, overseas, social, economic, territorial, and other. His classification is used in Table 2.3 which lists the departments comprising central government, giving details of expenditure and, hence, an indication of their importance within the total framework of government. It is important to stress that this is only an indication; in particular, the picture is somewhat confused by the fact that some functions are the responsibility of local government and therefore either funded directly at a local level or 'lost' within the DoE. It is misleading, for example, to assume that education expenditure represents only 4 per cent of public expenditure as this is one of the main responsibilities of local government.

Table 2.3 Main government departments in UK and expenditure

Department	Forecast expenditure 1992–93 £ billion	% of total
Central		
Cabinet Office	1.5	0.7
Treasury	4.9	2.5
Overseas and Defence		
Foreign and Commonwealth Office	3.4	1.8
Ministry of Defence	24.0	12.6
Social		
Department of Health	25.2	13.2
Department of Social Security	36.5	19.1
Department of Education	7.4	3.9
Economic		
Department of Trade and Industry	3.1	1.6
Department of Employment	3.0	1.6
Department of Transport	5.3	2.8
Department of the Environment	45.1	23.7
Ministry of Agriculture, Fisheries and Food	1.0	0.5
Territorial		
Scottish Office (and Forestry Commission)	10.5	5.5
Welsh Office	6.0	3.1
Northern Ireland Office	3.1	1.6
Other		
Home Office	5.7	2.9
Department of National Heritage	2.6	1.3
Lord Chancellor's and Law Officers' Department	2.3	1.2
Total	190.6	

Source: HM Treasury 1993

Functional departments

These refer to the government departments covering the major policy sectors, such as transport, education, agriculture, and so forth. Table 2.3 sub-divides these according to their main function following the divisions used by Hennessy (1990: 381–382). Some ministries are single-sector departments; others have multiple functions. For those concerned with rural policy the Department of the Environment represents a particularly confusing picture as it has responsibility for both housing and local government, as well as for environmental protection. There are, of course, many issues which cut across departmental boundaries. For example, MAFF includes food within its brief but health matters fall within the purview of the Department of Health; there are, inevitably, grey areas when issues of food and health are under consideration. There are also examples of cross-departmental responsibility.

For example, agricultural research is funded by MAFF, the territorial departments, the Office of Science and Technology and the Department of Education, with several other departments funding research with significance for agriculture. Some relatively modern areas of concern have never fitted easily within traditional departmental boundaries. Tourism, for instance, originally came under the aegis of the Department of Trade and Industry (DTI) when the Tourist Boards were established in 1969, was later transferred to the Department of Employment, and in 1992 was switched again to the new Department of National Heritage (DNH). The DNH also has responsibilities for sport and general recreation. Countryside recreation, however, is the concern of the Countryside Commission, under the aegis of the DoE. The DoE is also responsible for the Rural Development Commission, which includes some aspects of rural tourism within its remit.

For our purposes the Ministry of Agriculture, Fisheries and Food is one of the most important of government departments. The Ministry of Agriculture and Fisheries, as it then was, was founded in 1919 in succession to the Board of Agriculture. It has been a central government department ever since, the additional responsibility for food accruing to the ministry in 1955. Currently (April 1994) the ministry parliamentary team comprises the Minister in Cabinet and a Minister of State (both from the Commons) each of whom has a Parliamentary Private Secretary and two Parliamentary Secretaries (one from the Lords and one from the Commons). In addition there are two other government spokespersons in the House of Lords. However, it should not be forgotten that further responsibilities for agriculture are also held by the three territorial offices for Wales, Scotland and Northern Ireland. Thus in Cabinet there are three further Secretaries of State with direct responsibility for agriculture in addition to the Minister of Agriculture.

Figure 2.3 shows the central structure of the Ministry of Agriculture, at sub-ministerial level. The five deputy secretaries and twenty-one under-secretaries are all based in London and employ between them a further 7,500 London-based and regional staff. A further 4,000 are employed by the Intervention Board and the MAFF agencies (1994–1995 figures).

In addition to the central organisation in London, MAFF now has ten regional service centres in England, established in April 1992 to replace five regional offices and nineteen divisional offices. Their main responsibility is to handle the administration of grants, subsidies, licensing and other services. Each regional service centre has a panel:

> The panels are non-statutory bodies appointed by the Minister whose membership includes farmers, landowners, farmworkers, representatives of the food industry, environmentalists and consumers. Their main functions are to help maintain a communication link between the Ministry and its customers and to advise Ministers on the impact of policies at the local level.
>
> (MAFF 1993a: 3)

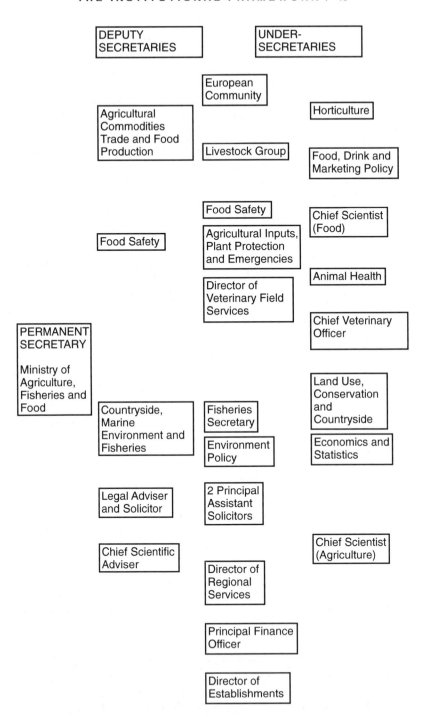

Figure 2.3 Ministry of Agriculture, Fisheries and Food, organisation chart, 1994

Source: *Vacher's Parliamentary Companion*

It says much about the centralism of the British political system that such regional panels serve the minister, rather than providing for any degree of regional or local accountability. Moreover, their activities are neither widely publicised nor well known.

In addition to the work carried out by core MAFF staff, much of the Ministry's work is sub-contracted to four agencies: the Veterinary Medicines Directorate (VMD) and the Central Veterinary Laboratory (CVL), both of which have been agencies since April 1990; and the Agricultural Development and Advisory Service (ADAS) and the Central Science Laboratory (CSL) which became agencies in April 1992. The VMD operates across the UK and is responsible to the agriculture and health ministers in all four countries; the CSL is responsible to MAFF; ADAS and the CVL to MAFF and the Welsh Office. Two further agencies are the Pesticides Safety Directorate (1993) and the Meat Hygiene Service (1995). The Agricultural Training Board, formerly under the Ministry's jurisdiction, was re-launched in 1994 as an independent company and registered charity (ATB-Landbase), although it is still heavily dependent upon MAFF funding.

Of the agencies, ADAS (responsible for consultancy and advisory services for farmers and a substantial research facility) is far the most significant to our purposes, employing some 2,500 staff in 1994. Notwithstanding the commercial objectives of the new agency, more than half of its income continues to be provided by central government (58 per cent in 1994). Regional ADAS offices remain housed in government buildings alongside MAFF regional staff, so that the agency status is easy to lose sight of 'on the ground'. A fifth agency, the Intervention Board, is a separate government body accountable to the four agriculture ministers of the UK.

Territorial ministries

These refer to the special arrangements that exist for the governance of Northern Ireland, Scotland and Wales, whereby several government functions are brought together within a single ministry exercising responsibilities for a particular territory. Thus, the Secretary of State for Wales is responsible for a diverse range of policies within the principality – agriculture, environment, industry and energy (and more besides) all coming together under a single ministry. MAFF's functions are assumed by the Scottish Office Agriculture and Fisheries Department (SOAFD), the Welsh Office Agriculture Department (WOAD) and the Department of Agriculture for Northern Ireland (DANI).

Inevitably, in practice many of the policies reflect very closely those developed for England by the functional departments. For example, mainstream agricultural policy varies very little between the four countries and is primarily developed by civil servants in the European Commission and in MAFF. Occasionally the reverse arrangement occurs as, for example, with forestry where the Scottish Office is the lead ministry developing policies with implications for the English MAFF. However it would be a mistake to see the territorial ministries merely as another form of functional department: they also adapt UK policies for implementation in the periphery; promote policies in response to national differences and needs; and lobby for resources not only for particular services but for the nation as a whole (Rhodes 1988: 143–144). The ability of territorial departments to promote successfully national/regional interests varies from sector to sector. Nor are the conditions identical in each of the three countries. For example in the case of agriculture, DANI and SOAFD have responsibilities for agricultural education which do not fall to either MAFF or WOAD. DANI and SOAFD are also important players in agricultural research, whereas WOAD's direct responsibilities for agricultural research are minimal.

Intermediate institutions

Functional departments require local or regional offices in order to function effectively, primarily through the provision of information and the implementation of central policies. In some cases, important decisions on resource allocation are made at this level, and some discretion in policy implementation is possible. However such possibilities are limited in the UK and it is not possible to detect a long-term trend towards the allocation of greater responsibility to the regional offices of central government. This is partly because of the sheer complexity of the arrangements, which make consistent regional political involvement and lobbying almost impossible. In the words of Rhodes, 'the resulting organisational pattern can be described as a maze only at the cost of some simplification' (Rhodes 1988: 153). Not only is there a wide variation in the number of regional intermediate institutions from department to department, but the boundaries show little consistency. So much so that even the Department of Environment, which was responsible for the Regional Economic Planning Councils, abolished in 1979, and then interdepartmental Regional Boards, itself never adopted the eight standard regional boundaries for England which the REPCs and RBs urged all departments to use. Indeed the DoE used different boundaries for each of its eight chief functions (Rhodes 1988) until integrated regional offices were established in 1994.

Non-departmental public bodies

Non-departmental public bodies provide an equally complex range of institutions encompassing public corporations, such as nationalised industries, various executive and advisory bodies, tribunals, and quangos (Rhodes 1988). All are related to a particular central department but are independent of that department in key respects. The independence is likely to be even greater at regional and local level, especially where a non-departmental public body has come to be responsible for a very significant aspect of government policy. This is the case with English Nature, the Countryside Commission, the Countryside Council for Wales, and Scottish Natural

Heritage whose decisive roles in the implementation of many aspects of the government's countryside policy are undertaken in relative isolation from the sponsoring departments. Depending on definition, these agencies can be seen as quangos, 'an area of modern government, which though murky and hard to define, is where much of the political and administrative "action" now undoubtedly takes place' (Hood 1982: 44). As Hennessy (1990: 440) points out, the DoE is the 'classic quangoid department operating through a network of statutory bodies, some of which are big spenders and employers'. In 1992, several of these formed the basis for the new Department of National Heritage, but many remain.

Agriculture is not without its non-departmental public bodies too. Amongst the most significant for any understanding of public policy in the agricultural sector are the Biotechnology and Biological Sciences Research Council (until 1994 the Agriculture and Food Research Council), the Agricultural Wages Board (AWB), and Food from Britain. Other bodies of note are the Agricultural Mortgage Corporation (AMC) and various marketing boards and commissions.

Local government

It is impossible in the space available here to do anything more than to mention briefly some of the functions of local authorities, and to point out that local government is not a unified system covering the whole of the UK. Budge *et al.* (1983) identifies eleven different local government systems covering the entire territory of the UK. Whilst some of these are of minor significance, such as the special arrangements for the Isles of Scilly, even the single most ubiquitous system, the triple structure of directly elected local government in non-metropolitan England, accounted in 1983 for just 49 per cent of the UK population.

Many local government functions are relevant to rural policy. The town and country planning system, operated through local planning authorities, is clearly of vital importance in terms of protecting rural land from many forms of development. Whilst local authorities have limited direct responsibilities for agriculture, county councils were, until 1992, responsible for agricultural education. And an increasing number of county councils and, to a lesser extent, district councils are taking an interest in wider countryside policies, including the provision of farm conservation advice to farmers (Winter 1995).

THE EUROPEAN DIMENSION

Finally, in this chapter we need to consider the European policy dimension. Again this is a large subject to be dealt with in just a few pages and much fuller discussions are available elsewhere (for example the excellent introduction by Nicoll and Salmon 1990). However it is important to indicate the nature and context of the European policy process, not least because so much

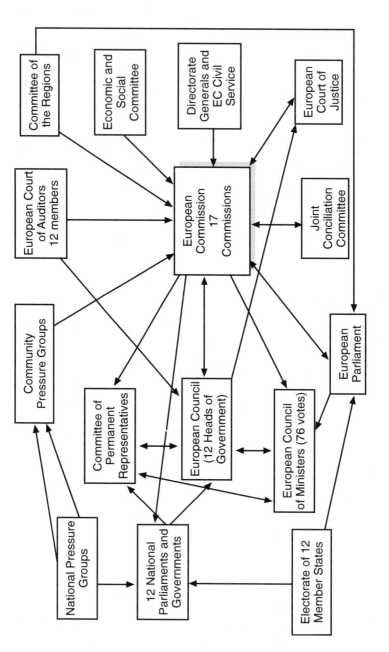

Figure 2.4 European policy making

of rural policy now derives from European legislation and policy initiatives. Figure 2.4 shows in diagrammatic form the structure of governance and policy making within the Community. The three main bodies are the European Commission, the Council, and the European Parliament.

The Commission's position is central to the European policy process, for although the Council has the final say in policy and has formally to approve all Commission decisions, it is the Commission which initiates policy and makes crucial decisions on how Council decisions are to be implemented. The Commission formulates either regulations (binding on all member states), directives (binding as to results but not means), or recommendations and opinions which are not binding, and is responsible for their enforcement. The Commission is led by seventeen Commissioners, who are, in effect, national political appointees. Each is assisted by a small group of specialist advisers, known as a Cabinet. But the more substantive policy work is conducted by the twenty-three Directorates General (DGs). The Commission is subject to many influences and to a number of in-built checks and balances. The most recent innovation, under the Maastricht Treaty, is the Committee of the Regions, a new advisory body set up to provide an input to the Commission representing the views and aspirations of various regions within the Community constituted at sub-national level.

The Council of Ministers is the prime legislature and comprises representatives of the governments of the member states. Formally there is only one Council but in reality its composition varies according to the topic under discussion so that, in effect, there are separate Councils for different topics such as agriculture.

The European Parliament, at first sight, has an impressive list of powers:

(i) it has extensive budget responsibilities; (ii) it can dismiss the Commission (i.e. the Commissioners) by a two-thirds majority of the votes; (iii) it may pose written and oral questions to the Commission and may call it to account in plenary or in committee debates; (iv) it can pass resolutions; (v) it is called upon to deliver opinions on draft legislation of the Commission.

(Meester and Van der Zee 1993: 140)

However, in reality its powers have been very weak in contrast to those of either the Council or the Commission. Some of this weakness is set to change under the terms of the Treaty on European Union, the Maastricht Treaty of 1991, which introduced major innovations and changes to the system of governance within the Community. The innovations include the extension of European powers to cover foreign and security policy, and home affairs and justice policy. These policies are to be agreed on an intergovernmental basis not, on the whole, through the existing institutions of the Community as outlined above. The Parliament has new powers of co-decision with the Council over a wide range of issues, including internal market legislation, environmental policy and consumer protection. This is an effort to increase democratic accountability within the Community and to reduce the subordination of the Parliament to the Council of Ministers. If agreement cannot be reached by the Parliament and Council, then a joint Conciliation Committee will be brought into play. In effect the Parliament will have a right of veto over those areas covered by this new approach. The Parliament is also to be given powers to veto the appointment of commissioners.

As these changes come to be fully implemented, the long-term consequences for the European policy process will be significant. With regard to rural policy, both the new Committee of the Regions and the enhanced role of the European Parliament will have implications for how rural policies are formulated and the manner in which the various interest groups will be obliged to operate. However it is possible to overestimate the importance of these changes not least because agricultural policy is such a complex matter that the policy community, at the heart of which lies the Commission, will remain hard for other groups and influences to penetrate.

Turning briefly to the example of agriculture, so as to consider the policy process in just a little more detail, DG VI, the Agriculture Directorate is one of the largest and most powerful of the twenty-three DGs. All its proposals are vetted by the Commission's Legal Service. In addition to approval by the Commissioner responsible for agricultural matters, a proposal must be approved by a majority of the other Commissioners. Then it is submitted to the Council of Ministers, where it is likely to be dealt with by the Agriculture Council (comprising national agriculture ministers), initially through its sub-committee, the Special Committee on Agriculture which consists of permanent officials from the member states. Only in instances where other areas of policy are involved is a proposal sent to the Committee of Permanent Representatives which is, except in agricultural matters, the Council's most important committee.

The Council is required to consult the European Parliament, which has its own Agricultural Committee; but the Parliament's influence has been severely constrained by its narrow budgetary powers. On most matters of CAP finance the European Parliament can only propose modifications and the Council has the final say. The Parliament is also constrained by Article 155 which provides for the Commission to set up management committees for specific agricultural products under powers conferred by Council; many of the details of legislation are dealt with at this level.

Contrasting the European and UK systems of governance

There are many important differences between the EC and British systems of policy making. For example, the EC system is inspired more by 'federal' notions than by the example of the British 'unitary state':

> The EC explicitly rests on a division of powers between the EC and member state levels. ... The EC also separates power between different

institutions on lines broadly analogous to those found in the American constitution, with the consequential tensions and conflicts between institutions that are typical of American politics. None of this is to suggest that the EC is yet or ever will necessarily be a federal system, but rather to make it clear that its institutional and political style is easier to understand in American terms than in conventional British terms.

(Wallace 1990: 153–154)

In contrast to America and to Britain, though, the Community was designed by politicians accustomed to coalition politics and to corporatist arrangements, whereby the interests of labour and employers are consulted and incorporated into decision making. This has not resulted in a developed closed corporatist system as such (Sargent 1985), but it has contributed to an open political system in which it is expected that interests will be articulated at various levels. The role of the Economic and Social Committee is the closest there is to a formal corporatist structure, the like of which does not exist in the UK, although the National Economic Development Office had some similarities in the 1960s and 1970s. The Economic and Social Committee has a formal advisory role and is representative of employers, workers and independents, appointed by the European Council on the basis of member states' nominations. Thus the European political system certainly appears to owe some of its structural characteristics to corporatist experience, even if it is not yet strongly corporatist in practice:

The power structure is diffuse and allows multiple opportunities for the exercise of influence and counter-influence. The process of building coalitions and log-rolling prevail rather than clear-cut ideological competition, which tends to be buried under the technical detail of legislative proposals.

(Wallace 1990: 156)

Another contrast between Britain and the EC system highlighted by Wallace (1990) is that lines between politicians and senior officials are much more blurred. Officials may publicly declare their political opinions and can stand for election. It is common for senior Commission staff to be recruited from the ranks of national politicians, and thus commissioners are, in a sense, political civil servants. Leon Brittan, a former cabinet minister under Mrs Thatcher, is an example. Moreover, in direct contrast to the UK system, it is the officials of the Commission who have had the sole right to initiate policy. Implementation of directives and regulations is much aided by the judgements made by the European Court of Justice which through its creative interpretation of law acts as a major creative force in policy making. There is no such parallel organisation in Britain. These comparisons of course beg the question as to how far national policy making is now superseded by European policy processes. Few commentators doubt that there is a process of Europeification (Andersen and Eliassen 1993), but it still has a long way to run:

> Decision-making is becoming increasingly Europeified in the sense that what happens at the EC level, now penetrates more and more areas of national policy-making. The member states are not, however, inclined to give up their central position.... National representatives sit on the committees which control the formation of EC policy.
>
> (From and Stava 1993: 55)

Even more importantly, considerable national discretion remains in many areas of policy making and implementation. Examples range from the UK's withdrawal from Economic Monetary Union and its decision not to implement the Social Chapter of the Maastricht Treaty, to numerous, indeed constant, less sensational decisions about how to implement directives nationally.

CONCLUSIONS

This chapter has focused on the key institutions and agencies relevant to the formulation of rural policy. It has demonstrated something of the complexity of the modern nation state. In the sections of the book that follow a number of the institutions and agencies mentioned here will be referred to as we examine in more detail how certain areas of policy have developed over time. But before commencing this detailed examination of agricultural and environmental policy, one further area of the policy process has to be analysed. We have referred once or twice, in passing, to issues of policy implementation and evaluation, and it is to that final stage of the policy process we must now turn in the next chapter.

3

POLICY IMPLEMENTATION AND EVALUATION

INTRODUCTION

This chapter tackles two important, but sometimes neglected, issues – the implementation and the evaluation of policy. Policy implementation is important for two main reasons. First, as indicated strongly in chapter 1, the policy process does not end with the delivery of a policy package which can be administered in a non-problematic and neutral way. Almost inevitably any policy output, whether it be legislation or grant schemes, policy guidance notes or court rulings, has to be interpreted by those charged with its implementation. The interpretation of policy is a necessary part of the policy process, one in which all manner of influence can be brought to bear. The decisions that have to be taken during the implementation stage are policy decisions and are as much a part of the policy process as the drafting of legislation or the formulation of policy statements.

Second, even when interpretation is relatively straightforward, policy may not always be implemented as those promoting the policy expect. Policy implementation is a challenging issue because of widespread perceptions, both lay and academic, that many policies and legislative provisions are not adequately implemented. Hogwood and Gunn (1984) cite two main instances of this aspect of policy failure: first, non-implementation, where policy is not put into effect at all; and second, unsuccessful implementation, where policy is carried out but fails to produce the desired outcome. In both cases the gap between the intentions of policy makers and the outcome may be referred to as *implementation deficit* (Weale 1992), *implementation deficiency* (From and Stava 1993) or the *implementation gap* (Marsh and Rhodes 1992a). The first section of the chapter examines this issue. The remainder of the chapter is devoted to policy evaluation. We consider evaluation from a two-fold perspective. First, there is a need to understand something of the politics of evaluation itself and why evaluation has become important, so much so that the 'rise of the evaluative state' has been coined as a term to describe recent trends (Neave 1988). Second, we turn to a consideration of the practical application of evaluation.

IMPLEMENTATION DEFICIT

It is a familiar predicament with human craftsmen, to be let down by the material in which they work; the clay crumbles under the modeller's hand, or it cracks in the heat of his oven. Nature, like the craftsman, seems to find her whole positive tendency in imposing a rhythm or a shape; only the shape does not always take, nor the rhythm always fulfil itself.

(Farrer 1994: 38)

The Anglican theologian Austin Farrer had profound issues of divine providence and evil in mind when he penned these words, but the words might equally well serve as a metaphor for the rather more mundane question of policy implementation. The intentions of those devising policy are not always easily translated into effective actions on the ground where the gap between the ideal and reality is often glaringly obvious. In examining this issue Weale draws a distinction between policy *outputs*, that is, 'the product of government activity in the form of regulations, laws, inspections and procedures' (Weale 1992: 154), and policy *outcomes*, that is, material changes that actually take place as a consequence of the policy outputs. The gap between output and outcome is the essence of the problem of implementation deficit. This can be seen, and is usually considered, at the national level, but it is worth noting at this point that implementation deficit is increasingly perceived as a problem at the European level due to the 'absence of institutionalized interdependencies between the decision-making level (the EC) and the implementing level (the member states)' (From and Stava 1993: 58). A helpful way to consider the issue is to examine a number of important preconditions for perfect policy implementation as set out by Hogwood and Gunn (1984: 199–206). These are considered in turn below, using some rural policy examples to illustrate the points being made.

That the circumstances external to the implementing agency do not impose crippling constraints

In the words of Hogwood and Gunn, 'some obstacles to implementation are outside the control of the administrators because they are external to the policy and the implementing agency' (1984: 199). For example, agricultural policies the world over have, at different times, faced intractable problems of implementation due to inclement weather, disease and/or the problems of farm structure. Other constraints may include the existence of powerful vested interests which effectively veto policies. For example, land reform policies in many countries have faced implacable opposition from large landowners, often backed by military leaders.

> That adequate time and sufficient
> resources are made available to the
> programme; and that the required
> combination of resources is actually
> available

Even when major physical and political constraints are not present, some policies fail due to lack of money or unrealistic expectations about what can be achieved in a specified period of time. For example, the opportunity for local authorities and national parks to engage in management agreements with landholders regarding public access or conservation was set out in both the Town and Country Planning Act 1932 and the National Parks and Access to the Countryside Act 1949 (Blacksell and Gilg 1981). However the resources allocated to this purpose were for many years insufficient, and it was not until after the 1981 Wildlife and Countryside Act that management agreements became a major (if costly) conservation mechanism (in the hands of the Nature Conservancy Council). A very similar story can be told with regard to the Control of Pollution Act 1974 which coincided with cutbacks in public funding, preventing Regional Water Authorities from taking on the staff necessary to implement the Act. It was not until a new agency, the National Rivers Authority, was born some fifteen years later that some of the 1974 policies could be effectively implemented.

> That the policy to be implemented is
> based upon a valid theory of cause
> and effect; and that the relationship
> between cause and effect is direct
> with few, if any, intervening links

Often the cause and effect assumptions of policy makers are implicit and unstated, and if policies fail it may be because the policy makers had an inadequate understanding of the problems they confronted. There are many examples of inadequate analysis prior to policy formulation. For example, the identification of the Somerset Levels as a suitable candidate for Environmentally Sensitive Areas status in 1986 presupposed that the ecological interest of the area was entirely a product of a particular farming system and that, therefore, the support of a particular farm management regime would be sufficient to safeguard the ecological value of the area. It is now widely accepted that continued ecological deterioration is a consequence of the activities of Internal Drainage Boards, activities which are not checked in any formal way by ESA status. Other examples might include policies for rural economic revitalisation directed at farmers when other sections of the rural business community might be better placed to stimulate the rural economy; or policies aimed at freeing up agricultural land from planning restrictions to solve the rural housing problem with insufficient examination of complex employment and housing linkages.

That dependency relationships are minimal

Pressman and Wildavsky (1973) argue that the greater the number of decisions required by different actors at different points in the implementation process, the more likely there is to be a policy failure. Thus multiple dependency relations between different agencies can cause significant policy blockages. Hogwood and Gunn suggest that a single implementing agency not dependent upon other agencies is required, whereas in fact,

> it is nowadays relatively rare for implementation of a public programme to involve only a government department on the one hand and a group of affected citizens on the other. Instead there is likely to be an intervening network of local authorities, boards and commissions, voluntary associations, and organised groups.
>
> (Hogwood and Gunn 1984: 203)

Here the contrast between agricultural policy, with a single ministry and an easily identified client group, and more diffuse rural development or environmental policies is striking. Rural development initiatives involving more than one government department, two or three tiers of local government and a diverse business community present considerable dependency relationship problems of the type described and help to explain some of the difficulties that have been experienced by agencies such as the Rural Development Commission or innovative county councils attempting to take a lead in this area over the last decade or so. Another good example of the complexity of administrative arrangements, particularly the proliferation of consultative arrangements, surrounds land use planning issues in designated areas such as the National Parks. In the absence of effective planning controls over forestry and agricultural developments in the parks, but in the light of the conservation policy goals that have been set, a complex system of dependency relations has developed involving a number of agencies and interest groups.

That there is understanding of, and agreement on, objectives

It is not uncommon for policies to lack clear objectives. Indeed rather the reverse:

> most research studies suggest that, in real life, the objectives of organisations or programmes are often difficult to identify or are couched in vague and evasive terms. Even 'official' objectives, where they exist, may not be compatible with one another, and the possibility of conflict or confusion is increased when professional or other groups proliferate their own 'unofficial' goals within a programme.
>
> (Hogwood and Gunn 1984: 204)

A good example of this sort of confusion, although not a topic central to the concerns of this book, is with regard to rural housing. After the relative policy clarity of public authority housing programmes, the transition brought about by the down-turn in council house building and sale of council houses after 1979 led to much concern about the housing situation in rural areas. Recognition of this concern led to an increased policy emphasis upon housing association provision. However the aims of government at national or local level have never been made clear. Definitional problems over such terms as 'local needs housing', 'social housing', and 'low-cost housing' mean that policy objectives are almost impossible to determine, still less to implement (for further discussion see Shucksmith 1990).

That tasks are fully specified in correct sequence

This precondition has to do with appropriate procedures for implementation, without which apparently feasible policies may flounder. The example of the Control of Pollution Act 1974 can again be used. The procedures for implementing the provisions of the Act with regard to agricultural pollution of water were not put in place for some years and required additional statutory provision.

That there is perfect communication and co-ordination

Policy objectives may be misunderstood by implementing agencies or by recipients, especially where they are not communicated well in the first instance. Like all of the preconditions examined here, perfect communication and co-ordination are unattainable. To some commentators, recognition of this has become the basis for a fundamental critique of the 'top-down' approach to policy implementation (e.g. Barrett and Fudge 1981; Barrett and Hill 1984). In other words the fundamental distinction between policy formulation and policy implementation is flawed. In reality imperfect communication means that policy inevitably evolves through all its stages and so-called *implementation studies* must acknowledge that policy continues to change even when in the hands of recipients.

It is, of course, often difficult to determine whether it is imperfect communication and co-ordination or wilful misunderstanding by those responsible for policy implementation which is at the root of apparent dissonance between policy objectives and outcomes. But there is certainly a case for seeing the implementation of milk quotas within member states following the decision of the Council of Ministers in 1984 as an example of this kind of policy failure. The details of the quota regulations were not available until *after* some of the key decisions regarding implementation had to be taken nationally. This, coupled with the opaque nature of some of the rules, meant that implementation details inevitably varied from state to state more than the policy makers had intended. In the case of the UK, arrangements for quota trading developed which were arguably in breach of regulations and certainly contrary to the intentions of Commission officials.

That those in authority can demand and obtain perfect compliance

If communication difficulties present one reason for implementation deficit, another is subversion. Those in authority rarely obtain perfect compliance. Subversion and reinterpretation of policy may arise from resistance within the implementing agency, inter-agency rivalry or the responses of policy recipients. The more radical the policy, the more strenuous the resistance is likely to be, as Mrs Thatcher discovered with regard to the replacement of the local rates by the community charge or poll tax. A less dramatic example might be the low-key resistance within agencies forced to charge for their services and become more market orientated, as occurred in ADAS and the Rural Development Commission from the mid-1980s. Employees of these two organisations in the regions initially resisted the policy change, often tacitly, because it seemed fundamentally to alter their role, implying a transition from public servants to commercial consultants.

IMPLEMENTATION IN PERSPECTIVE

The preconditions for successful policy implementation may be hard to achieve, but policies almost invariably have some impact and identifiable results. Complete policy failure is as rare as perfect policy implementation, certainly in modern capitalist societies, so what then do we make of the notion of implementation? Probably the best way to deal with the implementation issue is to treat it as part of the policy process itself, as indicated in the opening chapter. Thus for Barrett and Fudge: 'It is essential to look at implementation not solely in terms of putting policy into effect, but also in terms of observing what actually happens or gets done and seeking to understand how and why' (Barrett and Fudge 1981: 12). Key words for Barrett and Fudge are interaction and negotiation. Their aim is to move away from a focus on policy implementation as the transmission of policy to a concern for policy–action relationships, in which it is difficult to identify a sequential implementation process as such:

> Ideas about negotiation and bargaining between actors and agencies involved in the policy process lead to a redefinition of 'implementation'. Policy cannot be regarded as a 'fix', but more as a series of intentions around which bargaining takes place and which may be modified as each set of actors attempts to negotiate to maximise its own interests and priorities.
>
> (Barrett and Fudge 1981: 24–25)

The political processes by which policy is mediated, negotiated and modified during its formulation and legitimation do not stop when initial policy decisions have been made, but continue to influence policy through the behaviour of those responsible for its implementation and those affected by policy acting to protect or enhance their own interests. This view of implementation takes us away from the traditional focus on formal organisational hierarchies, communication and control mechanisms, and places more emphasis on:

(i) the multiplicity of actors and agencies involved and the variety of linkages between them;
(ii) their value systems, interests, relative autonomies and power bases;
(iii) the interactions taking place between them – in particular negotiation and bargaining behaviour.

(Barrett and Hill 1984: 220)

In subsequent chapters we will see that the distinction between policy making and policy implementation is often blurred in the manner indicated here.

POLICY EVALUATION

Evaluation has been defined as follows:

evaluation is an activity involving the systematic application of social science theory, methods and techniques to identify and assess the processes and impacts of governmental policies and programmes. It may be conducted retrospectively or prospectively, in secret or in the public domain, by governmental or non-governmental organizations.

(Pollitt 1993: 353)

One of the consequences of the 1980s Civil Service reforms, referred to in the last chapter, was the increased emphasis given to policy evaluation by Conservative governments committed to scrutinising government functions. The introduction of next-steps agencies and contract relationships for the provision of services in many sectors only served to increase the importance attached to evaluation as a means of ensuring value for money and a degree of public accountability.

Thus, during the last decade, policy evaluation has become a major concern in the UK, as government has sought to ensure that policies are more cost-effective. Many specific rural policies, particularly discrete grant schemes, have been evaluated as a matter of course. The results of these evaluations have, in some situations, been used to redirect policy. For example, evaluation of the MAFF Farm Diversification Scheme showed that tourist accommodation, despite being the most popular option for farmers joining the FDS, provided a lower rate of return on capital investment than any other enterprise. It was accordingly dropped from the scheme. This example alone should suffice to show immediately that evaluation is not a neutral administrative activity. It is, potentially, very much part of the policy process with implications for the further development of policy. Sometimes policy evaluations are quite explicit about the possibility of redirecting policy. This is especially the case where it is possible to evaluate two or more alternative policies with broadly the same objectives. For example, in 1994–1995 the author was responsible for an evaluation for MAFF of free farm conservation advice, which involved an evaluative comparison of provision by ADAS and the Farming and Wildlife Advisory Group.

Despite the importance now attached to evaluation it has, in institutional and organisational terms, 'never found a secure or permanent home near the heart of a (relatively centralised) state machine' (Pollitt 1993: 354). Heath's experiment with an independent-minded unit close to the heart of government, the Central Policy Review Staff, was used little by Wilson and Callaghan and abolished by Thatcher (Campbell 1993; Pollitt 1993). Organisations such as the National Audit Office, and, in Europe, the Court of Auditors, have come to some prominence for their evaluative work but it is work conducted from a strongly financial auditing perspective. Moreover, there are few examples of swift and clear parliamentary and departmental responses to NAO's assessments of cost-effective policy (Carter et al. 1992). Pollitt's indictment of the UK's record on policy evaluation is damning:

> What the UK has never done is construct one or more broad-scope public policy evaluation units, empowered by a mission of producing for the public domain timely evaluation of both new policies and existing programmes.... [During the mid-1990s] ... most evaluations will be of a narrow scope, managerialist variety and will be carried out by a fragmented array of specialist units, agencies and contracted consultancies, many of which will possess neither sufficient visible independence nor any particular commitment to democratic (as opposed to managerial) needs. After thirteen years of 'rolling back the state', it still seems that, as far as the biggest policy questions are concerned, we are expected to believe that 'nanny knows best'.
>
> (Pollitt 1993: 360–361)

Thus it is that policy evaluations in the UK range from uncommissioned works by think-tanks, such as the Institute of Public Policy Research, often with inadequate access to information, through evaluations put out to contract (often to private consultancies or universities) by government departments themselves, to internal mechanisms for policy monitoring and evaluation. The government has pressed departments to introduce formal and systematic systems of policy evaluation. Figure 3.1 shows the approach suggested by the Treasury in a guide for Civil Service managers as part of the Financial Management Initiative (see p. 36).

The policy evaluation glossary published as an annex to this guide provides a striking example of the economics thrust of the government's chosen style of evaluation (see also the example of policy appraisal and the environment and the hegemony of financial valuations now applied to environmental quality in another government guide: Department of the Environment 1991). Financial terms such as *additionality, compliance costs,*

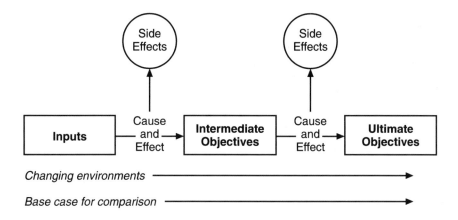

Figure 3.1 An evaluation framework

Source: HM Treasury 1988

cost-effectiveness analysis, externalities, gearing, investment appraisal and *unit costs* all denote a strong emphasis upon evaluation as a tool for improving the economic performance of government programmes and, indeed, upon the role of evaluation in introducing market disciplines to public sector operations.

However, in many ways this is a somewhat truncated form of evaluation for it assumes that the policy objectives are clear, and usually that they translate easily into financial terms. In practice, 'the transition from decision to action in the policy process is neither smooth nor obvious' (W.I. Jenkins 1978: 203). The issue becomes even more complicated when we consider that the aims and objectives of policies may be unclear, that they may partially succeed and partially fail, and that there may be unintended side-effects which negate the original stated aims. Different systems of policy evaluation have been devised which reflect the policy process approach recommended in this text. In the remainder of this chapter we will consider what form such evaluations might take. Table 3.1 shows the main stages of an alternative evaluation process. The stages are discussed using the example of the Farm Diversification Grant Scheme (FDGS). The first step is to establish broad policy goals, in this case revitalising rural economies, reducing surpluses and their cost, and supporting farm incomes. In establishing these broad goals we need to add to the list what might be termed the ulterior motives of government; for example, its perceived need to be seen to be genuinely concerned about an issue which, in reality, might be low in its priorities, or a desire to divert future criticism of declining support for agriculture.

Second, the specific objectives may be determined as grant aid for certain business ventures on farms. Targets for participation and level of spending may form a part of these objectives. Third, the identification of the agencies and their respective roles is important and is not as straightforward a matter

Table 3.1 Stages in retrospective policy evaluation

Task	Methods	Difficulties
Establish broad policy aims i.e. policy goals	Examine content of parliamentary acts and statutory instruments, EC directives, etc.; ministerial, governmental and EC statements; cabinet papers; interviews with key officials/politicians	Secrecy; conflicting goals; propaganda versus reality; ulterior motives; *post hoc* rationalisation
Identify specific objectives or desired outputs	As above with additional emphasis upon the policy refining carried out by government agencies	As above
Identify agencies and individuals responsible for implementation	As above with special emphasis upon interviews with key personnel	Multi-agency projects
Specify key stages in implementation process	As above with additional emphasis upon the experiences of recipients	*Post hoc* rationalisation can be a particular problem
Assess achievement	Define criteria for success; decide on means of measurement; collect data; analysis	Quality of the data; problem of 'other' influences; problem of side-effects

as it might at first appear. In the case of the FDGS, implementation might require the active involvement of MAFF, ADAS and local planning authorities. In addition, non-agricultural development agencies, such as the Rural Development Commission, might be involved.

The identification of key stages in policy implementation is a crucial procedure in evaluation, and extraordinarily it is one that is neglected in official programmes of evaluation. Five key elements, again designed with a scheme such as the FDGS in mind, deserve consideration:

1 Publicity
Did potential recipients know about the scheme?
Was there any bias in the publicity methods so that some recipients 'heard' before others?

2 Advice/extension
What advice, if any, was available to potential recipients? Was sufficient advice available?
Was the advice accurate or unduly optimistic or pessimistic?
Was there any bias in the advice so that some recipients were more or less likely to respond than others?

3 Participation in the scheme
Was the scheme easy to embark upon in terms of being administratively easy? Did financial hurdles exist which excluded target groups? Once started was the scheme managed and administered efficiently?

4 Monitoring and adaptation
What monitoring has taken place?
How and with what objectives?
Has the monitoring led to any modifications?
What do the modifications tell us about the changing nature of the policy, if anything?

5 End-point
Is there a logical end-point when the scheme is supposed to have attained its goal?

The fifth and final step is the assessment of outcome and achievements. In many policy evaluations this is far and away the most important stage and a number of techniques have been derived to consider this question. Five main categories of technique have been identified by Hogwood and Gunn (1984): before-and-after studies, modelling, experimental method, quasi-experimental method and retrospective cost-benefit analysis.

Before-and-after studies might appear to offer the most straightforward approach to assessing the achievements of a policy. In fact they present considerable difficulties. It is rare, for example, for policy makers to be far-sighted enough to establish the full information required prior to introducing a policy scheme. Even where information is available on circumstances prevailing prior to the introduction of a policy, difficult questions may remain, such as whether apparent policy achievements are exclusively the consequence of a particular policy and, if not, to what extent. Our example of the FDGS is a good one to consider in this context. No systematic national survey of diversification was carried out before the scheme although one has been undertaken since (McInerney and Turner 1991). The survey found that many farmers had diversified their operations using the FDGS, but it is very difficult to establish how many of these farmers would have done so anyway. Another problem with this approach is the issue of side-effects. A significant, if supplementary, aim of the FDGS was to stimulate the rural economy as a whole. Measuring this by before-and-after studies would have been very difficult and has not been attempted, yet it has to be accepted that while the FDGS might have achieved this aim it may also have had unintended side-effects on other non-farm businesses. For example, the increased competition arising from grant-aided farm accommodation may have resulted in a loss of business in the non-farm tourist accommodation sector.

In many ways *modelling* is a refinement of before-and-after studies. Models, mathematical or otherwise, attempt to incorporate systematically all factors which may have influenced a policy outcome. Lack of suitable data and erroneous assumptions about causal relationships provide a formidable challenge to those wishing to evaluate policy through modelling (Hogwood and Gunn 1984).

As the name implies, the *experimental method* relies on testing a policy on a particular group of people, ideally retaining a control group with identical characteristics to the experimental group on which the policy is not applied. The experimental method in policy evaluation works best where the influence of other factors or variables can be eliminated. Thus the method has some application in situations where groups of people can, to some extent, be isolated. Two factories run on broadly similar lines or two similar schools might offer opportunities for policy experimentation. However the heterogeneity of agricultural businesses and the diversity of rural land use present few opportunities for experimentation of this kind. Even where policies are applied in specific locations, for example Environmentally Sensitive Areas, it is usually precisely because these areas are so distinctive. Therefore any comparison with 'control' groups in other undesignated areas is likely to be a singularly fruitless exercise.

It is important to emphasise that the monitoring of a policy retrospectively, even with the use of control groups, cannot be construed as evaluation using the experimental method as this 'requires that programme delivery is focused entirely on evaluation considerations' (Hogwood and Gunn 1984: 231). However, well-designed programmes of this nature using control groups, the *quasi-experimental method*, do offer some scope for evaluation. For example, considerable advances have been made in examining the determinants of farmer behaviour over the last decade, enabling the use of 'natural' control groups when considering specific agricultural policy impacts.

Finally we turn to *retrospective cost-benefit analysis* as an extension to the methods already outlined. CBA offers the prospect of attributing a financial value to the costs and benefits associated with a particular policy. It does not, however, solve some of the conceptual or analytical problems of isolating costs and benefits, establishing causal relationships, and eliminating independent variables.

In practice, most formal policy evaluations will utilise a range of methods. Moreover, much comment on the impact and achievements of policy by politicians, policy agents and academics will utilise none of these methods. Intuition, ideology, received opinion, all figure largely in what may pass for policy evaluation. In the same way that implementation is as political a process as policy formulation, so evaluation is inevitably part of a continuing policy process too, a process in which political imperatives figure highly.

CONCLUSIONS

Implementation and evaluation studies have, for many years, been the poor relations of political science. However, in recent years interest in them has increased. According to Pressman and Wildavsky (1973), it is surprising that correct implementation ever takes place. By the same token, it might be argued that it is surprising that policy makers are willing to subject themselves to any kind of evaluation which will almost inevitably point up weaknesses in both formulation and implementation processes. Whilst the basis for such observations is clear from the observations made in this chapter, it has to be countered that the perception of failure is in the eye of the beholder. How policy implementation and the results of evaluation are perceived is itself part of a political process.

Implementation deficit cannot be observed in a neutral and value-free manner, as is apparent from the fierce political disputes between politicians regarding a myriad of policy outcomes. And evaluation, as an integral part of the policy process itself, feeds into politics. An evaluation showing policy failure can be utilised by those who wish policies to be 'cut' and public money saved *and* by those who seek more resources so that the job can be done properly in the future.

Part II

AGRICULTURAL POLICY

4

THE EMERGENCE OF AGRICULTURAL CORPORATISM

INTRODUCTION

In order to understand post-war and recent developments in agricultural policy an historical perspective is required. This is what this chapter provides. It shows how a nation which for long eschewed protection, commending instead free trade, gradually came to consider the merits of interventionist policies. It shows how wartime experiences, in particular, gave rise to new forms of agricultural politics alongside the new policies. The chapter starts with the repeal of the Corn Laws in 1846 and ends with a new framework for policy in place by the start of the Second World War. In a century the wheel had turned full circle, from agricultural protection to *laissez-faire*, to protection again. There are many strands to this story which is more than just an account of government fiscal policies. Indeed the climate which eventually gave rise to a fully fledged agricultural support system was born of increased state intervention in many non-fiscal areas, such as education, research, training and advice, marketing, and land ownership. The twentieth century has seen policies developed in all of these areas sometimes through direct government intervention and on other occasions through novel forms of self-government within the agricultural industry itself.

The balance between central government initiative and the devolution of power and responsibility to the farmers themselves is one of the key issues to be explored. At the outset of the twentieth century, farmer organisations had little political power and virtually no policy responsibility. By 1945, the National Farmers' Union was a partner of government, deeply embedded in all aspects of agricultural policy formation and implementation. Agricultural corporatism had arrived. Thereafter the corporatist deal would provide both the foundation for post-war agricultural success and a stumbling block to readjustment once success had been transposed into the crisis of surpluses and over-production.

The Corn Laws dated from the Middle Ages and became particularly restrictive of imports after 1773, although the more extreme protectionism of the earlier laws was removed in 1815 (Watson 1960; Woodward 1962). None the less the laws still presented a barrier to imports, one that was of particular concern in years of shortage. A succession of bad harvests after 1836 led to increases in bread prices and proponents of free trade gained much popular support for the case that imports of grain would have prevented this. But the fundamental issue at stake was not so much the specific matter of the price of corn as the gathering political confrontation between the traditional landed interest and the urban-based interest of industry and commerce. In 1846 the repeal of the Corn Laws was carried by Parliament amidst such great controversy that the Tory Party split and the Prime Minister, Sir Robert Peel, who had steered the measure through Parliament, was forced to resign (Barnes 1930; Chambers and Mingay 1966; Woodward 1962). Agricultural protection was at an end, and for the remainder of the century few mainstream politicians gave any serious thought to its reintroduction.

Historians have seen the repeal as a key symbolic victory for industrial and commercial politicians over traditional landed interests (Woodward 1962). The immediate consequences for agriculture were, in fact, limited and arable farming continued to flourish for another thirty years after repeal. However, agricultural depression came eventually and the last two decades of the century saw a rapid decline in the fortunes of high farming. The story of the depression was told most graphically by Lord Ernle (formerly Rowland Prothero) in his *English Farming Past and Present* (1912) and for many years this account was the only one available. It is a tale of relentless decline and misfortune. However, its author had his own particular interests. He was an agent for the Duke of Bedford and one of a small coterie of protectionists within the upper echelons of the Conservative Party who had gathered around Lord Milner (Cooper 1989).

More recent, and less partisan, estimations of the depression have stressed its regional variation and the fact that dairying, livestock production and horticulture were not nearly as adversely affected as arable farming (Orwin and Whetham 1964; Perry 1973; Saul 1969; M. Turner 1992). In some instances the depression gave opportunities to more progressive and entrepreneurial farmers who were prepared to adopt fresh methods and produce new commodities. However, even in sectors experiencing relative growth, considerable marketing and organisational difficulties were encountered, for example in dairying. In the long run it was such difficulties in sectors which, on the face of it, offered good prospects that gave added and crucial support for interventionist arguments. But this is somewhat to anticipate our story.

In the 1880s and 1890s intervention of any kind seemed a long way off, despite the rapid rise of agricultural protectionist policies in, for example,

France and Germany (Koning 1994; Lambi 1963; M.S. Smith 1980). A National Fair-Trade League, formed in 1881, foundered in 1892 after the Northcote Commission, appointed by Lord Salisbury, failed to reach a verdict on the issue (Zebel 1967). Michael Tracy adduces seven main reasons why the UK, alone among the major European powers, failed to adopt protectionist policies for agriculture during the late nineteenth century:

1   Britain's lead in industrial production favoured free trade;
2   The influence of economic theorists such as Ricardo and Adam Smith;
3   The political legacy of the anti-Corn Laws agitation;
4   The strength of the British navy;
5   The food production of British colonies;
6   The relative political weakness of the landowners as a result of democratic reforms;
7   The absence of a coherent and united agricultural pressure group as a result of divisions between landlord and tenant and between arable and livestock farmers.

(after Tracy 1982)

When eventually, at the turn of the century, the principles of free trade came under attack from members of the Conservative Party it was in response to a perceived need to promote preferential trading relations with the colonies of the British Empire rather than a renewed desire to protect home production *per se*, whether of agricultural or industrial commodities (although protection was of increasing importance as the campaign proceeded). Within Britain, the industrial manufacturing areas of the north and midlands were expected to benefit most from preferential trading with the colonies. The dream of men such as Joseph Chamberlain was that the Empire as a whole would become a trading entity competing against other nations and protecting itself from unfair competition through its own internal preferential arrangements (Brown 1943; Zebel 1967). An important additional element was the increasing need to raise revenues for public welfare and military expenditure (Cain and Hopkins 1993; Emy 1972). Faced with a Liberal Party committed to taxing the landed aristocracy in order to fund government spending, the Conservatives had few options: 'As defenders of the aristocracy and the status quo, the Conservatives faced with the need to improve Britain's defences, resisted direct tax increases. But they were forced in return to resort to tariffs' (Cain and Hopkins 1993: 203). Thus the tariff reform movement represented a coalition of industrial interests, Empire enthusiasts, and the interests of those with landed wealth: 'It offered a programme of "social imperialism" designed to unite property with labour in the cause of empire and to head off the formation of a mass party dedicated to socialism' (ibid.).

Despite the scaremongering by many free-trade propagandists that food prices (and hence returns to farmers) would rise under tariff reform, the implications for agriculture of Chamberlain's tariff reform campaign launched in May 1903 were, at best, uncertain and, at worst, potentially damaging to

British farming. Agriculture was one of the key sectors in which some of the colonies had a clear comparative advantage over Britain. Thus imperial unity might well be to the further detriment of home agriculture. By contrast, the colonies were seen as potential markets for Britain's industrial goods and it was British manufacturing that was expected to benefit most from tariff reform. Some free traders were quick to seize upon the possibility of negative implications for agriculture, prompting Chamberlain to consider the agricultural question rather more seriously and indeed to seek policies which would both benefit colonial agriculturalists and protect domestic agriculture (Marrison 1977). His dependence upon an alliance with landed interest made this move all the more pressing.

In October 1903 Chamberlain launched an unofficial Tariff Commission to examine all aspects of tariff reform. The historian A.J. Marrison (1986) has examined the workings of the Commission's Agricultural Committee in some detail. He shows that, despite suspicion among some farming leaders, many were prepared to assist the Committee in its deliberations, indicating some sympathy for tariff reform and protectionist principles. A.H.H. Matthews, the secretary of the Central Chamber of Agriculture (see pp. 82–83), was sympathetic as were key officials of the Board of Agriculture and many individual farmers. Towards the close of 1903 the Central Chamber came out in support of Chamberlain, as did the Farmers' Clubs (Marrison 1986). And there is some evidence of localised agricultural support (Brown 1943; Sykes 1979). But many farmers remained suspicious of policies originating from those whose prime interests were in protecting industrial manufacturing and the Empire.

Chamberlain had resigned from the Conservative Cabinet in September 1903 to devote himself fully to the tariff reform campaign. In many ways it was a campaign for the soul of the Conservative and Unionist Party, but hopes that the party would unite behind tariff reform were doomed to failure. The Prime Minister, Arthur Balfour, made various abortive attempts at compromise, but the Conservatives were torn apart on the issue (Blewett 1968; Ensor 1936). The 1906 general election dealt a devastating blow to Chamberlain and the cause of tariff reform. The Liberals won 377 seats leaving the Conservatives just 157, of which 79 were held by Chamberlainites (Blewett 1968); the Irish Nationalists took 83 seats and the newly emergent Labour movement 53. The Liberals remained solidly behind free trade, despite numerous internal divisions on other issues. The election demonstrated the huge distance that still had to be travelled before protectionist policies would be sympathetically regarded by the electorate. The election results showed little sign of support for tariff reform in rural areas, strongly suggesting that the Central Chamber of Agriculture and the Farmers' Clubs were out of step with the industry as a whole. Indeed by-election results, prior to the general election, in Hertfordshire, Dorset, Shropshire and Sussex, had indicated that tariff reform was not likely to be popular in the countryside. In the general election, the Liberals regained many agricultural seats in the south-west and East Anglia and won new rural seats in the traditionally Conservative south-east (Russell 1962).

The débâcle of the 1906 election may have left the Tories with fewer MPs and with a bitterly divided party but the younger Tory members with the greatest energy were almost all supporters of tariff reform, a point recognised by Balfour who remained the party leader despite losing his own parliamentary seat (Gollin 1965). Undaunted by either the election defeat or the ill health of Chamberlain, who suffered a stroke in July 1906, the Tariff Commission continued its work and the debate within the Conservative and Unionist Party moved in the direction of the tariff reformers. More than anything else the party united in opposition to Lloyd-George's famous 1909 budget, with tariff reform as a key element in the presentation of an alternative policy. Balfour had come to the conclusion that tariff reform represented the only alternative to 'the bottomless confusion of socialistic legislation' (*The Times*, 23 September 1909, quoted in Blewett 1968). Consequently the January 1910 election was fought by the Conservatives on a Chamberlainite programme, with 'free fooders' only a mere handful of Tory candidates. The election was lost. Once again the Liberals were able to play on fears of higher food prices resulting from tariff reform and, in many electors' minds these fears outweighed the potential advantages of expanded markets for industrial goods. In the second election campaign of 1910, in December, Balfour modified his programme by promising to put the tariff reform issue to a referendum and again he was defeated (Ensor 1936). The Unionists did win back some of the lost rural seats of 1906 but the results were devastating for the tariff reform cause because of the manner in which the Unionists had linked the issue with their opposition to the Budget, as well as to reform of the House of Lords:

> The merging of tariff reform with the anti-Budget crusade had tarred tariff reform irretrievably as the rich man's method of avoiding taxation.... 'Protection is indeed not merely dead but damned' crowed the Liberal *Morning Leader* and many Unionists were inclined to agree. The landed classes who had hoped that tariff reform might save them from the burdens of the Budget were disenchanted; Unionists who had shared Birmingham's optimism concerning the popularity of tariff reform were disillusioned; while the party strategists saw that in the very cases where tariff reform was to have won seats, in the industrial north and London, it had singularly failed.... Dissent spread in the Tory ranks.
>
> (Blewett 1968: 123)

After the loss of three successive elections Balfour resigned in 1911. His successor as leader, Bonar Law, who had previously been an ally of Chamberlain, was completely taken up with his passionate opposition to home rule in Ireland and despite his earlier convictions would not now countenance anything which seemed to be an electoral liability (Ensor 1936). He rapidly retreated from tariff reform and in January 1913 abandoned food duties as official party policy (Marrison 1986).

But a year later the country was at war and government intervention in agriculture became a necessity, at least in the short term. What are the main lessons of the tariff reform episode for rural policy? First, it showed that protectionist policies for agriculture were only likely to emerge as part of a much broader set of concerns. If agriculture was ever to receive protection again it would not be because of the inherent problems or desires of farmers themselves, at least not in the first instance. Second, it revealed a deep-rooted uncertainty and lack of effective organisation within the farming community itself.

If tariff reform provides one example of how turbulent national politics could shake the certainties of a rural world, another lies in the political assault on the assumed economic and moral superiority of the English tripartite system of capitalist agriculture.

THE DECLINE OF THE LANDLORDS

There is not room enough here to give anything but a very general and basic account of the changes which beset agricultural landowners in Britain during the closing decades of the last century and the first three decades of the twentieth century. A point at issue throughout this period, sometimes explicitly and often implicitly, was the need to maintain and establish the conditions for a prosperous agriculture in the context of an industrial and commercial nation and in the light of a declining landowning class. The period from 1875 through to 1939 was dominated by agricultural depression (with the exception of the war years) and by political debate regarding the role of the landowning class, with a growing assumption that the class could not survive. There are several major and exhaustive studies of patrician Britain which together provide a full account of the complex social, economic and political processes which brought about the decline of the British aristocracy (Beckett 1986; Cannadine 1990; F.M.L. Thompson 1963).

By the close of the Edwardian era, landlords found their economic power eroded by the agricultural depression and the declining importance of agriculture in the national economy; their political power eroded by local government and the democratic reform of national government; and their direct power over tenants diminished by successive pieces of agricultural holdings legislation. Many of the changes were due to the inevitable transitions consequent upon the emergence of mature industrial capitalism, but here we will concentrate more narrowly on the developments that can be seen as specifically contributing to the decline of the power of traditional agricultural landowners. The decline of the power of landlords was profound over a relatively long period of time but modest in its pace of change. It varied from region to region and, inevitably, smaller landowners more dependent upon agriculture suffered more deeply and declined more rapidly than large landowners well endowed with non-agricultural as well as farming interests. The decline was neither the direct result of a specific political movement nor, indeed, the intended consequence of all the actions which can now be seen as contributory factors. By contrast, the land reform *movement* challenged not only the landowners but the system of capitalist agriculture which had emerged from the progressive landlordism which had characterised English agrarian development. In the same way that tariff reformers dared to challenge the sacred doctrine of free trade so land reformers dared to suggest that the English system of land occupancy and farming was inferior to peasant proprietorship and/or land nationalisation. It is to the land reform movement that we now turn.

Land reform

In 1880, long before Lloyd-George launched his attack on landownership, Disraeli declared that alongside the governance of Ireland the other main political concern of the day was 'the principles upon which the landed property of this country should continue to be established' (quoted in Perkin 1989: 133). Perkin places the land reform movement on a par with the movement for electoral reform leading up to the Great Reform Act 1832 and the anti-Corn Law agitation of the 1840s and describes it as 'one final attack on the landed aristocracy' (ibid.: 48).

It is worth reflecting on just why the issues of land reform and nationalisation came to prominence. The rise of socialism at this time was certainly a contributory factor. Other more utopian critics of landownership had more in common with anarchist ideologues such as Peter Kropotkin (1898). Those in the 'back to the land' movement of the close of the century demonstrated sympathies for petty peasant methods of farming and simple communal living (J. Marsh 1982). But of greater significance, in terms of political impact, were the critics whose capitalist, indeed industrial credentials, were impeccable. Such critics were not interested in simple living on the land but in economic advancement. For example, some calls for the abolition of landlordism were based on a critique of urban land speculation and on the perceived economic inefficiencies of rent (S.B. Ward, personal communication, 1993). Some, notably free-trade Liberals such as John Bright and Joseph Chamberlain (whose criticisms of landownership continued even after his switch to the Tory party), attacked the distribution of landownership, but not the underlying principles of the landlord–tenant system. Others, such as John Stuart Mill, advocated peasant proprietorship on the grounds of justice and equity and a strong belief in the peasantry as a counter-balance to the industrial masses (C.J. Dewey 1974; Martin 1981). Even radical proponents of land nationalisation, such as Alfred Russel Wallace, or of land taxation such as Henry George, had weakly developed ideas concerning the role of the state. Wallace's vision of land nationalisation lay in the state as ground landlord with all citizens having the right of an occupying tenure on 1–5 acres (Bateman 1989). Later advocates of nationalisation were even more explicit in their desire to see an invigorated market-led agriculture. Thus in the 1930s and 1940s, the one-time Minister of Agriculture, Lord Addison, and the Oxford agricultural economist, C.S. Orwin, clearly wished to see land nationalisation in order to further advance capitalist agriculture, rather than as a point of socialist principle.

Despite this ferment of ideas, little in the way of tangible land reform measures emerged. The Liberal government elected in 1906, true to manifesto commitments, passed the Smallholdings and Allotments Act in 1908 (consolidating and greatly strengthening an 1892 Act of the same name) and the Small Landholders (Scotland) Act 1911 (Leneman 1989). These Acts were designed primarily to give farming opportunities for landless workers, but would-be smallholders faced often insurmountable difficulties in raising the necessary capital for the stocking and equipping of the holdings available from local authorities (Orwin and Darke 1935). The Land Settlement Act 1919, designed to give Great War soldiers a stake in the land, fared little better. Smallholdings and allotments barely provided even a modest contribution to structural change in agriculture. Whilst they became important in a few counties, but hardly on the scale of the social revolution dreamt of by many proponents of land reform, elsewhere their impact was cancelled out by the loss of erstwhile smallholdings through market forces. Between 1913 and 1935 in England the number of holdings of between five and twenty acres, the size category of holdings created by land settlement policies, decreased (Orwin and Darke 1935). By contrast, death duties, seen by some as a means to land reform, had a far greater effect, rising from 8 per cent in 1894 to 40 per cent (on estates worth more than £200,000) in 1920. The impact was increased by the fact that so many young men had pre-deceased their fathers (S.B. Ward, personal communication, 1993), contributing to the trend for estates to be broken up and sold to the emerging ranks of medium sized owner-occupiers. The structure of land occupancy did indeed change but hardly in favour of the peasant proprietor, still less of the landless workers. The class that benefited was the owner-occupying commercial farmer – the beneficiary of an agricultural decline that decimated the ranks of the landowning aristocracy.

Agricultural decline

Although the land reform movement undoubtedly did much to unsettle those holding traditional patrician values, its impact was far less significant than the sustained onslaught of market forces. Mention has already been made of the depression in British agriculture which lasted from the 1870s until 1939, with the exception of the period 1914–1920. The consequence for landowners was little short of catastrophic. Between the mid-1870s and the mid-1890s rents fell by 12 per cent in the pastoral north-west of England and 41 per cent in the arable south-east; and during the 1914–1918 war, while commodity prices increased, rent levels were frozen (Cannadine 1990). By the 1930s they were back at the level of the early nineteenth century. This, coupled with the increased death duties and a growing income tax burden, led to a rapid and dramatic sale of lands:

In the years immediately before and after the First World War, some six to eight million acres, one-quarter of the land of England, was sold by gentry and grandees. In Wales and Scotland, the figure was nearer one-third, and in Ireland it was even higher. Across the whole of the British Isles, the change between the late 1870s and the late 1930s was remarkable, as five hundred years of patrician landownership had effectively been halted and reversed in seventy.

(Cannadine 1990: 111)

In addition to the economic weakness of individual landed families resulting from agricultural decline, the failing fortunes of agriculture also meant that farming was marginalised in a manner unimaginable in countries such as France and Germany, where agriculture's role in the national economy remained more central. Self-sufficient in food in the first decade of the nineteenth century, by the first decade of the next the UK imported 40 per cent of its calorific food supplies, and less than 20 per cent of its wheat was grown at home (P.E. Dewey 1989; Whetham 1978). Gross agricultural output in the UK fell from £250 million in 1870–1876 to £208 million in 1894–1903 (Fletcher 1973). Agriculture's contribution to the national economy declined from 20 per cent of the Gross National Product in the late 1850s to just 6 per cent in the late 1890s (Perry 1973; Saul 1969).

Democratic reform

Economic decline was coupled with political decline. The Third Reform Act of 1884–1885 extended the vote to the majority of adult men (c. 60 per cent) and, through a redistribution of parliamentary seats, significantly reduced the imbalance between urban and rural interests (Cannadine 1990). 'The post-1885 electoral order resembled that of 1960 more than that of 1832', thus spelling 'the end of the historic House of Commons' (Blewett 1965: 27). Once defeated at the polls, particularly in the Liberal landslide of 1906, few grandees contested their Commons seats again: by the 1930s fewer than 10 per cent of Conservative MPs were of traditional landed stock (Cannadine 1990). The strength of the landowning aristocracy in the House of Lords remained but the upper house was also under attack, culminating in the curtailment of their powers embodied in the Parliament Act of 1911. Of course, landowners continued to be prominent in many governments of the twentieth century but slowly their significance declined and they were increasingly marginalised in junior government posts rather than holding the reins of power in one of the great offices of state (Cannadine 1990).

Landowners also found their local political power curtailed, a process shown by Lee (1963) in his seminal study of Cheshire to be inextricably linked to the emergence and progress of local government. The Local Government Act of 1888 created county councils in England and Wales (Stanyer 1967); a year later the measure was extended to Scotland, although Ireland had to wait a decade (Dunbabin 1963a). Agricultural matters were on the agenda of the county councils from the outset as a result of the responsibilities they inherited for the administration of regulations covering contagious animal diseases. In the early years, in particular, concerted efforts were made by the traditional ruling elites in some counties to ensure that traditional patterns of social control were maintained and that only limited public expenditure took place. The extent of the changes depended on local circumstances. The power of landowners lasted longer in those parts of lowland rural England less touched by industrialism and where large tenanted farms dominated, although there were exceptions even to this, such as the rapid decline in landlord power in Lincolnshire (Horn 1984). In Wales the loss of power was dramatic and sudden as the 'landed gentry who had dominated the countryside for centuries as justices of the peace were routed in an immense social revolution (Morgan 1981: 52).

Thus in most of Wales, industrial England and Gaelic Scotland local administration was transformed and became the preserve of the middle class. In the shire counties of England the transformation was somewhat slower, not least because some patrician families consciously refocused their attention from national to local government (Cannadine 1990). In 1911 over 20 per cent of the county councillors and aldermen in England qualified for entry in *Walford's County Families* (Horn 1984), and landowners or their agents were often elected unopposed. But even where landowning families involved themselves in county councils, they increasingly found that real power had either shifted to central government or, in the face of the growing complexity of local governance, resided with the salaried officials of local government. 'By the 1930s, the county councils were no longer the old rural oligarchy under a new name, but a professional hierarchy and structured bureaucracy which might – or might not – be sheltering behind a façade of patrician authority' (Cannadine 1990: 167). It was the increasing complexity, bureaucracy and professionalisation of local and national government, and also of those other traditional pursuits of the younger sons of the gentry (the Civil Service, the law, the church and the army) which so debilitated the gentry:

> With recruitment increasingly based on competition rather than connection, on merit rather than money, on ability rather than on social position, the traditional patrician preponderance was bound to be broken. Instead of being an outwork of the landed establishment, the great professions had become the almost exclusive preserve of the middle classes.
>
> (Cannadine 1990: 239)

The decline in political, administrative, economic and social power, both nationally and locally, meant that landowning interests could no longer be guaranteed a favourable hearing in government circles. To defend landownership the Country (originally Central) Landowners' Association was formed in Lincolnshire in 1907. In just three years it could claim a membership of 1,000, of which 100 were MPs (Newby 1987). Despite its rapid assimilation in the upper echelons of the Conservative Party the very need for the CLA to be formed provides evidence of the declining political fortunes of the landowning class.

Agricultural holdings reform

In addition to broad-based political and economic decline, a more specific set of developments concern the erosion of landlord power over tenants. By the end of the seventeenth century the erosion of erstwhile relatively secure feudal tenant rights of copyhold and freehold was virtually complete. The imposition of arbitrary fines had given rise to particularly insecure tenures and the creation of virtual leaseholds, in all but name, among many copyholders (Kerridge 1969; Hoskins 1938). However, the general legal framework remained substantially the same as in the Middle Ages (Denman 1958), and the surprising degree of adaptability within this feudal legacy actually provided the basis for capitalist development. The ability of landlords to establish 'market' rents, accompanied by general economic growth, led to a gradual differentiation of the peasantry, with the rise of the English yeoman, a tenant farmer committed to specialisation, accumulation and innovation (Brenner 1982). But, of course, such capitalistic endeavour was only possible with some degree of security of tenure. While landlords were willing to grant moderate security during the prosperous years in the middle of the nineteenth century, when circumstances changed during the last quarter of the century so the terms of tenancies became more contentious, with fixed-term tenancies sometimes being reduced to as little as one year.

But the issue which emerged in the closing decades of the nineteenth century as the central bone of contention was not security of tenure, but the rights of tenants to claim 'compensation for any unexhausted improvements made by him and remaining on his holding at the end of his tenancy' (McQuiston 1973). The Landlord and Tenant Act 1851 gave tenants rights, upon the termination of a tenancy, to remove (in certain circumstances) the buildings they had constructed. But further legislation on tenants' rights in 1875 was ineffectual as it rested upon the landowners' voluntary compliance. However the Agricultural Holdings Act 1883 provided a statutory framework for compensation and extended to one year the period of notice to quit, thus opening the way to further legislation to protect tenants. From 1883 the whole issue of farm tenurial arrangements shifted in focus from social relationships on particular estates to national statutory provision. Further provisions for protecting tenants' rights to compensation and guaranteeing freedom of cropping came in 1890, 1900, 1906, 1908, 1913, 1914, 1919, 1920, 1922 and in the Agricultural Holdings Act 1923 which Densham (1989) has referred to as the 'Magna Carta' of agriculture. The 1923 Act consolidated the various pieces of legislation and laid down a full code for tenants' rights to compensation. A relatively full measure of security of tenure, however, did not come until the Agriculture Act 1947 and the Agricultural Holdings Act

Table 4.1 Land tenure 1908–1989, Great Britain

	Rented/mainly rented		Owned/mainly owned	
	% area	*% holdings*	*% area*	*% holdings*
1908	88	88	12	12
1922	82	86	18	14
1950	62	60	38	40
1960	51	46	49	54
1970	45	42	55	58
1976	44	38	56	62
1980	42	34	58	66
1984	40	31	60	69
1987	38	28	62	72
1988	38	27	62	73
1989	37	26	63	74

Sources: *A Century of Agricultural Statistics*, MAFF, HMSO 1968; Agricultural statistics, annual, MAFF, HMSO.

1948, which provided full lifetime security of tenure even to existing short-term tenancies of two years or more. The decline in tenanted land following the 1923 and 1948 Acts (see Sturmey 1955) is shown in Table 4.1.

THE POLITICAL ORGANISATION OF COMMERCIAL AGRICULTURE: THE RISE OF THE NFU

The decline of the landlords left a political vacuum and socialist hopes that the farmworkers might fill it were doomed to failure (Danziger 1988; Dunbabin 1963b and 1968; Howkins 1991; Madden 1957; Newby 1977 and 1987). Instead by the early twentieth century the time was ripe for a very different form of union activity. Cox, Lowe and Winter (1991) have contrasted the success of the NFU in the twentieth century with the conspicuous failure of several earlier agricultural organisations. They point to the fact that some of these organisations were too broad in that they attempted to present a united agricultural case across the boundaries between landowners, tenants and workers, and hence were flawed in terms of the harsh political realities facing the industry. Others responded to the problem of breadth by avoiding any political involvement at all. A second group, by contrast, was too narrow in choosing to associate with a particular political interest or specific set of policies. In the first category were the Royal Agricultural Society of England (RASE) founded in 1839, the Central Chamber of Agriculture (CCA) formed in 1865 and the National Agricultural Union (NAU) which emerged from a conference organised by the CCA and the Lancashire Federation of Farmers' Associations in 1892.

The RASE, like many local agricultural societies and farmers' clubs founded in the nineteenth century, studiously avoided any political and policy involvement, preferring to promote the science and husbandry of agriculture. For several decades it was the nearest to a representative body for farmers and yet it rarely made any comment on policy issues. The CCA was more overtly political with an initial emphasis on tenant rights and then, in the 1870s and 1880s, a sharp swing towards the interests of landlords (Brooking 1977). It continued to claim representation of all sectors of agriculture in spite of its strong landlord bias. The CCA survived until 1959 but went into terminal decline as soon as the NFU and CLA emerged as respectively specific farmer and landowner bodies. The NAU, like the CCA, aimed to represent landowners, farmers and workers, and largely worked in tandem with the CCA. In fact it did little to improve tripartite relations and, somewhat curiously, the bulk of its 50,000 members in 1895 were farm-workers. The death of its leader, Lord Winchelsea, in 1898 led to a rapid decline in membership. Subsequently the NAU was taken over by propagan-dists for agricultural co-operation, being re-launched in 1901 as the Agricul-tural Organisation Society (Matthews 1915).

The second group included the Farmers' Alliance, founded in 1879 and disbanded in 1888, and the National Federation of Tenant Farmers' Clubs (NFTFC) which, although not founded until 1892, failed to survive into the next century. The Farmers' Alliance stood for Liberal principles of free trade, capitalist agriculture and the preservation of a reformed landlord–tenant system (Fisher 1978). The NFTFC was primarily concerned with promoting security of tenure and tenant rights (Mutch 1983). Whatever the weaknesses of the individual groups, the underlying reasons for their failures were much broader and had to do with the contrasting fortunes of arable and livestock farmers:

> Very clearly the basic conflict of interest, felt rather than formulated, between the two main groups of English farmers, and masked by the ambivalent rural–urban, protection–free trade division, prevented the formulation of any co-ordinated view, any single, forceful, agricultural policy, and effective co-operation. And in the absence of unanimity, with on the one hand a declining corn-growing interest nostalgically looking to the past and on the other hand thousands of livestock farmers – with the small hard-working family farmer in the majority, motivated by profit and with a keen eye to the main chance – quietly increasing their output with the aid of cheap, imported cereals, the vacuum left by these divergent interest was easily filled by the confused idealism of Free Land Leaguers, Georgists and anti-landlord radicals, bimetallists and all the advocates of peasant ownership, small-holdings and the return to the land.
>
> (Fletcher 1973: 52)

The formation of the Farmers' Alliance was one of the factors behind the establishment in 1879 of a Royal Commission on the Depressed State of the Agricultural Interest. Despite a membership of twenty-one, not one could be described as a small or even medium-sized farmer. Of the thirty-five farmers called to provide evidence, only one farmed less than 100 acres at a time when the average farm size in England was 85 acres (Fletcher 1973). The only clear outcome of the Commission was the formation of the Board of Agriculture in 1889, a government department with limited powers, an even more limited budget, and a staff recruited by patronage from the ranks of the landed elite. Up until the First World War, the Board acted primarily in the administration of legislation to prevent diseases and in a modest programme of research. A second Royal Commission was at work between 1894 and 1897. It fared little better than the first, with most evidence still coming from larger farmers. Protection was espoused by some arable farmers, but rejected by others and by livestock farmers. The essential problem of the depression was that of market adjustment (from corn to livestock) and this the Commission shied away from.

An organisation that could find a way of resolving the conflicts between farmers and of replacing 'confused idealism' with a coherent political force based on a single economic interest would have great potential. Notwithstanding the political salience of tariff reform and land reform, in the long run it was a much quieter set of developments during the Edwardian years that was of lasting significance to the politics of agriculture in Britain. The emergence and rapid growth of the National Farmers' Union is one of the most important factors in the story of twentieth-century British agriculture.

The first local farmers' union was formed in Lincolnshire in 1904 and the National Union in 1908. Newby claims that the Union was formed in response to the emergence of the CLA in Lincolnshire in 1907 and to the spread of trades union activity amongst the county's farmworkers. However, this is based on the erroneous view that the Lincolnshire farmers union did not commence until 1908 (Newby 1987: 163). Starting as it did in 1904, it pre-dated both the CLA and the revival of organised labour. This is an important point for although the Union was clearly opposed to both landowners and workers in some circumstances it was not a purely reactive force.

On the contrary, the conditions for the emergence of the NFU in the first decade of this century were ideal for other reasons. State intervention in agriculture, or certainly the possibility, was gaining fresh credence. The depression was less severe, with the emergence of new mass markets and corresponding marketing complexities which required a collective approach. This combination of factors made the emergence of the NFU possible and ensured that it would not become a single-issue organisation liable to collapse with changed circumstances. The genius of the NFU was to restrict its membership to bona fide farmers but at the same time to avoid taking an overt and antagonistic stance regarding the organisations of either land-owners or workers. True there were many clashes of interest with workers and landowners over the years but the Union was careful to treat these pragmatically and to avoid class interest being articulated as a point of principle. Thus the stimulus that led to the merger of extant county unions into a national union was a dispute not with landowners or labour but with the National Federation of Meat Traders' Association (Cox, Lowe and Winter 1991). This was significant, for the NFU was deeply concerned with ensuring a collective strength in the market place.

The context in which the NFU developed was an increasing governmental interest in a wide range of ideas on how agriculture might be helped, without recourse to direct price support policies. Whatever the rhetoric surrounding the continued advocacy of *laissez-faire* policies, gone were the days when the state was completely distant from the problems of a major industry. At the very least it sought to understand the nature of problems and to consider modest involvement in remedies. This partly stemmed from the continuing role of the Board of Agriculture, but in 1909 another agency, the Development Commission, was formed to assist the development of agriculture and 'rural industries', forestry, land drainage and reclamation, rural transport, harbours, inland navigation and fisheries. The Commission placed considerable emphasis upon research and education – indeed the origins of the Agriculture and Food Research Council (AFRC) can be traced to the work of the Commission rather than to the Board of Agriculture. Agriculture remained a depressed industry and part of the Commission's brief was to explore means for rejuvenating the rural economy, both through agricultural and non-agricultural means. The Commission was the brainchild of Lloyd-George, as Chancellor of the Exchequer, whose personal interest in rural affairs stemmed from his upbringing in a remote part of north Wales. The Commission was a rural example of a more managerialist and interventionist approach to the economy beginning to emerge at this time. Lloyd-George justified the move on the grounds that Britain spent less on the development of 'national resources' than its industrial competitors (Harris 1972; Minay 1990).

THE 1914–1918 WAR AND THE STIRRINGS OF AGRICULTURAL CORPORATISM

In 1914, Britain was unique among the combatant countries 'in having no arrangements for food production as part of her defence plans, and this policy of official *laissez-faire* was retained almost without modification for more than two years of war' (Harris 1982: 137). The story of how this policy came to be changed has been told in detail elsewhere (Barnett 1985; P.E. Dewey 1989). A determined group of officials and politicians, which had been deliberating on these issues since well before the war, gathered around Lord Selborne, the President of the Board of Agriculture, and applied considerable pressure first to Asquith's Liberal government and then to his coalition government, urging the need for interventionist agricultural policies to increase food production (Cooper 1989). A key figure was Lord Milner, a Unionist and self-made man, with a technocratic, almost Fabian, approach to the need for efficient government. One of Milner's acolytes was Rowland Prothero (later Lord Ernle), a land agent to the Duke of Bedford, whose appointment, in Lloyd-George's new coalition government, to the Presidency of the Board of Agriculture in December 1916 signalled an end to prevarication on the question of state involvement in agriculture.

The new policy was enunciated by Prothero, within days of taking office, and came into operation under emergency powers early in 1917, prior to the enactment of the Corn Production Act in August 1917. The Bill faced opposition from free-trade opinion and some Labour MPs, concerned that farmers would be disproportionately rewarded for fulfilling their wartime duty. But radical or 'constructive' Tories, such as Lord Milner, allied with Lloyd George Liberals and some Labour opinion in applying the necessary pressure. The policy contained three main strands: first guaranteed prices for wheat coupled with guaranteed wage levels for workers (but a freeze on rents); second, powers of compulsion over the cropping and stocking of land; and third, the reorganisation of existing county committees into County Agricultural Executive Committees to implement the powers of compulsion. Between 1914 and 1916 there was little increase in tillage but after 1916 two million acres were added to the total arable area in Britain, an increase of 20 per cent. Thus the ploughing campaign was largely successful, primarily due to the success of the committees, themselves usually led by farmers (Chapman and Seeliger 1991; P.E. Dewey 1989; Sheail 1974, 1976a).

The need to maintain and to increase food production for wartime purposes after 1916 led to the formation of a government committee to consider the implications for post-war policy. In most respects the Selborne Committee, 1916–1917 (Cmnd 9079, 1918), urged a continuation of wartime policy with the provision of guaranteed minimum prices for the chief arable crops. But this was not to be seen as an open-ended commitment to farmers. The Committee recommended that farmers and landowners should, in return for the benefits of guaranteed prices, provide higher output. This was to be regulated by assessors empowered to inspect and report on farming practice, and ultimately there were to be powers to terminate tenancies and manage estates in the national interest (Whetham 1978). With the additional concern to establish minimum agricultural wage rates and hours of work the Report was a far-reaching document, characteristic of wartime hopes for 'a land fit for heroes'. As Whetham rather graphically puts it, the Report contained 'a mixture of history selected to point a moral, condemnations of backslidings and lack of zeal, suggestions for reform, and visions of a new world which might emerge from Armageddon, free of poverty, squalor, insecurity, and ignorance' (Whetham 1978: 88). However the focus on guaranteed prices for grain crops reflected a continuing bias towards arable as opposed to livestock agriculture.

In the immediate aftermath of the war, Selborne principles held fast, and the Agriculture Act 1920 put in place a system of support for agriculture along similar lines to those of the Selborne Committee. But with post-war prices tumbling, thereby escalating the cost of guaranteed price support, the Agriculture Act was repealed only one year later. So as Whetham concludes,

> as the farmers began to harvest the crops of 1921, all that remained of agricultural policy was an official belief in the virtues of uncontrolled

prices, at a time of disordered currencies, world-wide deflation of prices, and persistent unemployment in Britain's major industries.

(Whetham 1978: 14; see also Whetham 1972)

But, politically, there remained an experience of 'partnership' in agriculture which had proved not entirely unpalatable to farmers and government alike. Government may have been guilty of a 'great betrayal' of agriculture but it had gained experience of an industry beset by price fluctuations and instability. The representative groups of agriculture, and particularly the NFU, had tasted, for the first time, a direct involvement in the agricultural policy world, particularly at the local level. This embryonic corporatism was a formative experience and not one that would easily be forgotten.

Not that the NFU leadership, nor even the farmworkers' unions, were entirely opposed to the repeal of the 1920 Act. The NFU, at national level, was anxious to see an end to controls over wages and was unhappy with the role of the state in the new arrangements:

This is not to say that the leadership of the NFU advocated a return to the form of individual competition which had dominated in pre-war times. The method by which it preferred to cope with economic change, though, was by collective action, not state paternalism.

(Cooper 1989: 57)

None the less the Union was not at all clear how such collective action might be organised (although a start had been made with attempts to organise the co-operative marketing of milk) and the collapse of the post-war settlement left the NFU in something of a policy vacuum. Moreover, the leadership was discomfited by the fact that many local branches took a different view and deplored the loss of guaranteed prices of the 1920 Act as a 'great betrayal' (Whetham 1972; Cox, Lowe and Winter 1991). In many parts of the country, the reaction was bitter:

Protests poured in from the county branches. Some even went as far as comparing the Government's 'breach of faith' with Germany's violation of Belgium in 1914, while others suggested that politicians could never be trusted in future and described the Coalition Government as a 'set of rogues'.

(Brooking 1977: 195)

For the farmworkers, support for repeal of the Act hinged on their close alliance with urban unions whose 'cheap food' policies demanded a free-trade regime in agriculture. Only slowly and painfully in the 1920s and 1930s did the farmworkers come to realise that their own interests were not necessarily served by such a policy (Newby 1977).

INTER-WAR AGRICULTURAL POLICY

A short-term erosion of state intervention in the immediate post-war period was not, of course, confined to agriculture (R. Lowe 1978). In the case of the repeal of the 1920 Act, the impact on the thinking of the wartime reformers was far-reaching. It forced upon them the recognition that an agricultural policy modelled on a return to the corn-led agriculture of Victorian high farming was doomed to failure. In time reformers came to espouse scientific and techno-logical approaches to agriculture that gave rise to the agricultural revolution of the post-1945 years (good examples of contemporary work include: Astor and Murray 1932, 1933; Astor and Rowntree 1939; Hall 1942; Stapledon 1935; Street 1937). But in the 1920s these were distant dreams at best, for the inter-war years were undeniably difficult ones for agriculture. By the outset of the Second World War British agriculture supplied less than one third of its domestic food requirements. During the twenty inter-war years the UK agricultural area had fallen by 2.5 million acres to just over 31.5 million acres, of which nearly 60 per cent was in permanent pasture (Murray 1955).

However it is entirely fallacious to consider that no policy developments took place until prompted by the shortages of the Second World War. On the contrary, the 1930s saw the establishment of the foundations of a peacetime interventionist agricultural policy representing a marked departure from the experience of the previous hundred years. The prospects for this in 1921 looked bleak indeed, and it is worth reiterating exactly why this was so. The post-war price guarantees and state control had proved both cumbersome and expensive and had been dismantled. Land reform was floundering through indifference and lack of cash. Chamberlainite tariff reform policies were no longer on the political agenda. The NFU was in some disarray over both strategy and policy in the aftermath of the great betrayal.

To these may be added two other factors. First, the impact of electoral reform – the franchise was extended in 1918 to cover 78 per cent of the adult population (S. Moore 1991) – meant that politics was increasingly urban-dominated and that a party with no clear agricultural roots at all, the Labour Party, was now a force that could not be ignored. Second, the increasing complexity of the state's involvement in the industrial capitalist economy amounted to a clear and dramatic transformation of society and polity from the Edwardian era. Thus Tomlinson outlines the rise of a 'managed economy', showing how by the 1950s the economy was managed in 'a manner inconceivable in 1900' (Tomlinson 1990: 9; see also Booth 1987). Middlemas (1979) talks of the emerging 'corporate bias' in politics, as the government increasingly drew upon the experiences of business and the unions in the management of the economy. More recently Runciman (1993) has identified a shift from one sub-type of capitalism to another with the 1914–1918 war as the watershed:

From the 1920s on, we are in a world of national parties, class voting, career politicians, and the real or imagined threat of Socialism as an alternative system which a working class now outnumbering the rest of the electorate might all too easily be persuaded to vote for. The continuing dominance, in the event, of the Conservative Party should not blind us to the concessions to working-class interests which it involved. . . .

(Runciman 1993: 60)

It was, paradoxically, these fundamental changes that ultimately provided the basis for a new-look agricultural policy, with agriculture becoming as much part of the managed economy as other sectors, and modernising influences predominating in the approaches of all political parties. The dominance of the new urban-industrial politics meant, not that agriculture was forgotten, but that it was treated to the same logic. Thus Andrew Cooper (1989) has demonstrated convincingly how during the 1920s the Conservatives threw off the legacy of what he terms 'agrarianism', the belief that many more people could be employed on the land through the promotion of a new class of yeoman farmers, the Tory version of land reform. With the shedding of such romantic notions, notwithstanding the ruralism that continued to pervade much Conservative rhetoric, the way was opened for pragmatic economic management policies aimed at improving agriculture's contribution to the economy as a whole:

> Social engineering, the major concern of the pre-war reformers, gradually became overshadowed by the pressing need to rectify the weakening competitive position of British industry in international markets. In an attempt to overcome structural deficiencies, stricter attention was paid to both promoting new types of manufacturing, and to reforming existing organisation and distribution with the help of selective state aid. This technical development, albeit clumsy and hesitant, had significant collateral effects on domestic agriculture. Above all, it signalled another step in the integration of agriculture into the wider, all embracing economic system.
>
> (Cooper 1989: 64)

When Stanley Baldwin assumed the Conservative premiership in 1923, his inclinations on agriculture were reformist and one of his first tasks when re-elected to power in November 1924, was to persuade Lord Bledisloe (formerly Charles Bathurst), a reformer, to become Parliamentary Secretary in the Ministry of Agriculture. Only a year earlier Bledisloe had reluctantly turned down an offer from Ramsay MacDonald to become the first Labour Minister of Agriculture. Bledisloe emphasised the productive potential of a reformed agriculture, in the main market-led, but with government intervention to support scientific advancement, marketing initiatives, and land reform designed to facilitate the rapid rise of owner-occupying small and medium-sized farms. Henceforth, in Conservative thinking, agriculture became associated less with keeping people on the land to balance industrial and commercial interests and more with ensuring an economic performance in the countryside that would serve those interests – intervention to oil the wheels of the free market rather than to provide for social reform.

Small policy steps were taken in the 1920s and policy deliberations within the ministry showed a continuing commitment to bring agriculture into the mainstream of policy development. A ministry publication emphasised that 'fluctuations in prices could only be diminished by some collective control over supplies' (Whetham 1978: 164, paraphrasing Ashby, Enfield and Lloyd 1925). While the government's response was to shy away from market controls, a subsidy for sugar beet production was introduced in 1925, resulting in a ten-fold increase in sugar beet production in five years and £1 million was earmarked for land drainage (Whetham 1978). The Agricultural Credits Act 1928 set aside £5 million for long-term loans to farmers, through the newly established Agricultural Mortgage Corporation, a policy entirely consistent with the new modernising policy thrust:

> The rationale was simple; the state, in exerting its legislative power through the medium of the banking institutions, could dictate the terms upon which loan facilities would be granted, thus forcing through technological and commercial modifications to its liking.
>
> (Cooper 1989: 73)

However none of these developments amounted to more than the tiniest dent in the underlying *laissez-faire* policies that pervaded after 1921. Bledisloe left government in 1928 and his disillusionment with the agricultural politics of his time remained with him: writing in 1942 he lamented the 'long-standing and persistent lack of vision on the part of our statesmen of all parties' combined with 'an easy going acquiescence in the myopic aims of a powerful urban plutocracy' (quoted in Cannadine 1990: 458).

The new protectionism

None the less, new policies were to emerge, in the light of developments at the close of the 1920s and during the 1930s. As in the hey-day of free trade, agriculture was caught very much in the slipstream of wider urban-industrial developments. The emergence of protectionism as a policy advocated by business interests, who had benefited in the war and now suffered under a free market (Blake 1955), no longer had the gloss of Chamberlainite social imperialism. 'What motivated their pro-protectionist activities was not passion but sound commercial principles' (Cooper 1989: 80).

The Empire Industries Association, formed in 1925, had 280 Tory MP supporters by 1929, and in urging new policies to protect and encourage domestic industries highlighted agriculture's role as a consumer of industrial products (fertilisers, electricity, etc.). Rural protectionism for business rather than for altruistic or emergency motives was born. In the 1930s, fear of war was also a factor encouraging agricultural protection. 'The importance of maintaining trade in order to protect industrial and Dominion interest became less important as concern with ensuring home production in case of war increased' (M.J. Smith 1990b: 74). Increasingly, as the depression deepened in the 1930s, a new slogan emerged in agricultural policy debate: *marketing*. Schemes to regulate and improve the efficiency of agricultural marketing had their origins in the second Labour government's Agricultural Marketing Acts of 1931 and 1933. Some success was recorded by the Potato Marketing Board, less by the Pigs Marketing Board, and triumphantly by the Milk Marketing Board. In many ways, the Milk Marketing Scheme of 1933 provides the single most important initiative in inter-war agricultural policy. The initial incentive, as far as government was concerned, had more to do with the interests of urban-industrial consumers than with those of agriculturalists, for milk was seen as a healthy food and it was a Labour government which introduced the enabling legislation in 1931. None the less, the market weakness of farmers also appealed to those on the co-operative wing of the Labour Party, as to some Tories with corporatist leanings. The formation of the Board has been seen as a good example of the emerging corporatism in the inter-war years as the government sought fresh ways to grapple with complex issues of economic management (Winter 1984).

The background to the establishment of the Milk Marketing Board has been described in full elsewhere and these paragraphs draw from that account (Cox, Lowe and Winter 1990b; see also Baker 1973; M.J.F. Goldsmith 1963). The Board was given powers over producers and prices, and was charged with the overall development of the industry. Eight broad areas were to be covered by the MMB under the first milk marketing scheme:

1) The discipline of individual producers by their own collective marketing organisation.

2) The determination of prices at which milk could be sold by producers.

3) The pooling of receipts from the different markets for payment to producers.

4) The encouragement of improvement in the quality of milk.

5) The development of services to assist producers in the production of milk.

6) The improvement of transport and marketing arrangements.

7) The improvement and development of markets for both liquid milk and milk for manufacture.

8) The acceptance of certain (rather weak) safeguards to ensure attention to the interests of Government and consumers.

(after MMB 1956)

The Board comprised fifteen elected producer members and three appointees of the Ministry of Agriculture, Fisheries and Food (MAFF), with no direct consumer representatives nor representatives of manufacturers, retailers, farmworkers, or the agricultural supply industry. A Consumers' Committee and a Committee of Investigation were established to safeguard wider interests. During the pre-war years the Board made steady progress, restoring order in the market and enabling total production in England and Wales to increase from 856 million gallons in 1933–1934 to 1,063 million gallons in 1937–1938 (Whetham 1978). The Board's progress convinced many sceptics within the NFU of its usefulness. The market support offered by the Board allowed an expansion of production amongst smaller farmers, especially in the more peripheral regions, who had hitherto been in the weakest bargaining position *vis-à-vis* the buyers. Thus in remote areas such as West Devon the 1930s heralded a transition from the traditional pattern of stock rearing with sales of surplus milk, to more intensive and specialised milk production. By the 1950s, over 90 per cent of farmers were selling milk to the Board and the regular monthly payment from the Board became the staple income for most small and medium-sized farmers (Winter 1986).

A fierce campaign to revoke the Board in 1935, spearheaded by the press baron Lord Beaverbrook and a small number of producer-retailers was a failure, culminating in 86.5 per cent of producer votes being cast to retain the Board (Pepperall 1950). A further threat surfaced in 1938 when a bill to establish a commission which would have severely limited the Board's powers came before Parliament. The Ministry and the NFU joined the MMB in saving the Board in a campaign which demonstrated the extent to which a tightly-knit and effective agricultural policy community had now emerged. Crucial to their argument was the special nature of agriculture and its need to be self-governing. In less than ten years an initiative born of urban-managerialist concern was being defended in terms of pure agricultural fundamentalism. The denunciation of the proposed commission by the Board's chairman, Thomas Baxter, was typically vigorous:

> Superimposed on the industry is to be a commission of persons whose qualities are to be that they know nothing about dairying and who, through a maze of advisory committees, are to be the authority charged with the direction of the industry. There was never conceived of a more hopeless form of organisation to run any industry than that which the Government now provide for in the Bill, which producers everywhere must strenuously oppose.
>
> (*The Times*, 25 November 1938: quoted in Giddings 1974: 185)

Under pressure from the Conservative Parliamentary Agriculture Committee, the bill was withdrawn (Giddings 1974). The failure of the bill illustrates the speed with which corporatist arrangements, often termed meso-corporatism, could become firmly established, especially when they covered particular, relatively uncontroversial, economic sectors.

The consequences of the campaign were far-reaching, for the Prime Minister Neville Chamberlain replaced the Minister of Agriculture, W.S. Morrison, with Sir Reginald Dorman-Smith, a former NFU President, who had entered Parliament only in 1935. The move from NFU politician to national politician was to be repeated by others on several occasions after the war, another illustration of the close-knit nature of the agricultural policy community. Dorman-Smith's parliamentary tenure was destined to be a short one as he relinquished his position in 1940, when he became Governor of Burma.

If market organisation represented one platform on which a new agricultural policy was being constructed, another was the 1930s' revival, in a very small way, of subsidies for certain products, chiefly wheat, sugar beet and fat cattle, as well as limited import protection. An important step for domestic agriculture was the introduction of the Wheat Act 1932, which introduced, for the first time in peacetime, the concept of deficiency payments: 'a processing tax was levied on every sack of flour milled in Great Britain, and the money used to pay a deficiency payment covering the difference between the actual price of wheat and a guaranteed price of 10s. per hundredweight' (Mowat 1955: 439). This was not an unlimited guarantee and deficiency payments would decline above specified levels of production, but a further safeguard to home producers was provided for by the imposition of a low duty on imported wheat. Small subsidies were also introduced for oats and barley.

An important feature of the Wheat Act was its mode of implementation. Foreshadowing the plethora of boards, commissions, agencies and quangos that so characterised the post-1945 administration of agriculture, the Wheat Act established two new bodies to administer the legislation. The Flour Millers' Corporation was obliged to buy and dispose of unsold wheat if ordered to do so by the Ministry and the Wheat Commission administered the Act as a whole (Mollett 1960). The Commission was a nominally independent agency, although all nineteen of its members were appointed by the Ministry. Between 1931 and 1934 the area of wheat grown in Britain increased from 1.2 to 1.9 million acres, although much of this involved a substitution of wheat for oats or barley (Mollett 1960). Agricultural production increased by one sixth between 1931 and 1937. The total financial support for inter-war agriculture, in terms of direct subsidy payments, amounted to only £104 million, and of that, 80 per cent related to wheat and sugar beet (Murray 1955).

These relatively novel agricultural *support* measures were buttressed by some traditional *protection* measures, such as the Import Duties Act 1932, which introduced a 10 per cent tariff on most imports including food. This had been preceded in 1931 by duties on some fruit and vegetables. However these developments were bound up with the issue of preferential treatment for the dominions. The Ottawa Agreements Acts 1932 enshrined assurances that new protectionist policies were not only designed to protect home producers. For example, the Act explained how meat policy was designed 'first, to secure the development of home production, and secondly, to give to the Dominions an expanding share of imports' (quoted in Hammond 1962: 163). But this did not stop treaties with other individual countries, for example the 1933 trade agreement with the Argentine, which, in effect, greatly limited the protection available to domestic producers. The complexity of trade agreements with the Argentine and with South Africa, Canada, Australia and New Zealand and many smaller dominions need not detain us here (but see, Drummond 1974), except to say that Britain's imperial legacy delayed a smooth transition to agricultural protectionism.

For all the evidence of a changing outlook on agriculture, the concentration of support on arable production could be seen as protecting old 'high farming' and inhibiting the shift to livestock products which was economically desirable (Self and Storing 1962: 19). In addition the measures were piecemeal, and certainly not designed specifically either to restructure agriculture or to tackle agricultural poverty and decline where it existed. Even before the demands posed by the Second World War, the need for new agricultural policies was becoming apparent:

> As the depression began to lift, a positive long-term policy for British agriculture was called for, aimed not at restoring a past structure but at stimulating the adjustment to changing demand, not at raising prices but at reducing costs, not at restricting output but at increasing the efficiency with which output was obtained from the resources of land, labour and capital.
>
> (Whetham 1978: 260)

The Agriculture Act 1937 introduced subsidies for lime and basic slag and deficiency payments for oats and barley and further measures to control disease. The measure heralded a deepening involvement in the agricultural industry as the threat of war approached.

THE NFU IN THE INTER-WAR PERIOD

Before leaving this period of history, it is important to pay some attention to the NFU during the inter-war period. Despite the importance of its local organisation for the implementation of policy during the 1914–1918 war, the Union's direct contribution to policy had been limited and it was not even invited to give evidence to the Selborne Committee (Cox, Lowe and Winter 1991). After the war, the NFU set about improving its efficiency and engaging more closely with government. However it was slow to develop a clear policy on agricultural support. At the 1919 AGM a resolution calling for 'the formation of a Committee to approach the Prime Minister with a view to ascertaining the policy of the Government towards agriculture' was carried but an amending resolution calling for guaranteed prices for meat, corn, potatoes, milk and cheese was lost. A compromise motion, instructing the executive committee to formulate a clear policy on agricultural policy gave the Union hierarchy the opportunity to develop a more positive and constructive role in policy formulation (Cox, Lowe and Winter 1991).

An early opportunity came with its leading role in the work of a Royal Commission on agricultural prices appointed in July 1919, from which landowners were excluded (Morgan 1979). Its evidence demonstrated clearly how it had not yet resolved its ambivalence of policy on protection:

> A considerable body of evidence given by farmers went to show that in the opinion of many of them no measure for assisting the farming

industry by means of guaranteed prices of cereals is necessary solely in the interests of farmers themselves. It was said by witnesses speaking on behalf of the NFUs, which represent altogether over 100,000 occupiers, that the farmers are prepared, if freed from control of their farming operations and permitted to make their own bargains in the labour and produce market, to carry on their industry in a manner satisfactory to themselves without guarantees from the State. In their opinion, it is for Parliament to decide whether the national requirements necessitate increased corn production and consequential restriction on their freedom of action as regards their system of cultivation.

(NFU 1919)

The Union's involvement in governance was boosted by the terms of the 1919 Ministry of Agriculture and Fisheries Act, which instituted a three-tier system of agricultural representation and regulation, comprising county council committees, national councils and a central advisory committee. In England, each County Council Agricultural Committee appointed two members to the Council of Agriculture for England. Other representatives were appointed by the minister. The Council was to act as a national forum for the discussion of agricultural issues and as a source of advice for the minister and the smaller Agricultural Advisory Committee. Soon after the passage of the Act the NFU scored a notable success by increasing from two to five its direct representation on the Agricultural Advisory Committee for England and Wales (Morgan 1979).

The NFU used the Council of Agriculture in two quite different ways. At the outset it urged its membership to be prepared to act as county nominees, in order to maximise the NFU's influence on the Council. Over four years this gave valuable experience of policy discussions at the national level to many individual members. Participation in the Council did much to secure for the Union a national identity, and it appeared the Council might become a permanent feature of the corporatist governance of agriculture. However, the NFU did not hold a monopoly of representation on the Council, nor could county council NFU members always be relied upon to follow the official NFU line. The Council increasingly showed signs of independence, and in 1924 there was an abrupt change of tactics by the NFU after the Council had 'adopted a report which claimed that this statutory body expresses the authoritative opinion of agriculturists' (NFU *Yearbook* 1925). This was a move which seriously alarmed the Union and it responded by recommending its members to boycott the Council. As a result, the Council dwindled into insignificance (Self and Storing 1962). Thus the Union used the Council to demonstrate successfully its own claim to representational monopoly, a lesson that was not to be forgotten when the time came for the details of subsequent corporatist arrangements to be worked out. None the less the 'dwindling' of the Council did not occur overnight, and in 1927 Sir George Courthope, chairman of the Conservative backbench Agriculture

Committee, used the Council as an opportunity to censure government for its agricultural policy failings, a move which led directly to the NFU adopting a more critical attitude to government and a more directly pro-protectionist policy (Boyce 1988).

The short-lived 1920 Agriculture Act was the first piece of agricultural legislation in which the NFU negotiated with government as the sole representative of the agricultural industry. Having received support from Milnerite Unionists in the past, the NFU for a short period seemed to identify more with the Liberal Party and Lloyd-George's onslaught on landlordism. Indeed, the 1920 Act was amended in the Lords by Unionist landowners critical of the NFU's new role (Morgan 1979).

After 1921, having fought off attempts by some government members to retain the Wages Board, the NFU 'steadily withdrew from any partnership with the State and carried out its activities from a position of political isolation (Brooking 1977: 196). An early attempt to launch a system of MP sponsorship ended in failure and links with the Conservative Party throughout the 1920s were marred by the different objectives of back-bench MPs and representatives of the industry (S. Moore 1991). Progress was limited, but despite this the NFU's membership did not decline dramatically (see Table 4.2) and nor did its organisational efficiency:

Any but the strongest and most determined organisation would have declined in similar circumstances. That the NFU did not decline says much for the administrative abilities of its leaders and the resilience and efficiency of its established procedures. On the other hand the very desperate nature of the NFU's situation helps explain why it fared as well as it did. The repeal of guaranteed prices caused the organisation to lose its innocence. After 1922 it was much more hard headed in its dealing with government and came to regard any single panacea,

Table 4.2 Membership of the National Farmers' Union of England and Wales

	Membership	Membership as % of eligible farmers (estimate)
Inception December 1908	10,000	4
Summer 1910	15,000	6
1913	20,000	9
early 1916	22,000	9
1918	60,000	25
early 1920s	100,000	41
mid-1920s	90,000	37
1935	120,000	48
1953	210,000	84
1957	194,000	90

Sources: Brooking 1977: 134–135; Self and Storing 1962: 40; National and agricultural censuses

whether State assistance or co-operation, as unrealistic. From that moment on it came to rely more on its own resources. That tendency helps explain how survival was later turned to real advantage despite the negative suspicion of State assistance bred by the repeal.

(Brooking 1977: 197)

The Union also used the opportunities provided by county council committees to good effect. Their responsibilities largely related to permissive powers in the provision of agricultural advice and education; statutory duties for controlling animal diseases and the quality of agricultural seeds and feeds; and the provision of smallholdings. Their greatest powers were derived from the work of the War Agricultural Executive Committees concerning the control and stimulation of production along the lines envisaged by the Selborne Committee. Here the 1919 Act, as was made quite explicit in parliamentary debate, envisaged the county committees assuming a role hitherto played by landowners in encouraging good farming. By the same token, such controls exerted on behalf of the state were seen as a necessary part of the social contract between farmers and state (Hansard, 11 November 1919, vol. 121, cols 289–290, 302). With such an explicitly contractual and corporatist approach it is not surprising that these responsibilities were lost to the county councils with the Corn Production Acts (Repeal) Act in 1921.

It is interesting to note that the composition of the committees was not to be determined solely by the county councils, with the Ministry having the right to nominate up to one third of the membership of the county committee and any sub-committees. The justification for this was to provide a check on the spending of taxpayers' money, particularly with regard to the provision of smallholdings. There was no thought, however, that the consumers of food should be involved and a suggestion by a Labour MP in debate that they might be directly represented on the central and local councils was roundly dismissed by the Minister of Agriculture, Arthur Boscawen. In practice the Ministers' appointees were usually agriculturalists, and often local NFU activists. In Devon, for example, the two longest-serving chairmen of the County Committee were William Tremlett and John Metherell, who each also held periods of office as county chairman of the Devon branch of the NFU (Cox, Lowe and Winter 1991).

Having lost the powers relating to the control of production, there was a risk that the council committees might become moribund. Again the NFU must take much of the credit that this did not occur. The committees emerged as the main instigators of agricultural education and advice. By 1939, 55 of the 60 English and Welsh counties had appointed county agricultural organisers, many of whom doubled as principals of farm institutes, and the total agricultural education and advisory staff in the counties numbered 468; by the 1930s 36 counties had started farm institutes or demonstration centres (Holmes 1988). Nationally there were in 1938 just 68 provincial advisory officers funded directly by central government, interestingly through the

Development Commission rather than the Ministry (Holmes 1988). However, by the 1930s central government grant aided 60 per cent of the county's costs in this area (J.L. Evans 1936; G.E. Jones 1963). The staff figures may be small compared to the 1,350 officers employed by the National Agricultural Advisory Service in the early 1960s (G.E. Jones 1963), and the additional hundreds employed in county colleges, but they nevertheless represented a significant commitment by the counties. Prior to the 1914–1918 war fewer than one half of the counties had made any agricultural education and advisory provision at all. In the 1920s it became almost universal – ironically expenditure was greatest in counties with substantial urban and industrial interests where the revenue from rates was highest (G.E. Jones 1963). A (national) survey of nearly 2,000 farmers undertaken in 1944 discovered that 17 per cent had consulted their county organisers prior to 1939 (D. Chapman 1944; Holmes 1988). This is a surprisingly high figure considering the low level of staffing and the fact that the service could offer little in terms of grant aid – although the lime and basic slag subsidies under the 1937 Land Fertility Scheme probably accounted for many of the contacts. All this gave local NFU activists a powerful stake in the administration of their own industry, responsibilities which stood them in good stead when stringent powers were again assumed by County War Agricultural Executive Committees at the outbreak of the Second World War.

During the 1930s the NFU again found itself drawn into policy deliberations, first of all by Neville Chamberlain as one of the leading economic thinkers of the Conservative Party. Chamberlain's corporatist approach led him to recognise the NFU as the legitimate representative of organised producers and to court the union in his own search for policy solutions, a process taken much further by Walter Elliot, Minister of Agriculture from 1932 to 1936 (Cooper 1989). Elliott was one of the most corporatist of Conservative ministers in the inter-war years:

> Government and interest groups were, according to Elliot, not to be separated. Their relationship was to be a symbiotic one. As opposed to the liberal view that society consisted of the sum of individuals, Elliot held the belief that society was composed of the sum of producer groups such as trade unionists, farmers, and employers. Concomitantly, Elliot acknowledged that consumers ... should receive less political attention than had been accorded to them in the past.
>
> (Cooper 1989: 164–165)

Elliott's role in establishing marketing boards was the chief example of what such corporatist policies might mean to agriculture, not least in the requirement of the leaders of the NFU to ensure compliance and acceptance of policy among their members. Here Elliot scored a remarkable success. A farmer himself and with a background in agricultural research he was the first minister of agriculture, arguably the first senior government minister in any position, to gain such a high degree of confidence from the leaders of the

NFU. The years of his ministry were thus a turning point in the relations between the state and the government.

CONCLUSIONS

This chapter has provided an overview of approximately 100 years of agricultural policy development. In 1840 agriculture, or at least *arable* agriculture, was protected by the Corn Laws, the result of the powers held by aristocratic landowners within the British polity. Agriculture was firmly under the control of landowners in a system where tenants and workers had little say in government. Repeal of the Corn Laws marked the first of a series of political and economic reverses for the landowners. By 1940, their role had greatly diminished. The NFU, representing working tenant and owner-occupying farmers, had replaced the landowners as the main representative of the agricultural industry in the policy process. The agricultural interest was exerted not so much by landed individuals in the Houses of Parliament, the traditional strongholds of the landed interest, but by the NFU with ministers and civil servants. The location for policy making had shifted from the Palace of Westminster to Whitehall.

The rationale for agricultural policy now had little to do with the agricultural interest *per se*, but was largely an outcome of wider imperatives for the management of the economy. However, these new imperatives presented opportunities for agricultural interests to assert themselves in fresh ways which might afford lasting protection for the industry, irrespective of the wider interests of the nation-state. The urgent need for the agricultural interest to work alongside government in economic management, at the same time as establishing a set of political arrangements which would be hard to unpick, was presented in particularly exaggerated form by the 1939–1945 war and it is to this we turn in the next chapter.

5

THE STATE AND THE FARMER: POST-WAR FARM POLICY

INTRODUCTION

The main purpose of this chapter is to explain the development of agricultural support policies in the post-war period, and to consider their operation prior to the radical upheaval caused by the crisis of the CAP from the early 1980s, which is dealt with in the next chapter. This is an important story, for contemporary accounts are inclined to consider agricultural policy solely in the light of the CAP. In reality the legacy of pre-European policy still remains and, in terms of agricultural production, the seeds of current difficulties were sown long before accession to the Common Market. The UK did not have protectionist and interventionist policies foisted upon it when it joined the Community on 1 January 1973; some were already in place.

At the outset of the chapter we examine the nature and characteristics of the agricultural policy community that emerged during the war and its immediate aftermath. This is followed by sections providing a factual account of the main policy developments in the period from 1945 to the early 1980s, with an intervening section on the issue of the monopoly of representation enjoyed by the NFU. It would be impossible to cover this period in full here, so just the main legislative innovations and policy shifts are indicated. A section examining the agricultural policy-making process within the EC, with particular emphasis upon the pan-European representative farmers' group COPA, brings us to the close of the period, the early 1980s, after which agricultural politics is transformed, as shown in the next chapter.

WAR AND THE EMERGENCE OF A POLICY COMMUNITY

It should be apparent from the last chapter that by the outbreak of war in 1939, the seeds for an interventionist agriculture, with the NFU as a major partner, had been sown and taken root. The NFU's involvement in policy discussions increased as the war went on and key boosts to farmer confidence were given in early assurances by government that the Union would be included in discussions to ensure a stable post-war policy for the industry (M.J. Smith 1990b). Government was experiencing the advantages of working closely with a representative farming organisation in the development of policy. When government appeared to farmers to be in breach of these early assurances, in the prices offered in the autumn of 1943, it also experienced the lobbying power of the NFU, now with approaching 200,000 members. Ministers and the Prime Minister were denounced for a breach of faith and, whilst the government did not retreat on the particular point at issue, the pressure brought about a commitment to a four-year plan for agriculture and a determination to develop a better machinery for agreeing prices in consultation with the Union (M.J. Smith 1990b).

One of the early consequences of the war for agriculture was the decision in 1939, under the terms of the Emergency Powers (Defence) Act, to establish War Agricultural Executive Committees (WAECs) (Murray 1955). The 'War Ags', as they became known, were given considerable powers to enforce standards of husbandry and estate management and to evict farmers if necessary. A crucial element in the government's thinking was the decision to establish new committees and not to build on the framework offered by the existing county council committees. This amounted to a clear endorsement of the progress made by the representative groups within agriculture and a rejection of local democratic policy making or accountability in agriculture. It is a decision that has never been reversed.

WAEC chairmen were selected by the minister and together they appointed committees comprising representatives of the different sectoral interests within the industry. In practice most of the chairmen selected were either leading farmers or landowners, usually office holders within either the NFU or CLA. NFU activists became the single most important voice within many of the committees, especially the sub-committees appointed at district level. However any suggestion of direct representation was played down:

> As the Executive Committee is nominated by, and directly responsible to, the Minister the policy was adopted from the first, and has been adhered to strictly, that no nominee, or even representative, of any organisation should serve on any Committee. Collaboration and co-operation with the various organisations is welcomed and is most desirable, but membership on Committees as representing such bodies would prove the very reverse.... (the Executive Committee) desires only to lead, inspire, and assist every occupier ... to produce the utmost from his farm.
>
> (Burrel 1942: 7)

The arrangement was 'primarily state-induced self-regulation through appeals to patriotic sentiment' (Cox, Lowe and Winter 1985: 134). In some ways the lack of a formal representative system was a great advantage to the NFU. As a consequence of good organisation and sheer force of numbers (for example compared to the CLA), it could avoid open conflict and became, almost by default, the main impetus behind the success of most of the committees. The committees' success, linked of course to the prices offered for agricultural commodities in a controlled market, meant that agricultural production increased dramatically. Between 1938–1939 and 1941–1942 gross output of British agriculture increased by two thirds and commodity prices doubled between 1939 and 1946 (Bowers 1985).

As the war wore on, attention turned to the future direction for agriculture. The NFU was at the forefront of discussions on the shape of post-war policy. In 1944 eleven organisations (the NFU, CLA, NUAW, TGWU, Royal Agricultural Society of England, Councils of Agriculture for England and Wales, Chartered Surveyors' Institution, Land Agents' Society, Land Union, Land Settlement Association) and a group of peers, combined to issue a declaration on post-war agricultural policy, which proposed increased food production through price support, grant-aided land improvement, credit facilities and an acceptance that, in return, the directive powers of the War Ags should be maintained in peacetime (Courthope 1944). The objectives were broadly in line with the objectives of all three main political parties and many members of the government. Indeed the Minister of Agriculture, Lord Hudson, had been instrumental in bringing the organisations together to consider future policy so as to bolster his position within Cabinet (Self and Storing 1962).

Consequently in the immediate aftermath of the Second World War, there was a strong degree of consensus that there should be no repeat of the 'betrayal' of agriculture that followed the 1914–1918 war. This was reflected in the emergence of a firm and coherent policy community, capable of defending and promoting the interests of farmers. What were the main features of this policy community?

First, its participants shared a deep economic interest within a well-defined policy area. Production issues are much more likely than consumption issues to give rise to such a shared sense of purpose and community, and the agricultural policy community that emerged in the aftermath of the war clearly represented such a community of common interest. The industry was seen to be of key importance by civil servants because of food shortages and the industry's importance for overall economic performance; to politicians because electoral prospects were so dependent on economic and food security factors; and to the key interest groups because of their livelihood.

A second feature was the strong belief by participants that more resources might usefully be applied within the sector. In other words a policy community came into existence with a shared set of priorities and, particularly, an agreed need for public expenditure. In these circumstances, MAFF was able to harness the NFU to assist its case for lobbying the Treasury and other cabinet ministers. In turn, the NFU adopted MAFF as the best means for vigorously advocating its case. It might be argued that those involved in a particular sector will inevitably share a belief in the application of more resources. However, as was shown in the last chapter, these conditions did not prevail in agriculture until the 1930s and 1940s. Government ministers and senior departmental civil servants do, almost inevitably, canvass support for their own sector whatever the circumstances, but the point here is that at certain times, the depth of this belief is greater than at others. Moreover, at certain points in history, external circumstances, in terms of the driving concerns motivating members of the core executive, are likely to favour, or at least condone, spending in certain sectors more than in others.

Third, the participants shared a common 'appreciation' of the issues to be tackled and of the political 'rules of the game' for tackling them. The primacy of food production as a central objective for the nation as a whole provided, in the immediate post-war period, a firm and clear focus for all members of the policy community. And the experiences of wartime administration and regulation, amended and consolidated by the Agriculture Act 1947, established the shared rules of engagement.

Fourth, there was an even deeper shared or common 'culture' within the community. Not only was there shared agreement on the need to devote resources, but also a shared understanding of the problems and priorities. The depth of this common culture helped to determine its internal success, its cohesion and its closure to other interests. In the case of agriculture it was an absolutely vital ingredient in the post-war formula. Some of the main features of commonly accepted beliefs are indicated in Figure 5.1. Some of them may appear to be in partial or complete opposition, but an important underlying feature of a common culture of this type is the way in which contradictions are reconciled internally, without external political conflict.

A fifth feature of a policy community is that it tends to be relatively stable with a continuity of membership and boundaries which are recognised by the members of the community. As stated on p. 28, there are few policy issues

- Belief in agricultural progress through the application of scientific principles, and therefore in the need for publicly funded research, development and technical advice.

- Belief in the potential of applying management economics to increase the efficiency of farm businesses, and therefore in the need for publicly funded education and extension.

- Belief in agriculture's import-saving role.

- Belief in the underlying 'moral' values of farming and country life, and their role in the nation's life – especially the virtues of family farming.

- Belief in the need to preserve agricultural land from development.

- Belief in Britain's mixed system of agricultural tenure and a rejection of land nationalization.

- Belief that environmental quality and food quality are 'natural' outcomes of an economically sound agriculture.

Figure 5.1 Some underlying beliefs typifying the common culture in the post-war agricultural policy community

which do not, in some measure, impinge upon the majority of people, it is impossible to envisage a coherent policy community comprising the entire electorate; or one where those concerned are constantly changing as would be the case for many consumer issues. In agriculture the stability and definition of boundaries has been provided by the pivotal importance of the NFU representing a discrete membership. Organisations such as the RASE were content to take their lead from the NFU and other organisations tended to be smaller specialist groups such as the CLA.

A final key feature is the closure of a policy community to other interests. Even Parliament may find itself substantially excluded from the details of policy making and, even more, from implementation. By the end of the war agriculture had become very much a technical matter, in which the interests of consumers, for example, were largely excluded.

THE POST-WAR FARM SETTLEMENT

Marsden *et al.* (1993) have identified the post-war period until 1980 as representing a new era for British agriculture in terms of the world political economy – the *Atlanticist food order*, in which the USA played a leading part in influencing UK decisions – contrasting with the *Imperial food order* that had dominated the political economy of agriculture from the 1860s to the 1930s. This phase was characterised by the 'intensive development of agriculture ... as part of a shift towards a mass consumption economy and greater dependency on domestic food supply, for strategic reasons linked to the UK's declining military and economic strength' (Marsden *et al.* 1993: 44).

Thus the newly elected Labour government in 1945 continued the policies of the wartime coalition. Indeed the pressures for food production increased after the war. Although Labour turned its back on land nationalisation (M.J. Smith 1989a), it did not entirely neglect the farmworkers. The Agricultural Wages (Regulation) Act 1947 established the Agricultural Wages Board (AWB) for England and Wales, whereby wage rates would be determined in an annual round of statutory backed bargaining. Although seen as a triumph by the NUAW, the AWB has been interpreted by others as a retreat from direct intervention by Labour to honour its pledges to raise agricultural wages (Flynn 1989). Certainly the awards in the 1950s were modest and real wages even declined between 1949 and 1955 (Newby 1977). The problem lay in the composition of the Board, which comprised five members from the NUAW, three from the TGWU, eight from the NFU and five appointed by the Ministry. The Minister's appointees became, in effect, referees and under Conservative governments, in particular, tended to favour cautious settlements. When the Conservatives threatened to abolish the Board in the early 1990s the NFU supported the farmworkers in lobbying for its retention.

But of far greater central concern to Labour than wages, was agricultural production and this meant the farmers. Early in the Parliament Labour indicated its desire to legislate and the Agricultural Act 1947 enshrined the dramatically altered role of agriculture in the economy and polity resulting from the experiences of the Second World War. The Act provided a commitment to agricultural support which was profoundly to affect farming in the 1950s, and formed the basis both symbolically and legally for subsequent legislation and policy initiatives. In introducing the Bill to the House of Commons, Labour's Minister of Agriculture, Tom Williams, cited the importance of the 1944 declaration and of his continued consultation with the main representative bodies of the industry.

The Bill was designed to provide a secure and guaranteed market for agricultural products, to assist agriculture's own adjustments to changing markets and production demands, and above all to provide for the nation and farmers alike a productive and expansionist agriculture. Government wanted secure food supplies, farmers a secure income; and so the idea of 'partnership' first conceived by the Selborne Committee (1918) came to fruition. The Bill was supported by the NFU although some county branches were suspicious that the Bill's failure to specify the *size* of the guaranteed market left a worrying loophole. After considering these concerns the Conservative Party decided not to oppose the Bill (Flynn 1986).

The objectives of the Agriculture Act 1947 were summarised as follows:

promoting and maintaining, by the provision of guaranteed prices and assured markets for the produce ... a stable and efficient agricultural industry, capable of producing such part of the nation's food and other agricultural produce as it is desirable to produce in the United Kingdom, and of producing it at minimum prices consistently with proper remuneration and living conditions for farmers and workers in agriculture and an adequate return on capital invested in the industry.
(Agriculture Act 1947 Section 1, HMSO)

The Act established a system of guaranteed prices for most major products, to be derived through the Annual Review and Determination of Guarantees, at which the representative bodies of the industry would have a statutory right of consultation. The consultation clause did not preclude the incorporation of other groups, nor did it name the NFU. However policy is more than just legislative provision; after several years of experience the Ministry had come to recognise the NFU as the body they could deal with, and the Union could legitimately claim to be the leading voice for farmers (see Table 4.2, p. 96).

The Annual Review became the main focus for policy making throughout the period up until entry into the EEC. The Act also allowed for special interim reviews if unexpected changes in the cost-structure of the industry occurred during the year, and these took place in 1951, 1955, 1956 and 1970 (Bowers 1985). In effect the price reviews meant the removal of agricultural policy formulation from either parliamentary or wider public scrutiny in what has been described as 'a closed policy community with some corporatist features' (M.J. Smith 1989b: 96). As a result, the NFU–MAFF axis became so strong that relations between ministry civil servants and NFU officers could be closer than between the civil servants and government ministers, as Smith found in an interview with a retired Permanent Secretary at MAFF:

> The Minister would go and argue the position worked out between MAFF and the NFU in Cabinet Committee and Cabinet. Therefore if there was a disagreed review it was not really a conflict between the Ministry and the Union but between the Union and the government.
>
> (M.J. Smith 1990b: 131)

In addition to the contribution of the Union and the Ministry, a key role was assumed by agricultural economists in university departments whose provision of farm financial data through the MAFF-sponsored Farm Management Survey was a central element in the review process (Bowers 1985).

The Act, whilst establishing a broad framework of policy objectives, took a sufficiently flexible line to allow changes of direction in policy according to changing conditions. Thus it remained the basis for policy until 1972–1973, during which time three major policy phases may be identified (Davey *et al.* 1976):

* from 1945 to the early 1950s
* from the early 1950s to the early 1960s
* from the early 1960s to 1972.

The period after 1972, by which time the UK had joined the EC, represents a fourth period prior to the crisis engulfing agricultural policy after 1980.

The first phase of post-war policy was marked by expansion of output regardless of cost in order to provide food in conditions of shortage, a deteriorating trade deficit, and in the context of continuing government controls over both wholesaling and retailing. Agriculture had a key role to play in 'saving dollars and thus ending Britain's need for Marshall Plan aid' (M.J. Smith 1990b: 103). The cost to the exchequer was considered of minor importance in the light of such conditions. Although the cost to government excited some attention at the time, in the light of CAP expenditure in later years, it was relatively small as genuine home demand rose with the economic expansion and full employment of the 1940s. Fixed prices meant that all farmers received identical commodity payments irrespective of contrasting costs or market acumen. Production incentives were based on a sufficiently high guaranteed fixed price and fear of supervision orders or eviction. M.J. Smith (1990b) has identified this expansionist period as crucial to the development of a tight and closed policy community.

Before considering the next two periods of policy we need to spend a little time on one remaining feature of the 1947 Act, for the partnership between farmers and the state found further expression in the requirement of the Act for farmers and landowners to comply with the jurisdiction of the County Agricultural Executive Committees (CAECs), which inherited the mantle of the WAECs.

The County and District Agricultural Executive Committees: birch in the cupboard or partnership in the field?

The most recent full-length book account of post-war agricultural policy (M.J. Smith 1990b) makes virtually no mention of the CAECs or of their importance. By contrast Self and Storing (1962) devote two chapters to the committees, in which they emphasise the importance of sanctions in the post-war deal (the birch in the cupboard) and the role the CAECs played in ensuring a successful outcome to the productivity drive (partnership in the field). Their significance has been further developed by Cox, Lowe and Winter (1985, 1986a) who see the CAECs as crucial to the implementation side of the corporatist deal.

The 1947 Act's requirement that farmers and landowners should comply with 'rules of good husbandry' and 'good estate management' (failure rendering guilty parties liable to supervision orders or eviction) was taken in parliamentary debate and by contemporary observers to be a central feature of the legislation:

> it represented one way of trying to relate state support to improved efficiency, and it called for close co-operation between government and agricultural interests providing a striking example of the principle of partnership in action.
>
> (Self and Storing 1962: 27)

The CAECs embodied this principle, providing a focus for a local sectoral coalition of state and farmers. The Minister of Agriculture appointed five members directly, one being a member of the local county council; and seven further members were selected from a list of nominees put forward by the three interests involved – three farmers' representatives, two workers' representatives and two landowners' representatives. In practice this meant a selection process operated by the NFU, the NUAW, the TGWU and the CLA. The committees thus combined state and industry interests, enabling the NFU to claim that agriculture was self-governing and government that all county representatives were solely responsible to the Minister:

> the committee are not delegates, subject to instruction and recall, but the subtlety of committee administration lies precisely in their dual role as Minister's agents and sectional representatives. . . . The Ministry did not expect committee members to cease to take an active part in the affairs of their organisations, and it saw no incongruity in a member holding the public balance in the morning and wielding the private sword in the afternoon.
>
> (Self and Storing 1962: 144–145)

Certainly there was official acceptance of the double responsibilities of the committees: 'being the agents of the Minister they have to accept responsibility for the local implementation of government policy. Being local bodies, they must also protect the farmers in their area from ministerial interference and unreasonableness' (PEP 1949).

The committees were involved in three main types of work. First they were responsible for the maintenance of standards of farming and land-holding and for imposing sanctions where necessary. Although the number of dispossessions and supervision orders was not high (see Table 5.1), the committees took their duties seriously and it is impossible to estimate the number of farms who made adjustments so as to avoid the threat of sanctions. The emphasis on this work diminished as memories of wartime food shortages receded in the 1950s. Under the Agriculture Act 1958 the powers of supervision and eviction were repealed altogether, not without a measure of opposition from the NFU which feared that the price for such a freedom might be a future reduction in the scale of support (H.T. Williams 1960).

Second the committees provided the general administration of subsidies, grants and regulations. Much of the routine work was undertaken by officials of the National Agricultural Advisory Service (NAAS), established within the Ministry but working under the committees' auspices. The committees were, however, in a strong position to intervene in more delicate cases and provide 'local' policy initiatives on the more complex administration of regulations, for example the Milk and Dairies Regulations designed to improve hygiene in dairying. A 'local corporatism' is perhaps a useful way to describe a situation where policy is administered through a committee and a state advisory service working closely together:

> In their administration of grants and regulations the committees could mediate between the Ministry's demand for national uniformity and strict compliance with regulations and farmers' preference for individual exceptions and lax administration. In the case of the milk regulations, the committees were usually able to maintain a steady pressure on substandard producers, while at the same time restraining their often impatient technical officers; but more often they tended, as might be expected, to be too soft-hearted. It was extremely difficult, for example, to persuade them that any farm was not an economic unit, and they were reluctant to discriminate between the needs of different farmers. An example is the marginal production scheme, which was introduced in wartime to help farmers who had poor land or other production difficulties and which was continued until 1959. Some county committees found it easier to use up most of the money in grants for various specified operations carried on by any farmer rather than to investigate the awkward question of which farmers most warranted this assistance.
>
> (Self and Storing 1962: 147)

The third function of the committees was the provision of technical advice, again in close connection with NAAS. The district committee acted as sponsor for freshly recruited young advisory officers, introducing them to

Table 5.1 Supervision and dispossession: action taken under the Agriculture Act 1947, Part 2

Action	1948	1949	1950	1951	1952	1953	1954	1955	1956	1957	Total
Estate management supervision orders	25	73	159	163	186	122	53	18	12	0	811
Husbandry supervision orders	530	838	724	608	812	421	183	51	30	3	4,200
Dispossessions on grounds of bad estate management[1]	0	0	1	9	1	5	5	1	1	0	23
Dispossessions on grounds of bad husbandry[2]	1	23	64	69	112	82	19	6	0	1	377

[1] Certificates enabling the Minister to purchase the land compulsorily
[2] Orders terminating an occupier's interest or occupation
Source: Self and Storing 1962: 238

farmers, publicising meetings, and finding farms for demonstrations. Committee members instructed the officers about local agricultural conditions and warned of particular farmers' idiosyncrasies and prejudices (Self and Storing 1962).

Problems with the powers vested in the CAECs resulted in two committees of inquiry in the 1950s, the Ryan Committee of 1951 and the Arton Wilson Committee of 1956. Both identified inefficiencies in the closeness of the county committees' administrative and regulatory functions to their technical advisory function. The Arton Wilson Committee, in particular, took its lead from the desire of the Conservative government of the day to remove more of the vestiges of wartime economic intervention. What the recommendations amounted to was the desire to hive off the administrative functions to a higher level, and reduce the county committees' role to that of providing moral support for a rapidly growing and technically sophisticated advisory sector. The Arton Wilson Committee, in particular, did not mince its words on how the committees should adapt:

> With the recent transition to 'freer economy' in agriculture we consider that the last link between War AECs, and the modern Committees has been severed. It would be as well if this were recognised by the country, the industry and perhaps by more Committee members themselves. . . . the work which mainly occupied CAECs has already come to an end.
> (Cmnd 9732, 1956: paras 25, 27)

In addition to the careful deliberations on administrative matters by the Arton Wilson Committee the Conservative government was influenced by a number of cases of malpractice associated with the Ministry's powers to intervene in the land market. The most notorious, the Crichel Down case, led to the resignation of the Minister of Agriculture, Sir Thomas Dugdale, and almost certainly convinced the government of the need to reduce the powers of the Ministry and the CAECs (Chester 1954; Self and Storing 1962). The Agriculture Act 1958 removed the CAECs' powers of sanction, but stopped short of abolishing them. However, the new streamlined advisory committees were disbanded under the Agricultural Miscellaneous Provisions Act 1972 (to be replaced by Ministry Advisory Committees at a regional rather than county level), a Conservative measure designed to reduce bureaucracy. The new regional committees, which still exist, have very limited functions indeed, and most farmers are probably not even aware of their existence.

Meanwhile, a year before the demise of the county committees NAAS was replaced by the Agricultural Development and Advisory Service (ADAS), amalgamating NAAS and three other groups of advisory experts: the Ministry of Agriculture's veterinary officers, the Land Drainage Department, and the Agricultural Land Service (Beresford 1975). The formation of ADAS aroused suspicion within the NFU as it substantially blurred the distinction between an independent advisory service for farmers and an executive arm of government. The demise of both the CAECs and NAAS, whilst largely unlamented and unnoticed outside NFU circles at the time, might in retrospect be seen as a fundamental shift in the nature of the policy arrangements for agriculture, a shift somewhat lamented by some more recent commentators anxious to improve the administration of more complex farming and conservation policies (Pye-Smith and North 1984; Winter 1985a, 1985b).

AGRICULTURAL POLICY, EARLY 1950s TO EARLY 1960s

The sterling crisis at last eased in the early 1950s at the same time as a general improvement in world food supplies. 'When world prices were high the cost of guaranteeing agricultural produce was fairly cheap. The return of plentiful supplies of food to the world market caused home prices to fall and the cost of support to the Exchequer to rise' (M.J. Smith 1990b: 119). An early policy change, and one consistent with the market-orientated policies of the Conservative government elected in 1951, was to end the system of fixed prices. There was a gradual shift after 1953 from fixed prices payable to all producers to a system of minimum support prices, or deficiency payments:

> In this way, government guarantees to producers would operate only when the free market price of a commodity fell below the minimum price negotiated at the annual review. From the farmer's point of view, the main difference was that the guarantee was no longer related to the individual producer, but commodity by commodity to the industry as a whole. He picked up a deficiency payment if the market price fell below the minimum price, but the deficiency payment was the same for the weak seller as for the strong. The weak seller might make less than the minimum price; the stronger seller made more.
>
> (Beresford 1975: 26)

This was a policy designed to limit exchequer expenditure by encouraging efficient home production. Curiously, M.J. Smith sees this as 'only a change in the mechanism of price support' (1990b: 119), pointing out that the increased dependency upon world market prices meant less control for government and the prospect of escalating costs. In reality it was a fundamental policy shift, notwithstanding the political rhetoric of consensus and continuity, for it presaged the cheap food policy of the 1950s and 1960s at a time when other European countries had returned to tariffs and import levies (Bowler 1979).

In order to limit expenditure, there was renewed attention to the efficiency of home production. By the early 1950s, the across-the-board policy was, in any case, proving insufficiently sensitive to the structural diversity of the agricultural industry. Certain sectors of production, for political, social or economic reasons, were seen to be deserving of special treatment. More importantly, the largesse handed out to farmers in the immediate post-war years, whilst it encouraged production, did little to encourage long-term structural change in the industry. Farm sizes, for instance, remained unchanged so that the government was under constant pressure from the NFU to set prices that would give a good living to farms that, it might be argued, were barely viable. Measures to increase on-farm efficiency, although not at this stage to encourage the amalgamation of farms, were required. In 1951, the Livestock Rearing Act extended to lower ground the grants for capital investment that had already been made available to the upland farmers in the Hill Farming Act 1946; and in 1952 fertiliser subsidies, ploughing grants and beef calf subsidies were all introduced (Bowler 1979). Not only did these measures point to the need for the selective encourage- ment of certain sectors of food production, they also pointed to the increasing awareness that a goal of agricultural support must be improving agriculture's own efficiency in the use of resources as well as simply providing price support.

If capital investment and more intensive production methods provided one means of increasing efficiency, improved marketing offered another. The Agricultural Marketing Act 1949, whilst restricting some of the monopolistic powers of marketing boards, reaffirmed their centrality in the post-war policy settlement.

However, 'unless guaranteed prices could be brought down, the "eff-iciency drive" served simply to make the public expenditure problem worse' (Bowers 1985: 70). Consequently milk and wheat guarantees were cut in 1957 and sheep guarantees in 1959. The legislative centrepiece of this period was the Agriculture Act 1957 which attempted to combine both the new-felt need to limit expenditure on agricultural support and the continuing commitment to the principles of the 1947 Act. Thus for the Conservative Minister of Agriculture, Heathcoat Amory, introducing the Bill in Parliament it was:

a logical development of the Agriculture Act, 1947, which, with the support of all the parties, has for the past ten years been the statutory instrument through which the national support for agriculture has been provided.... agricultural development must concentrate on ever-improved efficiency and reducing unit costs of production.... farmers must have two things: confidence and capital.

(Hansard, 25 March 1957, vol. 567, cols 817–818)

Much of the Act was concerned with refining the arithmetic procedures to be utilised at the annual review. It set limits on the permitted reduction of the total value of the price review award (not more than 2½ per cent in any year) and on the variation in single commodity prices (not more than 4 per cent in any year). The award to agriculture was also, henceforth, to formally reflect the expectation that the industry would absorb increased production costs of £25 million each year (£30 million after 1966) through increased efficiency (Bowler 1979). Efficiency was also accorded high priority by the extension of farm improvement grants, hitherto confined to the hills, to all farms. Subsequent developments aimed at the improvement of agricultural efficiency included the introduction of a Small Farmer Scheme to assist the development of viable small farms, and the development of a nation-wide advisory service, the National Agricultural Advisory Service (NAAS). Another key policy change was the abolition in the Agriculture Act 1958 of the powers of sanction held by the agricultural executive committees.

In an attempt to limit the cost of support, standard quantities for domestic products were introduced. Under this policy the government set limits on the level of production they were prepared to support; beyond a certain level the total of deficiency payments for a product was to be reduced. The measure was introduced by Amory for milk and potatoes prior to the 1957 Act; it was extended to barley and pigs in 1961, eggs in 1963, and to cereals in 1964 (Bowers 1985; Bowler 1979; Wilson 1977). However, the details of standard quantities and the level of deficiency payments remained a topic for determination at the annual review, and the NFU had become increasingly adept at driving a hard bargain. The government's drive to shift the focus of support did nevertheless meet with considerable success. By 1959–1960 grants accounted for nearly 40 per cent of the agricultural support budget (nearly 50 per cent by 1970: Bowler 1979; Bowers 1985). Despite the obvious strengths of the farming lobby, it did not have everything its own way during this period. It is sometimes assumed that corporatist relationships give in-built advantages to interest groups. This is not always the case and in the 1950s the NFU found itself locked into a relationship that did not always yield the results it was looking for.

AGRICULTURAL POLICY, EARLY 1960s TO 1972

The third phase of agricultural policy, from the early 1960s to entry into the EEC in 1972–1973, marks a period of agricultural protection in the face of growing world food surpluses. The changes were presaged by a notable victory for the NFU in 1960 when it took an initiative outside the normal procedures of the annual review, by securing a meeting between the NFU President, Harold Woolley, and the Prime Minister Harold Macmillan. This resulted in talks between the agriculture ministers and the NFU leaders outside the normal review process – the resulting white paper was considerably more favourable to farming than the annual reviews of the time (Bowers 1985; Cmnd 1249, 1960). One of the suggestions in the white paper was that consideration should be given to international negotiations to control markets 'to the mutual benefit of all'. Consequently, despite a world market in the 1960s potentially of great benefit to British consumers, protection came to the fore in British agricultural policy, with the Labour governments of 1964–1966 and 1966–1970 placing great emphasis on the import-saving role of British agriculture. Wilson was particularly susceptible to import-saving arguments because of his obsession with the balance of payments problem. Thus standard quantities for wheat and barley were removed in 1968 and 1969, and the vestiges of a standard quantity policy for milk (administered 'informally' by the Milk Marketing Board after 1963) gradually disappeared.

A more immediate outcome of the NFU's campaigning was the introduction of import controls in 1963 and in the annual review of 1964:

> the British government, almost unnoticed, breached one of the basic
> principles of British trading policy since the repeal of the Corn Laws
> – that there should be an open door for imports of cheap foodstuffs,
> particularly from the Dominions. The sharp rise in costs convinced
> policy makers that deficiency payments and an 'open door' for imports
> could not co-exist in an era of large surpluses on world commodity
> markets. Typically, the 'open door' was partially closed, while de-
> ficiency payments were only slightly modified.
>
> (Wilson 1977: 14)

Minimum import prices allowed for payments to foreign suppliers to make
their prices *higher* than the world market price. This was done in order to
maintain British farm prices and, at the same time, to avoid physical
restrictions or tariffs on imports which might lead to damaging retaliation
against British exports.

Import saving to ease the UK balance of payments difficulties became an
often repeated rallying cry in the 1960s and has been identified by M.J. Smith
(1990b) as one of the main myths propagated by the NFU to maintain
agricultural support. As Smith documents, the NFU position found ready
support from, *inter alia*, ministers, civil servants, the House of Commons
Select Committee on Agriculture, and the National Economic Development
Council (NEDC). Opposition seemed confined to the ranks of academic
economists (see Hallet 1959; Josling 1970; McCrone 1962; Nash 1965),
whose claims that the resources devoted to agriculture might be better
deployed elsewhere in the economy, or that agricultural intensification too
often led to increased imports of such things as fertiliser and machinery, fell
on deaf ears.

Between 1960 and 1970, net farming income rose from £392 million to
£561 million, fractionally ahead of the rate of inflation; productivity
continued to rise but total production did not rise as fast as many hoped
(Beresford 1975). Notwithstanding the growth of protectionism during this
period, the 1960s were not happy years for ministers of agriculture or for the
NFU. Agricultural discontent and restlessness threatened the corporatist
'behind-the-scenes' style of policy making in an unprecedented manner.
Between 1964 and 1965 there was even talk of 'industrial action' by the NFU,
and relations between Labour's Minister of Agriculture, Fred Peart, and the
NFU were decidedly cool. Both sides were discovering the problems of
corporatist policy making, particularly for such a diverse industry as
agriculture. For the big-farm leadership of the NFU, one of the problems
was that Peart was attempting to redress some of the imbalance between large
and small farmers. He was not content to confine agricultural policy to price
guarantees but was keen to extend the Small Farmer Scheme and to do more
to help smaller producers with support for marketing and business advice.

Later in the decade it was the turn of the smaller farmers to feel under threat, and 1969 and 1970 witnessed scenes of militancy on a scale not seen in British agriculture before or since. At the heart of the problem was a dispute about the figures on net farm income used by the government in its calculations for fixing the annual review. A steady increase in farm income for the industry as a whole was, not surprisingly, used by the government to justify restraint in the determination of its award to the industry. But many farmers felt that the official figures did not tally with the experiences of their own particular business and accused the government of massaging the figures or relying on figures from unrepresentative farms (Day 1968). There was probably some truth in the latter accusation as the figures essentially derived from the self-selected participants of the Farm Management Survey, conducted on behalf of the Ministry by university agricultural economics departments. These farms tend to be somewhat larger and more efficient than the average.

Agricultural discontent at this time is scarcely touched upon in accounts of agricultural policy development in these years, where the emphasis tends to be upon the 'closed' politics of agriculture (e.g. M.J. Smith 1990b) or upon the policy changes presaging entry to the EEC and the new era of protectionism (but for discussion of discontent during the early 1970s, see Grant 1978). In retrospect, farmers appear to have had relatively little cause for serious alarm. The cost–price squeeze was more than compensated for by technical advance and incomes held up relatively well. However the pace of technological change was rapid and this inevitably led to casualties. Economies of scale really began to bite in British agriculture and the average size of farms, so long stable, began its inexorable rise. Many farmers felt insecure in the face of uncertainty over whether or not the UK would enter the Common Market. Perhaps, too, they compared themselves unfavourably with other sections of the community as they contemplated a fast-changing economy and society. Another factor was undoubtedly the new Labour government after 1964. There was not so much an antipathy towards Labour's agricultural record, but rather the feeling that Wilsonite labourism was so thoroughly urban and modern in its preoccupations that the farmers were being left out. The rhetoric of the early years of the Wilson era with its emphasis upon economic planning, the expansion of education, technological advance, the 'white heat of the scientific revolution', must have meant little to many farmers – likewise the preoccupation with sterling in the later years (see Pimlott 1992).

THE NFU AND MONOPOLY OF REPRESENTATION

Before turning to the issue of European agricultural policy, it is important to make some further comments on the role of the NFU during the period under discussion in this chapter. It is important to recognise the extent of the Union's involvement in policy and wider aspects of the industry; so much so that membership came to be seen as almost 'compulsory' for many farmers. There are a number of ways in which membership levels may be kept high by interest groups and the potential for competing organisations severely reduced. The first is the provision to members of important fringe benefits. The insurance and legal services of the NFU have always been an important element in retaining members and in financing the work of branch secretaries. Second, the Union has, at different times in its history, involved itself in commercial activities. For example, at one time it held a 75 per cent stake in Europe's largest meat company, the Fatstock Marketing Corporation. But in contrast to the insurance service, these endeavours have not been particularly satisfactory and have done little to make membership essential. Finally, the NFU has been a provider of technical information through its publications and branch meetings. Again this service can do much to attract and retain members whose business interests demand this kind of service. Thus while membership can in no way be described as compulsory as some pure corporatist theorists suggest it should be, the dominance of the NFU as a politically representative body for farmers provides a scenario of 'no alternative' to many. 'In other words, risk avoidance and insurance can loom large in calculations about membership, and an association may not have to offer excessive benefits to make membership on balance a strong pull' (Williamson 1989: 78).

As Williamson points out, 'organized interests representing functional interests show a tendency towards a position of monopoly ... with unprivileged groups excluded from the policy process' (ibid.: 68). On this count, it is fair to say that the relationship between government and the NFU (taken to include for this purpose the Scottish and Northern Ireland unions with which the NFU enjoys close links) was close during the period in question and that few organisations, if any, were able to challenge the monopoly established by the NFU. Some caveats need to be applied to this so that a complete picture may be painted.

In the sphere of marketing policy, the NFU has not formally enjoyed a monopolistic position owing to the existence of various statutory marketing boards and commissions. However, in terms of the debate on monopolistic representation these could be viewed as merely another opportunity for the NFU to exercise its monopoly position. For example, the Milk Marketing Board was essentially a corporatist structure outside the terms of the 1947 Act. In reality the NFU heavily influenced the candidatures of prospective Board members. Conversely, the Union's stance on milk price negotiations depended heavily on the evidence supplied by the MMB. The two worked in close concert if not total agreement.

The Country Landowners' Association (CLA) might potentially be seen as competing with the NFU to represent agricultural interests. In practice open competition has been rare. On most issues of agricultural policy, particularly commodity issues, the CLA has been content to let the NFU take the lead. Similarly the NFU has been happy to see the CLA develop its own specialisms, many of which also benefit NFU members. For example the CLA's expertise on legal aspects of landownership, taxation and public rights of way is widely recognised. On one issue, agricultural holdings legislation, there has been periodic conflict and the absence of unanimity on this means that, in effect, farm tenure is an issue situated somewhat outside the scope of the normal corporatist arrangements.

There have, of course, been some explicit challenges to NFU hegemony, and it is worth spending some time considering the nature of these challenges and whether or not they diminished the NFU's monopoly of representation. We consider here the best known case – the Farmers' Union of Wales (FUW) (drawing on Cox, Lowe and Winter 1987, 1990a; see also Murdoch 1992). The formation of the FUW in 1955 was occasioned principally by the dissatisfaction of hill sheep farmers and small marginal dairy farmers with an NFU perceived to be dominated by larger lowland arable farmers. The first president (Ivor T. Davies) and general secretary (J.B. Evans) had been chairman and county secretary respectively of the Carmarthenshire branch of the NFU. In the early years the Union had a considerable struggle to establish itself against the hostility of the NFU and the indifference of MAFF. But support from predominantly Welsh-speaking smaller hill farmers grew, and indirectly the rise of political Welsh Nationalism in the 1960s assisted the cause. The early struggles, like the opposition to central English bureaucracies, nurtured the Union's at times beleaguered mentality, neatly linking threats to Wales and Welsh agriculture with threats to the Union itself. The words of one-time president Glyngwyn Roberts, writing in the FUW's journal *Y Tir*, sum up the position:

> Born out of the soil of Wales – without capital, staff or office facilities – we can be proud of the growth of our Union. But there can be no room for complacency. Because we refused to conform, we were born into controversy.... The opposition in the early years was constant and intense. The only result of this was to send our roots deeper and we face the future knowing that all efforts to crush and suppress our name have failed.

In contrast to the NFU's comfortable accommodation with mainstream, and particularly Conservative, English political thinking, the FUW espoused a populist ideology, defending family farming as economically, politically and morally superior to large-scale capitalist agriculture (see Ionescu and Gellner 1969). The protest activities of agrarian populism may go beyond the narrow defence of farming independence. In particular they can be linked to a cultural nationalism, reflecting a sense of the decline of a traditional way of life. Thus, addressing the FUW Annual General Meeting in 1970, Plaid Cymru's leader Gwynfor Evans declared, 'in fighting for a prosperous agriculture based on the family farm we are in fact fighting for the traditional culture of Wales'. But alongside these ideological strands, has been an increasing political pragmatism. For example, the Union vigorously supported retention of the unified English and Welsh Milk Marketing Board during the threat to its existence in the mid-1970s after entry into Europe, not least because of its good representation on the Board. Labour's John Silkin, also popular with the FUW for according it full consultative status with government, was given a hero's welcome at the 1978 FUW AGM for his part in saving the Board.

Thus the Union had to wait more than twenty years to achieve consultative status within the annual review, albeit at a time when relations between the NFU and the Labour government were particularly sour. This means that its challenge to the monopoly of the NFU must be seen as serious and genuine. However its regional identity prevents serious damage to the NFU's monopoly as it cannot be seen as a competitor over the whole range of the NFU's territory, either geographically or in policy terms. It is probably fair to say that for all the irritation that the FUW has posed to the NFU, it has scarcely dented the underlying monopolistic position of the NFU.

AGRICULTURAL POLICY, EARLY 1970s TO 1980

The fourth phase of agricultural policy concerns the entry of Britain into the EEC and the adaptation of its agricultural policy to fit the requirements of the Common Agricultural Policy. Before considering precisely how this affected UK policy after 1973, it is worth giving some consideration to the origins and development of the CAP itself, prior to UK membership. The 1957 Treaty of Rome, signed by the original six members of the EEC, included Article 39 which laid the foundations for the CAP in the following terms:

> The Common agricultural policy shall have as its objectives: (a) to increase agricultural productivity by promoting technical progress and by ensuring the rational development of agricultural production and the optimum utilisation of the factors of production, in particular, labour; (b) thus to ensure a fair standard of living for the agricultural community, in particular by increasing the individual earnings of persons engaged in agriculture; (c) to stabilise markets; (d) to assure the availability of supplies; (e) to ensure that supplies reach consumers at reasonable prices.
>
> (quoted in B.E. Hill 1984: 19)

In 1958, the first agricultural commissioner, Dr Sicco Mansholt, chaired a conference at Stressa to examine how such a policy might be developed. In addition to national government representatives this was attended by representatives of the farming organisations, a clear recognition of the importance of the agricultural lobby from the very outset of the CAP. This was hardly surprising in view of the fact that over 20 per cent of the Six's working population were engaged in agriculture at this time. The conference, building upon Article 39, agreed the following objectives:

(a) to increase agricultural trade within the Community and with other countries;

(b) to maintain a balance between structural and market policies;

(c) to avoid surpluses, and to give scope to the comparative advantages of the regions;

(d) to eliminate subsidies which would distort competition;

(e) to improve returns to capital and labour;

(f) to preserve the family structure of farming;

(g) to encourage rural industries which, by providing new job oppor-tunities, would assist the removal of surplus labour, and to provide special aid to disadvantaged regions.

(Butterwick and Rolfe 1968: 6–7)

Between 1961 and 1963 the details of the policy were put into place, covering most of the major agricultural commodities, but common pricing did not come into place on any commodities until the late 1960s (B.E. Hill 1984). On the face of it the objectives of Article 39 and of the Stressa Conference were broadly similar to the UK's own policy objectives. But considerable stumbling blocks were encountered when the UK first attempted to negotiate entry to the Community between 1960 and 1962. The UK's historic openness to imports of agricultural commodities from Commonwealth countries posed a particular problem (Neville-Rolfe 1990). The 1970–1972 negotiations were little easier but they were successful. The UK obtained concessions on continued trade with New Zealand and with sugar-producing countries. It also secured a five-year period of transition to the higher agricultural prices of the Community (Neville-Rolfe 1990). However, the only major element of British policy to continue virtually unaltered in the Community was its policy for the support of hill farmers, adopted for the Community as a whole in 1975 as the 'Less Favoured Area Directive' (Neville-Rolfe 1990: 177).

The policy that Britain had to comply with upon entry to the EC contained three main elements (B.E. Hill 1984; Neville-Rolfe 1990):

- *common prices*
- *community preference*
- *common financing.*

In the terms of the Commission itself, 'the single market, Community preference and financial solidarity ... have become the golden rule of the CAP' (Commission of the European Communities 1979: 12).

In theory there is a common policy of pricing throughout the EC, based on separate price regimes for most major products, decided annually by the Council of Ministers (Marsh and Swanney 1980). In practice the use of monetary compensatory amounts and floating exchange rates seriously undermined the principle of common pricing (Fennell 1979), so that the 'single market has become a myth, a pretence permitted through the medium of MCAs' (B.E. Hill 1984: 91). MCAs, in particular, operated as a system of border taxes or subsidies, benefiting those whose currencies had appreciated and penalising those with depreciating currencies, such as the UK in the 1970s (Grant 1981). Thus, by 1979 West German farm prices were 45 per cent higher than those of the UK and seven price zones operated rather than a single common price (Swinbank 1978).

The pricing policy of the EEC presents a bewildering array of terms – target price, guide price, normal price, basic price, intervention price, withdrawal price, minimum price, production aid, deficiency payment, threshold price, sluice-gate price, reference price, variable levy, supplementary levy, customs duty and restitutions. Many of these pricing policies are variations on a common theme and represent modifications of basic principles for different agricultural commodities. Thus there are basically two types of pricing mechanism – *internal* and *external*. Internally, the Community sets target prices annually as the desired or normal prices for most commodities. These are not guaranteed prices as such; rather 'frontier protection and internal support measures derive from them' (Fennell 1979: 107). A minimum guaranteed price is provided by the intervention price. When the market price for a commodity slips to a certain level the Community buys into intervention sufficient quantities of the product to ensure that prices do not fall below the intervention level. This is financed through the Guarantee section of FEOGA (European Agricultural Guarantee and Guidance Fund). Stocks are released on to the market when the prices have again risen. Externally threshold prices are set annually as prices which must be reached by imports, so fulfilling the principle of community preference. These are maintained by variable levies which are a charge on imported goods.

The third principle, common financing, has proved important symbolically but is as practically complex as common pricing. As B.E. Hill (1984) points out, there is no economic reason why the CAP could not be financed through national budgets, but the use of a common fund became very much an article of faith for the Commission during the 1970s. The result, because the CAP consumes such a disproportionate share of the EC budget, is that those countries with agricultural exports are subsidised by those with agricultural imports, irrespective of their wealth. The benefits accrued by Denmark and the Netherlands and the losses sustained by Italy and the UK have only reinforced existing inequalities of GNP (B.E. Hill 1984; see Table 5.2). One of the major consequences of this aspect of the CAP, as we shall see in chapter 6, has been seriously to discourage some countries from contemplating radical reform of the CAP.

Although the Guarantee section of FEOGA has dominated the Guidance section, a variety of structural policies have been funded with Guidance funds. Thus, capital grants schemes continued with Britain's membership of the EEC, but were more closely linked into overall farm development schemes, the aim being to establish efficient farms and, especially in the light of the European 'small farmer problem', efficient agricultural structure. The Modernisation of Farms Directive (Directive 72/159/EEC) led in Britain to the Farm and Horticultural Development Scheme (later the Agricultural and Horticultural Development Scheme) with its comprehensive fixed term plans (usually five years). Another major directive, of continuing importance to

Table 5.2 Budgetary contributions and receipts 1979 (million ECU)

| | Contributions from own resources | Receipts | | | | | Net contributions |
		FEOGA	Social fund	Regional fund	Collection costs	Total	
Germany	4,407.2	2,592.1	61.4	46.0	197.5	2,897.0	1,510.2
France	2,886.5	2,549.4	93.7	103.6	98.9	2,845.6	40.9
Italy	1,793.2	1,341.3	156.3	143.7	96.2	1,829.7	−36.5
Netherlands	1,344.1	1,559.5	11.1	8.6	84.9	1,664.1	−320.0
Belgium	966.5	806.5	7.8	3.1	58.8	876.2	90.3
Luxembourg	19.4	14.2	0.3	0.3	0.4	15.2	4.2
UK	2,513.5	589.7	201.9	165.7	168.6	1,140.9	1,372.6
Ireland	104.6	602.1	38.8	32.9	6.3	746.2	−641.6
Denmark	337.4	783.2	24.5	9.1	14.9	831.7	−494.3
Total	14,372.4	10,837.9	595.7	513.1	726.6	12,846.6	1,525.8[1]

Source: B.E. Hill 1984: 94

[1]The excess of contributions over receipts for the Nine reflects administration costs, overseas aid, etc.

Britain, is the Mountain and Hill Farming in Certain Less Favoured Areas Directive (Directive 75/268/EEC), which has had a significant impact throughout the upland areas of Britain.

The policy mechanisms within the EEC contrasted with Britain's post-war policy instruments, and, more significantly, boosted expansionist policies which had appeared to be faltering in the late 1960s. The relinquishment of a cheap food policy signalled a renewed period of agricultural prosperity. Whilst the problem of surpluses was recognised at the European level from the mid-1970s, this did not prevent Britain from continuing to encourage expansion in an attempt to reduce the budgetary and balance of payments costs of entry into the EEC (Bowers 1982; Cmnds 6020, 1975; and 7458, 1979).

Despite the expansionist framework, the early years of UK membership were characterised by a return to the agricultural discontent that had marked parts of the previous decade. This was largely due to an unprecedented crisis in the beef sector, the consequence of inept handling of the transfer from UK deficiency payments to EC guide and intervention prices. In 1973, the Conservative government ignored NFU advice to delay the switch-over (Grant 1978). The Labour government, elected in 1974, inherited escalating prices but was determined to avoid fuelling food inflation still further (not least because the Labour leadership wished to avoid anything that might jeopardise public confidence in the EC in the run-up to the membership referendum). They promptly opted out of taking beef into intervention. Prices fell, despite an emergency package of EC measures permitting the UK to introduce a headage premium. Demonstrations against imports of Irish beef took place at Holyhead, Fishguard and Birkenhead, prompting clashes with the police. The NFU steered a delicate course, neither condoning nor condemning the actions of its more militant members, and managing to continue a dialogue with government, which ultimately was persuaded to challenge the European Commission to amend its policies:

> agricultural politics has become more crisis-ridden in recent years, not just, of course, because Britain is now a member of the European Community; one must also take the example of successful industrial militancy into account. These outbreaks of militancy are not always easy for the NFU leadership to control if they are to retain both the confidence of their own members and the trust of government. The fact that the unity of the NFU has been maintained in such difficult circumstances is a reflection of the political skill of the organisation's leadership.
>
> (Grant 1978: 104–105)

AGRICULTURAL POLICY MAKING IN THE EUROPEAN COMMUNITY

Clearly another consequence for agriculture of entry into the EC was the impact upon agricultural policy making. One of the issues in both the 1960–1962 and the 1970–1972 negotiations over EC membership was the NFU's insistence that the annual review should be retained and, if possible, extended to the European arena. In 1963, the six countries agreed to regular consultations with COPA (the umbrella body for European farmers' organisations) and with this the NFU had to be content (Neville-Rolfe 1990).

European agricultural policy making has been considered in detail by a number of writers, and it is to their work we now turn. In chapter 2 we considered the policy process within the EC in general and Figure 2.4 (p. 50) showed in diagrammatic form the chief institutions formally involved in the policy process. To these must now be added the pressure groups which seek to influence policy. In the case of agriculture four warrant particular attention:

- the Comité des Organisations Professionelles Agricole (COPA), the umbrella body for European farmers' organisations;
- the Comité Général de la Co-opération Agricole (COGECA), the umbrella body for co-operatives;
- the Commission des Industries Agro-Alimentaires (CIAA) within the Union des Industries de la Communauté Européenne (UNICE), representing agricultural and food manufacturing interests;
- the Consumer Consultative Committee with representatives from the Bureau Européen des Unions de Consommateurs (BEUC), the European Trade Union Confederation (ETUC), the Comité des Organisations Familiales auprès des Communautés Européennes (COFACE) and the European Community of Consumer Co-operatives (EURO-COOP).

(Averyt 1977; Snyder 1985)

It is within these groups that the national pressure groups operate, so, for example, the NFU is an active constituent member of COPA striving to ensure that COPA's policies reflect, as closely as possible, its own. Inevitably compromises have to be struck and on occasions the NFU has been dissatisfied with aspects of COPA policy. In such instances it can lobby the European Commission or other EC agencies directly. Thus the NFU's Brussels office allows it to be in close contact both with COPA and with the Commission itself. In addition, it can revert to national policy lobbying, attempting to ensure that the MAFF in its consultative role *vis-à-vis* the EC puts forward views with which the NFU is in accord. Thus the 'promotion of a so-called "national interest" in Community bargaining is therefore the outcome of a prior battle in each national capital to determine what domestic interest should receive priority' (Wallace 1983: 69). The continuing importance of national political processes should not be underestimated when considering the impact of the UK's entry into the Community.

The example of COPA

As with the NFU in the UK, COPA is often presented as the model of a successful pressure group within the EC system (e.g. Moyer and Josling 1990), although since Averyt's pioneering research, published in 1977, there has been a surprising lack of detailed empirical research on this aspect of the agricultural policy process. COPA was founded in September 1958, following the Stressa conference, comprising fourteen farmer organisations from the six member states. Another French union joined in 1967 and membership increased to 22 when Denmark, Ireland and the UK joined the Community. The accession of Spain, Portugal and Greece increased the number of constituent members of COPA to a total of thirty (1994). Within the UK, membership is confined to the NFU, the NFU of Scotland and the Ulster Farmers' Union.

Interestingly, the FUW has been excluded from membership despite its recognition by the UK government in 1978 (although it is a member of an alternative European small farmer grouping). An organisation seeking COPA membership requires the unanimous support of all other members, and the NFU has blocked the FUW's attempts to join on the grounds that it is a regional organisation rather than a national one. In the light of the NFU's own regional complexion (it makes no claim to represent farmers in either Scotland or Northern Ireland), not to mention the existence of other regional groupings within member states (such as separate bodies representing Walloon and Flemish farmers in Belgium) this is clearly an exclusionary tactic by the NFU rather than merely a constitutional matter.

COPA's main focus of power is its presidium, which originally comprised one permanent representative from each member organisation. This meant that countries with only a single representative organisation, such as Germany, were at an electoral disadvantage within COPA, compared to countries, such as France, with several farmer unions. In 1973 this system was changed to a weighted voting system, which remains the case. Not surprisingly there are question marks concerning the representativeness of COPA:

> The elites that have the greatest resources to dominate national farm groups are also the ones that dominate COPA. The skills needed to work through COPA in order to influence EC policy accentuate the diversity of political resources among various groups of farmers. National farm policy in France or Germany is complex enough, but EC farm policy is so Byzantine that only farmers with large amounts of resources, especially time, knowledge, and wealth, are able to understand the decision-making process sufficiently to be able to influence

the output. COPA leaders thus represent an elite of an elite. They are drawn from large-scale agriculture. . . .

(Averyt 1977: 75)

This has led, on many occasions, to attempts by national farmer groups to subvert the process by lobbying ministers and even Commission officials directly. However COPA's level of resources and its corporatist relations within the EC make such opposition hard to sustain.

Members of the presidium have very high levels of access to commissioners, and middle-ranking COPA officials have good contacts with Commission officials. Presidium meetings are often attended by Commission officials as the presidium offers the Commission an immediate access to specialist information and knowledge. Thus farming organisations are potentially in a powerful position to influence policy debate, especially during the early stages of policy deliberation. COPA's influence rests on four main bases identified by Bowler (1987) as:

- the willingness of national affiliates to agree a common position on a given issue, including the use of qualified majority voting;
- the establishment and resourcing of active working groups carrying out and disseminating research on policy issues;
- assured representation within the decision-making machinery of the EC through advisory committees and the Economic and Social Committee;
- the presentation of a uniform view at national level to the ministers who make up the Council of Ministers.

Against this seemingly powerful position, it has to be said that COPA's close relations with the Commission mean that the Commission also expects to gain legitimacy for its own decisions. Another caveat regarding COPA's power is the lack of contact with the Council of Ministers, individual ministers often looking to national farmers groups rather than to COPA (Averyt 1977). The same point is made by Neville-Rolfe (1984) who argues that as agriculture has become more politically sensitive an issue in the Community so the focus for pressure group influence has shifted from Brussels to the lobbying of national ministers within member states. By contrast, Gardner argues that 'it is probable that the power which the agricultural pressure groups wield within the Community was and is a major factor which has subverted and diverted both the policy formulation and decision-making process' (1987: 170). Some more recent commentators on the decision-making process have given only the briefest of mentions to the influence of pressure groups as they attempt to unravel the complexities of the formal decision-making procedures within the institutions of the Community, in itself no mean task (e.g. Meester and Van der Zee 1993).

There is something of an unresolved research problem here. It is quite clear that in the 1960s and 1970s, the climate for successful co-operation within COPA was ideal:

Increasing expenditures on agriculture created a positive-sum game which encouraged the farm organizations to work together. The size of the budgetary pie was growing so new benefits for one farm group did not have to come at the expense of another. Co-operation seemed logical in that it created the maximum pressure on policy-makers to enhance the share of agriculture, which could then be divided among the various farm interests.

(Moyer and Josling 1990: 44)

Moreover, the European Commission itself was open to the demands of COPA in so much as agriculture was, from the outset of the Community, a major policy sector, a flagship in the emergent Community's struggles to establish its legitimacy:

The general weakness of the EC's public legitimacy as an authoritative political institution made it absolutely imperative for the Commission officials to establish a new corporatist relationship with farm organizations at the Community level. The farm lobby had every reason to oblige, given that Brussels is the primary source of benefits. The development of the CAP did not undercut and may even have strengthened the corporatist relationship between national agricultural ministries and farmers.

(Moyer and Josling 1990: 45)

According to Moyer and Josling, the incomplete integration of Community political institutions served to strengthen the power of COPA within this supra-national corporatist arrangement. Within an individual member state, key decisions on agricultural policy are likely to be brought before cabinet or its equivalent, where a minister of agriculture may or may not be strong. However, in the Community, decisions are likely to reside with the council of agricultural ministers. There is little central co-ordination of policy and only in cases of crisis do heads of state intervene. In the case of the UK, though, this mode of policy formulation resulted in agriculture becoming more of a central policy concern. As Britain is a net contributor to the Community's finances and agriculture takes such a significant share of the money, agriculture was seen by successive UK governments in the 1970s and 1980s as an area where total EC expenditure might be curtailed. Consequently, domestic agricultural policy came closer to the centre of the political stage than it had done for many years.

As agriculture became more politically sensitive as a result of the surplus crises of the 1980s, so the role of COPA has changed. It is no longer so central to policy genesis or to the lobbying of individual member states regarding CAP reform; but its influence on the form that policy proposals take has some remaining significance. According to a national representative of one of the member bodies of COPA, the organisation is increasingly handicapped by the difficulty of forging a consensus (personal communication 1993). Whilst consensus was relatively easy to attain during the expansionist period of EC agriculture, the crisis of the 1980s presented significant problems to COPA. Consequently, its policy statements were often reduced to a blandness or level of generality of no great help in policy making or, alternatively, to over-ambitious attempts to attain impossible goals.

The situation is aggravated by one of the consequences of majority voting in the Council of Ministers, namely an even greater importance to the role of the Commission in policy generation. The Commission is in a better position to achieve a clearer view than the Council and has less immediate need of COPA support than in the past. Under the rules of unanimous voting, COPA only had to find one government prepared to back its position in the Council of Ministers to allow it to bring considerable pressure to bear on Commission officials as they prepared policy. That is no longer the case, and there is an increasing tendency for national lobby groups to adopt direct lines of communication with Commission officials rather than attempt to seek an acceptable policy through COPA. When asked about the value of COPA membership, the official mentioned above responded that it was very useful as a source of information on the views and experiences of other lobby groups and member states. Valuable though information exchange might be, his comments provide scant evidence of a highly successful pan-European pressure group.

COPA's argument for retaining a unanimous voting system is that it does not wish to do anything which might marginalise or isolate minority farming groups within its membership. This is understandable for the prospect of a fragmentation of interests, even a disintegration of COPA itself, might be very real if a majority voting system were introduced. After all there would be little incentive for any union constantly holding minority views to remain within the organisation in such a context, a curious and rather paradoxical threat for a pan-European organisation during a period of deepening European integration.

CONCLUSIONS

This chapter has spanned nearly forty years of agricultural policy in the UK, a period which saw deepening levels of state and supra-state intervention in the industry. In this process, the NFU played a crucial role in the policy formulation process and in facilitating a relatively smooth implementation of policy. Despite differences of emphasis, which at times led to surprisingly strained relationships, the state and the farming lobby were essentially at one in wanting a prosperous and expanding agriculture. In this objective, policy was successful for this was a period of unprecedented expansion in agriculture. For example, UK wheat production increased from 3 million tonnes in 1960 to 8½ million in 1980. Barley production expanded from 4 to 10 million tonnes during the same period. There were more modest, but equally important, gains in the livestock sector.

Non-agricultural interest groups, representing food consumer or environ-mental interests, remained largely marginal to the policy process which can be characterised as closed and insular. As late as 1980, the NFU could still be seen as the model of a successful pressure group. But the seeds of future change had been liberally sown. The technological revolution in the industry was making rapid inroads into the numbers of farmers with long-term consequences for NFU membership figures. The impact of expansion on the natural environment was becoming an increasing cause of concern, but even that concern was surpassed by the worry about the net cost of European agricultural support to the UK government and taxpayer.

If the 1960s and 1970s had been the decades of a production and technological revolution for agriculture, the 1980s and 1990s were set to be decades of a political revolution.

6

THE CRISIS OF THE COMMON AGRICULTURAL POLICY

INTRODUCTION

On 31 March 1984 the EC Council of Agriculture Ministers introduced milk quotas with immediate, indeed retrospective, effect. This was high political drama and appeared, at first sight, to represent a fundamental shift from the ponderous execution of incremental policy shifts that had characterised the Commission's attempts to reform the CAP thus far. Farmers waking to such news on April Fool's Day might have been disposed to pause until after noon to check the veracity of the news reports. Not since the inception of the CAP had there been such a dramatic policy shift and the apparent lack of preparation by either the government or the industry's representatives gave an added potency to the deep symbolic importance attached to the imposition of quotas.

Thereafter, the remainder of the decade and the early years of the 1990s were a time of almost continuous policy review, debate and further rounds of reform. However many commentators have continued to lament the slow pace of reform and even quotas have been demonstrated to have been far less draconian than they first appeared. Writing three years after their imposition, a highly respected commentator on the CAP could find:

> no substantive evidence that preparations are in hand which would bring about a metamorphosis in the policy over the next decade. The existing pressures will remain; new ones are being added, but short of some cataclysmic occurrence bearing in on the CAP from outside, its crab-like progression can be expected to continue.
>
> (Fennell 1987: 63)

Whether the reforms that have subsequently taken place, and those that may yet be in hand in the wake of a long-awaited final agreement to the Uruguay GATT round in December 1993, amount to anything more than crab-like progression is open to debate. Few would argue, though, against the proposition that the pace of change has increased rapidly in the 1990s. At the time of writing the reform process is far from complete and there remain many voices sceptical about whether the cumulative effect of the policy switches will be adequate to deal with the problems facing the European Union.

This chapter examines the emerging crisis in European agriculture in the period after 1980 and the policy measures taken. The first section considers some of the main political and economic aspects of the crisis. Then the path to reform is considered, starting with an overview of the barriers to change and concluding with an examination of the main reform proposals and policy initiatives up until 1994.

THE GATHERING CRISIS OF THE CAP

The main, but not the sole, cause of the crisis of the CAP lies in the increasing surpluses and their cost to the EC budget. Being far from self-sufficient in food, the UK's accession to the Community in 1972 postponed the crisis for some years. As Table 6.1 shows, surpluses had already emerged in the six original members of the Community by 1972. The respite offered by the UK's accession proved short-lived and, by 1980, the Table shows that again the Community was over-producing significantly in several key commodities, particularly wine, butter, wheat, barley and sugar.

Thus in the UK alone, the cost of storing and handling food surpluses nearly quadrupled between 1980 and 1984. In the EC, total expenditure on storage grew from one billion ECUs in 1973 to a peak of seven billion in 1986 in real terms (1992 values) (Thomson 1994). Some of the trends are shown in Figure 6.1. A consequence of the cost of storing surplus food was not only that the cost of the CAP increased still further, despite the recognition since the early 1970s that it should be reduced, but that the CAP's proportionate share of the EC's budget also rose. By 1984, 69.8 per cent of the Community's entire budget was spent on agriculture. In the Community as a whole total gross FEOGA expenditure rose from just over 11 billion ECUs in 1980 to nearly 20 billion in 1985, increasing to 35 billion in 1993. Even in real terms, the trend was strongly upwards as shown in Table 6.2 and Figure 6.2. But, as many commentators have shown, these direct costs tell only part of the story. In addition to the direct expenditure through the public purse, the CAP also costs consumers in so much as it raises food prices above the levels they would otherwise be if world market prices prevailed. Table 6.3 gives an indication of the extent of the distortion, although it is important to point out that free trade would result in a reduction in EC production and consequently the possibility of an increase in world prices:

> the big question concerns the extent of the world price rise which extra imports would induce. No precise answer is possible; indeed estimation would be a mammoth undertaking – including, for example, how much the higher prices would cause other importers to reduce imports, which depend *inter alia* on how much their own agricultural industries would respond to the stimulus of higher prices.
>
> (B.E. Hill 1984: 88)

Estimating the costs to consumers of the CAP in 1980, Buckwell *et al.* (1982)

Table 6.1 Self-sufficiency ratios in the European Community for major agricultural commodities

	Wheat	Barley	Maize	Sugar	Potatoes	Wine	Butter	Beef	Pork	Eggs
1956–60 (6)	90	84	64	104	101	89	101	92	100	90
1972 (6)	111	110	68	122	101	95	124	81	99	99
1972 (9)	99	102	58	100	100	93	106	84	100	99
1974 (9)	107	103	59	92	100	115	93	100	100	100
1976 (9)	101	103	53	105	98	98	107	99	99	100
1978 (9)	102	112	58	125	101	93	118	95	100	101
1978 (10)	103	111	56	123	101	95	118	94	100	101
1980 (10)	114	111	62	125	101	112	120	103	100	101

Source: B.E. Hill 1984, based on: Statistical Office of the EC, Yearbook of Agricultural Statistics, various years
Note: Figures in parentheses indicate the size of the community to which the data refer

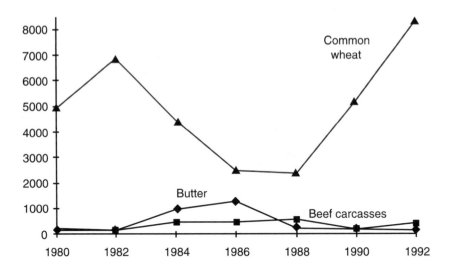

Figure 6.1 EC surplus stocks in storage (thousand tonnes)

Source: *Agricultural Situation in the Community*, annual report, Luxembourg: Commission of the European Communities Office for Official Publications

Table 6.2 CAP budget expenditures, million ECU, 1992 value

	Expenditure for storage	*Export refunds*	*Other EAGGF guarantee*	*Total EAGGF guarantee*
1973	1,181	3,839	7,815	12,835
1974	1,503	1,769	7,893	11,164
1975	2,315	2,640	6,860	11,815
1976	2,939	3,922	8,359	15,221
1977	2,668	6,326	7,378	16,372
1978	4,426	7,902	7,040	19,368
1979	2,934	9,654	8,605	21,193
1980	2,994	10,094	7,860	20,948
1981	2,745	8,312	7,694	18,752
1982	2,833	7,425	9,076	19,335
1983	4,215	8,478	8,344	21,037
1984	5,048	9,465	8,796	23,309
1985	6,402	9,160	11,259	26,820
1986	7,189	9,368	11,392	27,949
1987	4,628	11,493	12,174	28,295
1988	5,539	11,836	13,762	31,137
1989	3,211	11,118	13,253	27,581
1990	4,493	8,469	14,385	27,346
1991	5,846	10,519	15,519	31,884
1992	5,554	9,348	16,334	31,236

Source: For the years 1973–85: Koester and Tangermann, 1990. For the years 1986–92: *The Agricultural Situation in the Community*, various issues

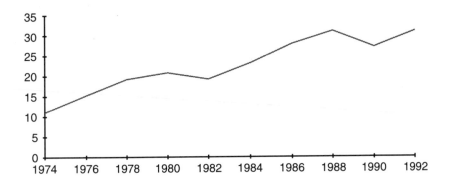

Figure 6.2 EC agricultural guarantee expenditure (billion ECUs 1992 value)

Source: Thomson 1994

concluded that they were about double the budgetary costs of the CAP itself, an estimate broadly in line with earlier work by Morris (1980; Dilnot and Morris 1982). Others consider the cost to be far higher. Roberts and Tie (1982) estimate consumer transfers to be six times the budgetary costs. Whatever the extent of consumer transfers, all commentators concur that these costs exceed the budgetary costs of the CAP itself.

In addition to the concerns over costs to the taxpayer and consumer, a growing concern of the late 1970s and early 1980s was with the manner in which this support was distributed. As long as the money could be seen as flowing into the agricultural industry itself, some political and economic justification for the costs borne by the wider community could be made. Farmers could be applauded for their key strategic importance to the nation's security and for making a valuable contribution to the country's balance of payments. Family farmers could still be paraded as a vital social and economic ingredient within otherwise declining rural communities and as an important group in society as a whole. In the UK such arguments were deployed, not only by the farming lobby but within government as late as the early years of the Thatcher government.

But evidence was mounting that much of the support for farmers was either siphoned off into inflated land values and rents or into the supply sector (Traill 1982; R.W. Howarth 1985). Even more devastating for the reputation of the industry were suggestions that the channelling of support into capital projects contributed directly to the decline in the agricultural labour force, hardly consistent with the social elements of the CAP (Wagstaff 1983). Traill (1982) estimated that a 1 per cent increase in support prices led directly to a 1 per cent reduction in the level of employment of hired labour over a period of six years.

The evidence of farm income data appeared to present an even more damning indictment of expensive support policies, especially in the period following 1976. Figure 6.3 shows how the volume of gross output in UK

Table 6.3 Agricultural prices in the EC as a percentage of world prices

	1967–8	1968–9	1969–70	1970–1	1971–2	1972–3	1973–4	1974–5	1975–6	1976–7	1977–8	1978–9	1979–80	1980–1
Soft wheat	185	195	214	189	209	153	79	107	124	204	216	193	163	146
Hard wheat	200	214	230	231	254	181	116	120	145	236	218	216	159	138
Rice	117	138	186	210	205	115	60	81	137	179	128	157	131	100
Barley	160	197	203	146	185	137	96	107	117	147	206	225	161	134
Maize	160	178	159	141	176	143	98	106	128	163	203	201	190	147
Sugar	438	456	298	203	151	127	66	41	109	176	255	276	134	85
Beef	175	169	147	140	133	112	110	162	196	192	196	199	204	190
Pork	147	153	137	134	131	147	131	109	113	125	137	155	152	135
Butter	397	504	613	481	171	249	320	316	320	401	388	403	411	286
Olive oil	166	173	167	165	153	125	96	113	207	192	211	200	187	214
Oil seeds	200	203	155	131	147	131	77	80	127	121	153	161	185	168

Source: B.E. Hill 1984: 87, based on: Statistical Office of the EC, Yearbook of Agricultural Statistics, various years
Notes: After the first two years pork prices are on a calendar year basis; 1969/70 shown relates to 1969

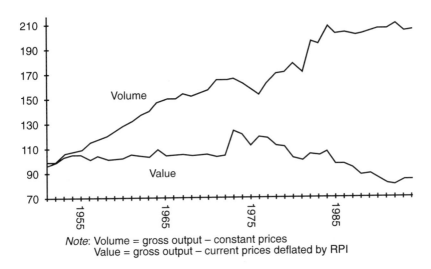

Note: Volume = gross output – constant prices
Value = gross output – current prices deflated by RPI

Figure 6.3 Volume and value of UK agricultural output (1952 = 100)

Source: Data provided by Berkeley Hill

agriculture has not been matched by a sustained increase in the value of that output, and it is this which lies at the heart of the farm income problem. The prices farmers have received for their products have declined in real terms despite increasing levels of support. UK farming income as a percentage of gross output declined from over 30 per cent in the 1940s to less than 20 per cent in the 1980s (Figure 6.4): 'When fluctuations in the value of output are superimposed on this narrowing margin, the implication for greater income

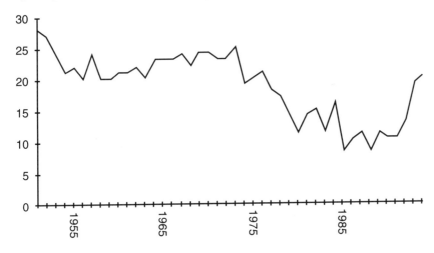

Figure 6.4 Farming income as a percentage of gross output

Source: Data provided by Berkeley Hill

instability is self-evident' (B. Hill 1990: 141).

These figures should not be taken to imply that all farmers and their families have suffered from a sustained period of declining living standards. Indeed one of the difficulties confronting those seeking to reform the CAP, as the extent of its problems became apparent in the late 1970s, was the recognition that while a sector of smaller farmers totally dependent upon agriculture faced real difficulties, many others appeared to thrive. For those who had been able to expand their businesses rapidly, the loss of income was more than made up for by increases in business scale, often brought about by the purchase of adjoining farms. The cost–price squeeze had forced the pace of farm amalgamations in most countries in Europe as shown in Table 6.4. Between 1975 and 1985, for example, the number of agricultural holdings in the nine member states of the EC (that is, prior to the accession of Greece, Portugal and Spain) declined from 5.1 million to 4.3 million and the persons working on the holdings from 12.7 to 10.5 million (Bryden *et al.* 1992).

Figures on net farm income do not take any account of other sources of income. Some farmers may have income from pensions and investments; others have additional business interests or income from employment. Many farm households contain spouses or siblings in full-time work off the farm making highly significant contributions to household income. This makes it very difficult to judge the success or otherwise of the CAP in fulfilling its pledges to provide a 'fair standard of living'. It might even be argued that by encouraging adjustments in the industry which compel farm households to diversify their income sources, the EC has fulfilled its pledges to farmers; although this is hardly what those who drafted Article 39 of the Treaty of Rome had in mind! This is not quite such a flippant comment as it might at first appear – the evidence of a recent study across western Europe shows that levels of farm household income correlate positively with the level of economic activity in the wider local economy (Bryden *et al.* 1992). Areas of stagnant economic performance tend to be areas of low agricultural household income and vice versa. Berkeley Hill (1990; see also B. Hill 1989 and Hill and Ray 1987) has estimated that for all farmers (including spouses) in the UK between 1977–1978 and 1986–1987, nearly 50 per cent of income came from non-farming sources. Hill concedes that these figures, based on Inland Revenue figures, include all with any self-employment income from farming activities whatsoever, and thus include many very small holdings and hobby farmers (which would include, for example, the author of this volume, a full-time academic with a five-acre holding yielding a couple of hundred pounds a year). This, of course, is the reverse problem to that posed by the standard measures of farming income which draw from the unrepresentative large-farm sample covered by MAFF's Farm Management Survey. As Hill has lamented:

> without a much more satisfactory approach to the measurement of income it is impossible to say how many have incomes which, according to some agreed criterion, fall below the threshold of poverty

Table 6.4 The structure of agricultural holdings: selected EU countries. Number of holdings (thousands)

Hectares	Year	Denmark	Germany	France	Ireland	Italy	Holland	UK
> 5	1975	17.6	318.9	361.8	34.4	1,987.2	54.5	42.0
	1987	2.5	231.2	236.0	34.9	2,150.1	44.0	50.1
5–10	1975	25.4	174.2	183.2	37.7	373.7	30.7	31.6
	1987	14.0	118.4	107.2	32.9	333.0	21.6	30.1
10–20	1975	36.8	209.8	272.7	70.6	179.2	44.0	45.6
	1987	21.8	148.5	174.7	63.3	171.3	29.3	37.1
20–30	1975	23.1	108.2	184.1	35.8	50.9	19.3	35.6
	1987	15.8	89.0	134.5	36.7	53.0	18.1	26.0
30–50	1975	19.3	70.4	174.6	29.8	35.7	10.9	42.9
	1987	18.1	77.2	164.6	29.6	38.6	13.9	35.8
50–100	1975	8.2	22.5	106.6	15.9	23.6	2.9	45.6
	1987	11.7	35.1	124.4	16.0	24.7	4.6	42.4
100 Ha +	1975	1.9	3.8	32.3	3.7	13.9	0.4	37.3
	1987	3.1	5.6	40.3	3.5	13.2	0.5	38.6
Total	1975	132.3	907.9	1,315.1	228.0	2,664.2	162.6	280.6
	1987	86.9	705.1	981.8	217.0	2,784.1	132.0	260.1

Source: Statistical Office of the EC, Yearbook of Agricultural Statistics, 1991

and justify help from the rest of society. This is an important and major gap in our knowledge about the UK agricultural industry.

(Hill and Ray 1987: 314)

And, of course, the ignorance applies to the rest of the European Union as well. The available evidence suggests that non-agricultural sources of income are hugely significant in many parts of the Union. Utilising data from over 6,000 farm households in 24 case-study areas in western Europe, it has been estimated that in 1987 just 40 per cent of farm households were monoactive in agriculture. By 1991 the figure had fallen to 37 per cent, with only one third of households committing 95 per cent or more of their available labour capacity to agriculture (Bryden *et al.* 1992).

Political responses to the crisis

With mounting evidence of this nature, the old arguments for the support of agriculture crumbled. An unlikely, and entirely informal, alliance of new right thinkers, environmentalists and the newly emergent food lobby combined to challenge the post-war policy consensus on agriculture.

The new right of the Thatcher revolution scorned state support of any private sector industry, and drew no distinction between agriculture and other industries. While Richard Body (1982, 1984, 1987), a maverick Conservative MP from Lincolnshire (in 1993 his disillusionment with the EC led to rumours that he might even resign his seat to fight as an independent Conservative), drew most of the publicity for his daring polemics against an industry from which he himself came, it was the quieter and more rigorous analysis emanating from right-wing think tanks (for example R.W. Howarth 1985) which probably did more, ultimately, to influence the opinion of ministers and of civil servants. The emergence of a more assertive environmentalism prepared to challenge the record of agriculture (Lowe *et al.* 1986; Pye-Smith and Rose 1984; Shoard 1980) also had some influence among decision makers, especially when the critique was linked to economic analysis as in the important contribution from Bowers and Cheshire (1983). The environmentalists were joined by a growing food lobby who combined concerns for agriculture and health with implications of the CAP for food distribution and equity in world trade (Clutterbuck and Lang 1982; Wheelock 1986). The effect of this new alliance was startling. In the space of four years between 1980 and 1984 agriculture was effectively re-politicised and the cosy corporatist settlement of the post-war era subjected to an offensive, the like of which was completely unfamiliar to the participants. Whilst corporatism was not exactly broken, the context in which policy making took place had at last begun to reshape. Whilst the core of the agricultural policy community remained intact, around the edges on issues of environmental and food policy a diverse and fragmented issue network was becoming an increasingly more appropriate metaphor.

In addition to the crisis of the CAP, there were several other contributory factors which served to make these years all the more dramatic in the UK. One was the impact of the passage of the Wildlife and Countryside Bill through Parliament between 1980 and 1981. This is dealt with further in chapter 9, but suffice it to say here that the wide publicity given to the Bill served to heighten public debate on environmental issues with agriculture very much taking centre stage.

Another issue for the new right was government itself and in particular, as seen in chapter 2, the work of government departments. At the very time when agriculture's place in the economy came under attack, so its own Ministry faced the rigours of a Rayner scrutiny, considerable uncertainty about its own future, and suspicion from the anti-corporatists of the new right, several of whom were private advisers to Number 10. It was hardly a propitious time for either senior MAFF officials or ministers to take the policy offensive. Instead, the Ministry adopted a low profile which caused increasing irritation, not only amongst some of the critics of agricultural policy looking for new directions, but also in the farming lobby itself. Once they realised that the problems besetting the CAP were no temporary blip but the heralding of a long-term structural crisis in the industry, then both the NFU and the CLA anxiously sought fresh ways of safeguarding the interests of their membership. The Ministry, at least in the early years of the decade, was not entirely responsive to such overtures.

BARRIERS TO REFORM

Despite the pressing demands for reform the obstacles were enormous. The complexity of the problems rendered the Commission, national governments and civil servants particularly prone to succumb to the lobbying tactics of the pressure groups:

> According to private conversations between involved officials of the Commission and the author over the period 1978–84, it is clear that in the process of formulating agriculture policy, Commission officials responsible took a great deal of note of what they were told by COPA representatives, in particular: (i) that immediate future productivity increases would not be anywhere near as great as independent analysts were consistently warning and (ii) that there would be a substantial expansion in international demand for agricultural products which would automatically reduce the budgetary pressure to control CAP surpluses.
>
> (Gardner 1987: 170)

Such misinformation certainly delayed the onset of the reform process in the early 1980s. But probably the most serious problem to face would-be reformers was the issue of national self-interest. As has already been discussed, the CAP involves a transfer of income from consumers to farmers. However, because not all countries have the same ratio of consumers to farmers this inevitably means that one of the consequences of the CAP is that member states with a low level of self-sufficiency in foodstuffs (that is, a high ratio of consumers to farmers) transfer income to countries with a larger share of farmers, often net food exporters (Pearce 1981). Two countries, in particular, have fared badly under the budgetary transfer arrangements – Germany and the UK. In 1980, for instance, West Germany made a net contribution to the EEC budget of 2,025 million ECUs and the UK contributed 730 million. All other countries gained as a result of their agricultural exports. When trade is taken into account, though, Germany and the UK are joined by Italy, Belgium and Luxembourg in experiencing negative transfers.

As a consequence, the majority of member states had no pressing reason to support radical reform measures. Indeed an increase in agricultural spending would be an advantage for most countries as it would further increase the flow of resources from a minority of members including the UK. Moreover, as long as this remained the case the imperative for agricultural expansion seemed almost unassailable:

> Common financing not only encourages some countries to favour higher CAP prices, it also encourages them to use national subsidies in order further to expand their agricultural industries. Indeed, the present system persuades *all* member countries that agricultural expansion is desirable. Taking the UK as an example, if food production could be expanded until no imports were necessary, the country's contribution to the Community would be reduced because there would be no agricultural trade transfers and budgetary transfers would no longer include levies on agricultural imports.... In essence, each country hopes that its expansion will reduce its CAP losses or increase its CAP gains more than its share of the additional costs of extra surplus disposal.
>
> (B.E. Hill 1984: 99)

Similarly, Harvey (1982) concluded that national interest not only rendered reform unlikely but that the continued emphasis by the European Commission on common policy, paradoxically, only served to reinforce and to increase the importance of national interests. In such circumstances the imposition of milk quotas, or indeed any other measure to stabilise or reduce production would seem to be irrational in terms of the individual interests of the majority of member states, even though it might seem eminently sensible from the perspective of the Community as whole. How then did any processes of reform come about? The UK's position here is critical – the net budgetary contributions rankled with the newly elected Thatcher administration. Moreover the solution of pumping more public money into a heavily supported agricultural industry in an effort to increase self-sufficiency was anathema to some members of a government committed to principles of free trade, the market economy and monetarist controls on public spending. Not surprisingly, then, a prime mover in pressing for reform was the UK government, in particular, Mrs Thatcher herself.

With characteristic belligerence she insisted that the UK would not accept an agricultural price settlement for the 1980/81 year without a revised formula to reduce the UK's budgetary burden. In this she was successful and in May 1980 the Council of Foreign Ministers agreed a three-year formula to reduce the UK's budget contributions (B.E. Hill 1984). However, the UK's policy position was based on worries about EC budget contributions more than a desire to return her own agriculture to the free market. Agriculture's 'special place in the pantheon of traditional Tory values' (Cox, Lowe and Winter 1989: 129) and the fact that as late as 1981 44 per cent of peers in the House of Lords were, or had been, farmers or landowners (Baldwin 1985; Shell 1992) contributed to a slow transition of the policy agenda. The first Minister of Agriculture in the Thatcher administration was Peter Walker, on the interventionist left of the Conservative party and with a firm commitment to agriculture. M.J. Smith (1992) has seen his appointment as signalling an area of government, alongside defence spending, where the free market philosophy did not yet prevail. He characterises Conservative agricultural policy up until 1983, when Michael Jopling became Minister, as contradictory. On the one hand, Peter Walker continued an expansionist policy, even referring favourably to the previous Labour government's 1979 White Paper *Farming and the Nation* (Cmnd 7458); on the other hand, Treasury and Foreign Office ministers and the Prime Minister herself increasingly sought ways to limit agricultural spending.

In fact, the contradictions are more apparent than real, for it was perfectly consistent for UK agriculture to continue expanding at the same time as attempts were made to curtail expenditure in the Community as a whole. Indeed, one of the arguments put over to justify an apparent volte face in policy when milk quotas were introduced, was precisely that it was better for UK agriculture to have expanded in the years immediately prior to quotas than to have stood still. The traditional links between the NFU and the Conservative Party also influenced the expansionist mood prior to quotas, and there are those who have questioned the extent of the UK's commitment to reform right up until the mid-1980s:

> The extreme specialisation of EEC agricultural policy mechanisms and the arcane CAP jargon has … consistently allowed UK ministers and, more specifically their civil service advisers, to obscure the true path of British agriculture policy from certainly the national press and the House of Commons, but probably also from the Cabinet. Although British agriculture ministers have frequently appeared to favour CAP reform, the longer-term trend of British policy has been towards acquiescence in an unreformed and therefore expansionist CAP.
>
> (Gardner 1987: 177)

REFORM MEASURES

It is now time to take a closer look at the reform measures themselves. The settlement to ease the UK's budgetary problems, known as the '30 May Mandate', involved a complicated formula worked out to accommodate the demands made by Margaret Thatcher. In addition, the Commission itself was required to set to work on a longer-lasting basis for reform, which resulted in the publication of three papers (Commission of the European Communities 1980, 1981, 1982). The first, *Reflections on the Common Agricultural Policy*, was essentially a thoroughgoing review of the CAP judged against the objectives set out in Article 39 of the Treaty of Rome. In terms of labour productivity, the provision of a fair standard of living for the agricultural population, market stability, and reasonable prices to consumers, the evidence seemed to add up to a damning indictment of the CAP (with success confined to the assurance of regular supplies and on that, of course, the CAP had been successful to the point of acute embarrassment). However, despite the weight of its own evidence, the Commission sought comfort in many points of detail and not only avoided an attack on the underlying principles of the CAP but also concluded that the objectives of the CAP had been broadly achieved. It did, however, focus attention on the problem of surpluses and the disparity between rich and poor farmers. The second paper, the *Report on the Mandate* accepted the somewhat sanguine conclusions of the *Reflections* paper and denied a need for radical reform: 'it is neither possible nor desirable to jettison the mechanisms of the CAP but on the other hand adjustments are both possible and necessary' (Commission of the European Communities 1981: 11). None the less the *Report* suggested adjustments which amounted to a significant departure from the principle of supporting any quantity of production irrespective of the market for the produce: 'it is neither economically sensible nor financially possible to give producers a full guarantee for products in structural surplus' (ibid.: 12).

The third paper, *A New Impetus for the Common Policies*, expanded on some of the points made in the *Report on the Mandate*. So what did the reform initiative amount to? In terms of concrete progress, the answer has to be 'very little'. But in terms of a recognition of the kind of measures that might be needed the 1980–1981 initiative was of considerable significance. There was agreement in principle on the following requirements:

- the need to reduce the level of guaranteed prices, especially where particular member states were supporting their domestic agricultures beyond the level agreed by the Community;
- the need to increase exports from the EC;
- the need to improve structural policies so as to support poorer farmers and regions without recourse to across-the-board price increases.

However, agreement within the Commission did not lead to approval by the Council of Ministers where the vested interests of the farm lobby, not to mention the budgetary advantages of the status quo for all countries except the UK and Germany, meant that in both 1981 and 1982 substantial price increases were approved. Even Germany, despite its budgetary position, was opposed to reform because of its own strong farming union. The only gesture towards a serious curtailment of the escalating cost of the CAP was the continued application of a co-responsibility for milk, originally introduced in 1977 (for a detailed account of the introduction of the co-responsibility levy see Tsinisizelis 1985). Co-responsibility levies involve a financial penalty incurred by producers when production reaches a certain level, the idea being that the producers themselves share some of the cost of storing the surplus produce.

In both 1980 and 1981 the Commission proposed a supplementary levy to apply to milk supplied above a reference level (Tsinisizelis 1985). The Council of Ministers rejected the proposals, and in 1982/83 a temporary decline in surplus stocks coupled with sharply declining farm incomes even spurred on the Council to adopt a 10.5 per cent increase in milk prices for 1982 and a 2.33 per cent increase the following year. Writing soon after these events Brian Hill's summing-up was bleak indeed: 'In the face of the collapse of the Commission's reform proposals and the lack of a permanent solution to the UK's "unacceptable" budget situation, the Mandate is obviously dead' (B.E. Hill 1984: 137). Despite the failure of these reform attempts, indeed because of it, the pressure for change from the UK was unabated. By mid-1983 it was apparent just how mistaken the easing of the reform initiative had been for there was a dramatic escalation of dairy stocks in store within the Community, and reform of the dairy sector became the most pressing matter in the policy debate on the CAP in the autumn of 1983.

In the political context associated with the Thatcher government's insistence on budgetary discipline in the Community and Chancellor Kohl's presidency of the European Council, in June 1983 the Council instructed the Commission to produce new measures to limit agricultural spending, encouraging it to take concrete steps compatible with market conditions to ensure effective control of agricultural expenditures (Petit *et al.* 1987):

A general sense of crisis prevailed. FEOGA costs skyrocketed, after a level period between 1980 and 1982, caused by the appreciation of the US dollar. Unless something was done, the increased revenue agreed in 1981 would have become exhausted at some time before the end of 1984.... Reform proposals could expect considerable opposition from the farm lobby in that farm incomes were declining in a number of member countries. Adding to the pressures facing it, the EC had to give a definite answer to the membership applications of Spain and Portugal. This was difficult as long as the budget crisis went unresolved.... The need to take action created an important impetus for policy-makers, allowing them to resist the influence of the EC farm lobby, particularly the dairy interests. The trade-off was a stark one – either the dairy programme was changed or the CAP would collapse.

(Moyer and Josling 1990: 67–8)

The result, on 31 March 1984, was the imposition of milk quotas, or strictly speaking 'a super-levy at farm level on all milk delivered to dairies over an annual amount determined by national quotas' (Harvey and Thomson 1985: 13). Essentially there were four stages in the political debate leading up to this move:

- the elaboration and publication of Commission proposals on 29 July 1983;
- discussion in national governments and in the Council of Ministers prior to the Athens summit in December 1983;
- bargaining prior to the Brussels summit of March 1984;
- final negotiations.

(after Petit *et al.* 1987)

In truth, the Commission's options were few. Savage price restraint would be politically unacceptable. The co-responsibility levy had already been shown to be ineffective, but helpfully provided a legal basis for the new initiative (Snyder 1985; Tsinisizelis 1985).

Quotas had been first mooted publicly in a Commission report of September 1978, although the report had favoured price restraint (Tsinisizelis 1985), so the MAFF and the MMB had long been aware of quotas as a possible development, although they did not, it seems, anticipate the speed with which change would occur. During 1983 the MMB made studies of quota systems in Austria, Canada, Switzerland and Queensland, but the first meeting between the Board and the Ministry to discuss quotas was not held until November 1983. The NFU also held meetings with Commission and Ministry officials and conducted internal discussions. However, unlike the three Scottish Milk Marketing Boards, which publicly declared their support for a quota system, the NFU, the Ministry and the MMB avoided public debate on the issue and in England the Scottish move was largely ignored. As Cox, Lowe and Winter explain:

> The reason for the lack of publicity given to the possibility of quotas must lie largely in the delicate stage in negotiations in which the three main actors – MAFF, the NFU and the MMB – found themselves. Having come to accept the likelihood, if not the inevitability, of quotas they did not expect imposition to occur so rapidly. Moreover, each of the three bodies had its own reasons for wishing to delay 'going public'. The Ministry, for example, was publicly committed to price restraint as the chief plank of CAP reform. This was in keeping with its long-standing 'big farm efficiency' approach, and was an appropriate stance for it to take under a free-market Conservative government. Just four years earlier, it had spelt out clearly and unequivocally its opposition to quotas. In evidence to the House of Commons Agriculture Committee in 1980 Ministry officials had declared that a sudden reduction of production was physically impossible, that excess production would still have to be disposed of, that quotas would counteract

market forces and weaken resistance to price rises, and that administration would present very considerable problems.... [The Ministry] was placed in an embarrassing situation when they became first a possibility and then a reality.

(Cox, Lowe and Winter 1990b: 98)

The imposition of quotas marked something of a turning point in the agricultural crisis of the 1980s, particularly for the UK government's approach to the issue. Hitherto the uneasy coalition of environmental and monetarist interests seemed to be on the point of radically reducing agricultural price support and imposing stricter controls on agricultural activities. The furore caused by quotas prompted several environmental pressure groups to re-examine their embryonic commitment to reducing agricultural support expenditure. For the Ministry of Agriculture it contributed to the emergence of a new assertiveness. The quotas option marked a renewed commitment within the European Community to interventionist measures. The UK had failed to convince the other member states, crucially the Federal Republic of Germany, that a return to the free market was either desirable or feasible. Therefore, although price restraint continued to lie at the forefront of the UK government's long-term strategy for agriculture, and much of its short-term rhetoric, its specific policies now contained strongly interventionist elements.

The NFU and the MMB had not spelt out their opposition in such detail as the Ministry. Indeed the NFU had begun to accept the inevitability of some form of limitation on milk production in the light of the Commission's July 1983 proposals:

It did not want to be seen to be disagreeing with the MMB, however, and it wanted time to be able to 'sell' a package to its members and to explain to them the concessions which the Union had won in the light of inevitable restrictions on their freedom. The uncharacteristic speed with which the Council of Ministers acted prevented all this and the Union and the Ministry were left in the unenviable position of having to explain a retrospectively applied policy which, at first sight, seemed to pose intolerable burdens on UK dairy farming and threaten its very survival.

(Cox, Lowe and Winter 1990b: 98–99)

In addition to the agreement on quotas, other commodities were subjected to one of the more severe price packages ever agreed in the history of the CAP including, for the first time, a production limit on a range of commodities, beyond which the intervention price would fall (Fennell 1985). None the less these thresholds were not set at a level likely to have a serious impact upon the burgeoning surpluses. Despite the widespread public mistrust of the Commission which emerged in the UK during the 1980s, not least in the context of the continuing crisis of the CAP, the blame for the slow progress of reform really lay elsewhere, with the politicians, as the House of

Lords Select Committee on the European Communities declared in 1987:

> The Committee consider that the blame for the state of Community finances rests not with the Commission, but with the Council of Ministers. The Committee have been impressed by the Commission's initiative in proposing solutions to the Community's problems: if its advice had been heeded over recent years, particularly in connection with farm prices, the present situation would not be so serious. The Ministers meeting in the various Councils have, in contrast, shown indecisiveness and lack of co-ordination.
>
> (House of Lords 1987: 13)

In the face of a continuing dilution of its policies, the Commission had little option but to resort to a campaign of education and persuasion – and so to a further series of documents designed by the Commission to stimulate yet more discussion, resulting in the uncertain 'crab-like progression' (Fennell 1987) that characterised the CAP throughout the 1980s. 1985 saw the publication of *Perspectives for the Common Agricultural Policy* (Commission of the European Communities 1985) and the following years brought fresh proposals for the curtailment of production. In July 1986 a co-responsibility levy for cereals was introduced and in February 1988 budget stabilisers were initiated. Designed to impose strict limits on the growth of CAP spending, stabilisers were heralded as yet another breakthrough, ranking alongside milk quotas in significance. In reality the mechanisms for achieving the sought-after stable budgets were not as well developed as for milk quotas. In effect stabilisers merely introduced a new type of price restraint with intervention prices set to fall when production reached a certain level. Their impact on farm incomes was never sufficient to curtail production levels seriously. 1988 also saw the introduction of a five-year scheme of voluntary set-aside, whereby farmers could choose to set aside some of their arable acreage in return for compensatory payments.

The next significant step was the publication by the Commission of *The Future of Rural Society* (Commission of the European Communities 1988). This discussion paper marked yet another new departure in Commission thinking, in so much as it represented an acknowledgement that agriculture could no longer be the sole or even the major focus for EC rural policies. Henceforth at least part of the focus would be on rural development, symbolised by the reorganisation of the former division for agricultural structures into two rural development divisions (Bryden 1991). Moreover, the document's analysis was not based solely on the identification of problems of agricultural adjustment. Instead the paper recognised that rural areas were subject to many forces of change and recognised three in particular: 'the pressure of modern development', 'rural decline', and 'depopulation and the abandonment of some land'. In practice there remained a strong, if not exclusive, agricultural element to the policies that emerged from this approach. The new emphasis on the *rural* economy led to a focus on new uses for *agricultural* land and buildings through schemes for farm diversification and alternative land uses, launched in 1988.

THE EFFECTIVENESS OF THE 1980s REFORMS

Before moving on to examine the new round of reforms in the 1990s, it is important to examine the outcome of the various measures passed in the 1980s. Here we examine the impact of milk quotas, set-aside, and farm diversification policies, principally within the UK. The explicitly environmental policies, such as ESAs and farm woodlands, are considered in chapter 9.

Milk quotas

The response of UK farmers to the imposition of quotas was to adapt their businesses so as to minimise the damaging consequences, and this was done with a degree of alacrity that surprised commentators and indeed the farmers' representatives themselves, whose dire warnings of impending disaster for the dairy sector proved groundless. In truth, quotas soon came to be seen by farmers for what they were – a valuable capital asset and a protection against the rigours of market competition. Quotas were set at a level that allowed most producers to take steps to implement necessary changes in farm management, sometimes even resulting in improvements in profitability. Moreover, the special cases procedure, for all its inadequacies, ensured some adjustments in quota allocations for a number of needy producers. Similarly, the Outgoers Scheme, introduced at the same time as quotas, offered relatively generous compensation terms to those who decided to quit the sector.

However none of these ameliorative arrangements was available for either hired farmworkers or workers in the ancillary sectors. In order to cope with quotas, farmers were encouraged to limit their inputs of bought-in feed and to take a careful look at all their costs, including labour. Thus farmworkers, the manufacturers of compound feed and milking machinery and, of course, workers in milk factories all suffered. The number of full-time hired farmworkers in the dairy sector declined by 27 per cent between 1983/84 and 1988/89, accounting for just over half of all hired job losses in agriculture during this period (MMB 1989). A study of producers' reactions to quotas in Devon showed an average reduction in use of purchased concentrate feeds of 27 per cent (Halliday 1987, 1988). Nationally, the number employed in the compound feed sector declined from 16,000 to 12,500 between 1983 and 1987 (BBC Radio, 'Farming Today', 23 November 1987; quoted in Cox, Lowe and Winter 1990b).

But for farmers themselves the situation was less grave. Indeed the long-standing structural adjustment in the dairy sector actually slowed for a time, partly as a result of the tying of quotas to particular parcels of land. Figure 6.5 demonstrates the rate of structural change in the industry before and since

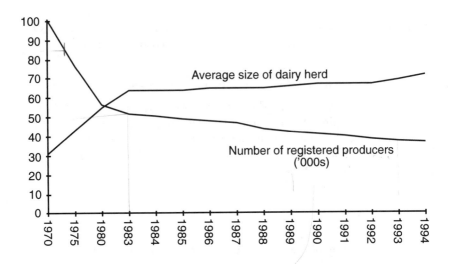

Figure 6.5 Structural change in UK dairy farming 1970–1993

Source: *UK Dairy Facts and Figures*, annual publication, Thames Ditton: Milk Marketing Board

quotas. The role of quotas in reducing production levels within the EU as a whole has been moderately successful, and quotas were retained as a key element in the 1992 reform package. Butter stocks continued to grow after 1984, peaking in 1986 at 1.3 million tonnes. By 1989 they had fallen to just 32,000 tonnes, only to increase again in the 1990s (standing at 160,000 tonnes in 1993).

There are three commonly perceived problems with quotas. The first concerns the continued high level of support payments made to farmers within the quota system. After the imposition of quotas milk prices continued to be set at a level unlikely to deter production, so rendering quotas – with all the associated administrative costs – an apparently semi-permanent feature of European agriculture. With its inevitable distortion of world markets in dairy products this was hardly a policy designed to appease non-EU countries in the complicated GATT negotiations on agriculture that commenced in 1986.

The second problem is the difficulty of setting quotas at a correct level for the market. If quotas are too tough then the EU could theoretically run the risk of a shortfall in production. Alternatively – and not surprisingly this has tended to occur rather than a shortfall in production – the vested interests of the agricultural industry work towards quota levels that are rather too generous, so resulting in a further accumulation of surpluses, paid for by the taxpayer.

A third problem concerns the potential distortion to the process of structural adjustment in the industry by the tying of quotas to the existing pattern of dairy farming in 1984. Strong interests in the UK resisted this aspect of quotas from the outset, insisting that quotas should be transferable in themselves. This is now accepted policy within the EU, although the allocation of quotas to member states still raises questions about the ability of the system to allow for long-term changes in the production pattern across Europe as a whole.

Set aside

In 1987, Regulation 1760/87 introduced a voluntary scheme for individual farmers to opt for a 20 per cent reduction in output of cereals (also beef and wine). A year later this was replaced by Regulation 1094/88, which continued to apply to beef (with minimal take-up) and wine as well as to cereals. The complex process of debate in establishing the set-aside scheme is illustrated in Table 6.5. The principal difference between the two regulations for the cereal sector was the decision to switch from a reduction in cereal production to the removal of land from production through set-aside. A farmer choosing to enter the scheme, for it remained voluntary, had to set aside a minimum of 20 per cent of his or her farm for between three and five years.

The impact upon levels of arable production across Europe was not great. For one thing, 'flexibility exists within Regulation 1094/88 regarding its interpretation, rates of compensation and those regions where, because of natural conditions or the threat of depopulation, the scheme should not be applied' (Ilbery 1990: 69). Even in those countries where the scheme was applied with what appeared to be a reasonably high level of commitment and expectation the results were disappointing, with a very modest proportion of land coming out of production: just 2 per cent in the EC as a whole.

Ilbery has identified four main problems that limited the effect of the first set-aside scheme on levels of cereal production: *slippage, selectivity effect, monitoring* and *low rates of compensation*. Slippage refers to the fact that 'the fall in production is proportionately lower than the amount of land set aside' (Ilbery 1990: 157). Although farmers involved in the scheme had to agree not to increase their arable area on the remainder of their farms there was nothing to prevent them setting aside the more marginal land where yields were already low. Slippage might also occur through the intensification of production on the remaining arable area on the farm. In terms of national objectives for the scheme, there might also be slippage because of the decisions taken by farmers not participating in the scheme. Thus most farmers are able to cite cases of a farmer setting aside land at exactly the same time as a next-door neighbour increased the area of arable cultivation.

The selectivity effect refers to the fact that a scheme of this nature inevitably attracts farmers who might have made the desired change in any case. In a dynamic economy of many producers at different stages in the business or family cycle it is inevitable that a proportion of those who adopted set-aside received subsidies that probably did not need to be paid to guarantee the removal of land from arable production. The monitoring problem refers to the difficulties of policing such a scheme especially given the potential for using the land released from arable production for other

Table 6.5 The set-aside debate

	European Commission's considerations	European Commission's proposals	European Parliament's recommendations and observations	National positions expressed through COPA/Council of Agricultural Ministers	Council decisions
Duration of programme	Commission considers an annual or pluriannual scheme. Pluriannual would allow results to be better assessed.	Five years 1988–93 with option of rescinding contract after three years.	Set-aside must be part of a comprehensive programme. Concern that EC's trading partners (especially the USA) would not be curtailing their production.	French government expresses major reservations about the scheme, fearing loss of valuable earnings from cereal exports.	Five years 1988–93 with option of rescinding contract after three years.
Compulsory or optional?	A compulsory scheme binding on all farmers seen as too costly and administratively difficult.	Compulsory for all member states; voluntary for farmers.	Agree that the scheme must be voluntary for all farmers in the EC. All member states must implement it.	Danish government refuses to implement set-aside scheme despite Commission threats. FR Germany hints at favouring a compulsory scheme for all EC arable farmers.	Binding on all member states. Voluntary participation by farmers. Portugal exempt and certain areas in France, Spain and Italy.
General or specific?	A specific programme is desirable though its application on a yearly basis is seen to be unrealistic.	General scheme for all arable land in the EC.	Agreed with the Commission's proposals for a general set-aside policy though need for special exclusion provisions for certain regions.	Member states in favour of a general scheme and agree with the Commission.	General (all arable land).

EC financed? Nationally financed? Both?	A specific programme would mean EC incurring costs for cadastral survey. Experience of table wine and olive oil sectors range against this option.	EC and national governments to share costs. EC monies for Guidance fund. Commission view is that national governments will ensure better monitoring of the scheme if they are jointly financing it.	Concern that rationalisation of CAP taking place. Level of subsidy should not be too high that member states cannot afford it. Suggest 50% Guidance, 50% Guarantee.	Spain, Italy, Greece argue against the use of Guidance funds. Commission proposes 50% Guidance, 50% Guarantee.	EC and nationally financed. 50% from Guidance, 50% Guarantee funds. EC contribution 60% of the first 300 ECU, 25% of 300–600 ECU.
Nature, level and modulation of subsidies?	Premiums must be at a level to genuinely encourage set-aside though not so high that they are considerably above normal returns from farming the land.	Commission proposes level of 100–600 ECU/ha within which member states set their own level. Different types of set-aside contract available including one rotation.	Absence of a uniform policy could result in significant financial flows to certain countries.	All states prefer flexibility in the payments scheme, thus able to demonstrate their strength of support for the policy.	Fixed by member states between 100 and 600 ECU/ha.
Size of parcels, reference period?	Have to ensure that parcels set aside are of a minimum size for administrative purposes and that the land set aside is arable.	Minimum area 1 ha. 20% of a farmer's arable land must be set aside for at least five years, though provisions to terminate the contract after three years.	Concern that the set-aside programme may increase land prices and hamper structural reform in agriculture.	FR German government most enthusiastic about the scheme though concerned about possible future production quotas using base years during the set-aside period.	Minimum parcel 1 ha. 20% of arable land must be set aside. Special dispensation for those setting aside over 30% of their arable land.
Effect upon levels of production?	Commission notes EC cereal production increasing by 2–3%/yr. 1988 cereal harvest highest on record at 163 mt.	Commission estimates 1 million ha being taken out of production, equivalent to a 2.3 mt cut in production. This estimate is based on land with a low yield.	Concern that the US Scheme had only a slight effect upon levels of production. Parliament urges Commission to give careful thought to the merits of the scheme.	FR German government perceives set-aside as likely to have major effect upon production levels.	No specific targets for lowering production of cereals in the Community.

Table 6.5 Continued

	European Commission's considerations	European Commission's proposals	European Parliament's recommendations and observations	National positions expressed through COPA/Council of Agricultural Ministers	Council decisions
Intensification effect?	The financial success of the programme will not depend on how much production is reduced on the idled land but on how much production on all hectares is reduced.	Commission encourages farmers to set aside over 30% of their arable land by offering reductions in co-responsibility levies.	Concern that farmers may increase production on parcels not set aside. The EC programme could therefore act as a subsidy for intensification.	All states concerned about possible increases in land prices caused by set-aside and therefore higher intensity of land use.	No special provisions.
Over-compensation effect?	Costs of operating the policy may be greater than the costs of disposing of the surplus production.	Commission estimates that the cost of the programme would be 400 million ECU.	In general financial terms set-aside is of very little importance – a view also endorsed by the Economic and Social Committee.	National variations in set-aside take-up could result in financial transfers between member states.	No special provisions apart from fixed budgetary allocation for the programme.
Structural effect?	Idled acreage is likely to have below average yields. Likely that all land where gross margin is equal to or below the level of premium will be set aside.	Commission recognises that it is likely to be less productive land which is set aside by farmers.	Parliament recognises this fact and urges Commission to rethink regional and environmental consequences.	Member states declare that it will be difficult to determine how much production has been reduced by the programme.	No special provisions.

Environmental effect?	Commission aware that set-aside in some areas will aggravate situation. Farmers need to upkeep the set-aside land.	Member states must ensure that the land set aside is not allowed to deteriorate in condition.	Concern that areas already experiencing environmental degradation will continue to suffer.	French, Spanish and Italian governments concerned about the possible further degradation of environmentally fragile regions.	Spain is given exemption for 29.5% of its arable land (Castille, Aragon, Estremadure, Andalusia).
Regional effect?	Recognition that such a scheme may not contribute to economic cohesion and may be inconsistent with the principles of the single market.	Commission establishes criteria for regions in the Community which can be exempt from the set-aside scheme.	Concern that some regions may suffer through accelerated depopulation. Negative effect upon agricultural workforce requirements. Parliament urges new measures for such areas.	French, Spanish and Italian governments concerned about the acceleration of rural depopulation in certain regions as a result of the scheme.	France is given exemption for 2% of its arable land in the south and west prone to fire hazard. Italy is given exemption for 0.05% of its arable land in the province of Trente where depopulation rates are high. Portugal is entirely exempt from the programme.

Source: A. Jones 1991: 110–111

Table 6.6 Arable set-aside (five-year scheme) in the European
Community: 1988–1993 (hectares)

	1988–1993 total	*% of arable area*
Belgium	877	1.2
Denmark	8,156	0.3
West Germany	415,413	5.6
Greece	213	0.0
Spain	87,712	0.6
France	225,015	1.3
Ireland	1,643	0.2
Italy	764,890	8.6
Luxembourg	94	0.2
Netherlands	15,379	1.7
Portugal	0	0.0
UK	109,239	1.7
EC Total	1,628,631	2.4

Source: Commission of the European Communities, 1993 and 1994

forms of agriculture, in particular grazing (some of which was legal under the scheme and some not). Low rates of compensation mean that farmers on good land, producing the highest yields, are unlikely to be attracted to participate.

In an evaluation of the scheme in England and Wales undertaken for MAFF, based on 259 on-farm interviews (representing 20 per cent of those joining the scheme in 1988/89), several of the problems highlighted by Ilbery surfaced. Farmers tended to be older and several regarded the scheme as a 'pension'; it was estimated that 15 per cent of the land set aside would have gone out of cropping anyhow (Ansell and Tranter 1992). Taking into account the slippage factors identified in the survey, the scheme appears to have resulted in a reduction of arable production in England and Wales of 0.5 per cent for wheat, 0.9 per cent for barley, 0.9 per cent for oats, 0.4 per cent for oil seed, 1.7 per cent for beans and 0.2 per cent for peas in 1988/89 compared to a reduction of less than 2 per cent in the UK arable area, as shown in Table 6.6. It should be pointed out that the slippage effect with regard to increasing intensity in the remainder of the farm was measured only in terms of the opinions of the farmers interviewed; nor was any attempt made to measure arable output and acreage on non-participating farms. Not surprisingly, the scheme failed to restrain significantly the growth of cereal surpluses within the European Union.

Farm diversification

Farm diversification provides a good example of how the contrasting, and at times conflicting, strands of opinion on agriculture within the Conservative Party could be reconciled around a specific set of policies within a particular sector. Diversification and the new land-uses initiative provide an ideological bridge between environmental and economic objectives, and also a means of ideological resolution of the contradictory objectives of economic liberalism and social conservatism within the Conservative Party. For a significant strand of Tory opinion, entry into the European Community promised the re-assertion of the traditional Conservative values of collective security and national purpose (Norton and Aughey 1981). Moreover, social conservatism, in the tradition of Macmillan and Heath, demanded a degree of economic planning and intervention, as exemplified in the origins of the CAP. Thus whereas UK monetarism and Thatcher's experiment in rolling back the state may have contributed to a degree of re-orientation of the CAP, they scarcely dented its underlying principles of market intervention and public sector support for agricultural producers.

Although the key symbols of the diversification debate – *alternative land uses, diversification, value-added, agro-forestry* – are very 'new', the style is reminiscent of the heady days of the 1940s and 1950s when the aim was to 'get agriculture moving' (Lowe and Winter 1987; Shucksmith and Winter 1990). The message that comes through is that land abundance presents new market opportunities, and that the role of government is merely to ease the transition for individual landholders and farmers. For social conservatism there is an apparently interventionist commitment to support agriculture and rural life during a period of change. For liberal conservatism there is the rhetoric of a transition to the market economy.

The opportunity for grant aiding small-scale tourist and craft enterprises was provided under Article 16 of the original Less Favoured Areas Directive in 1975 (75/268/EEC) but this was not implemented in the UK until 1985 when 25 per cent grants for tourism and craft work developments on farms were introduced in the LFAs. As the name implies, the LFAs covered only areas of particular agricultural handicap, especially the uplands. But the development of tourism grants in these areas occurred at the same time as the UK confronted the implications of the Structures Regulation of 1985 (Reg 797/85). From this emerged the Farm Diversification Grant Scheme (FDGS) (Statutory Instrument 1987 No. 1949) which commenced on 1 January 1988 and applied throughout Great Britain until the end of 1992. Before examining the working of the FDGS in further detail, it is worth just running through a range of other measures in the 1980s, which serve to emphasise the importance the Thatcher governments attached to farm diversification and rural economic revitalisation. For instance, in 1980 County Councils lost their power to direct district councils to refuse a planning application that did not conform to the structure plan, and a series of Department of the Environment circulars (22/80, 23/81, 22/84) pressed for the structure plan system to be simplified and streamlined to guarantee ample land for development. This was followed in 1985 by a White Paper entitled *Lifting the Burden* (DoE Circular 14/85), which indicated a presumption in favour of allowing development unless 'that development would cause demonstrable harm to interests of acknowledged importance'. A further circular in March

1986 (2/86), *Development by Small Businesses*, gave explicit encouragement to the conversion of redundant farm buildings.

Also in 1986, faced with falling agricultural incomes and mounting concern over the future for the countryside, the government set up an interdepartmental working party on Alternative Land Use and the Rural Economy (ALURE), involving officials from various departments with an interest in the countryside, including MAFF, the Treasury, and the Departments of Trade and Industry, Employment, and the Environment (Cloke and MacLaughlin 1989). In February 1987, amidst a storm of controversy, the DoE and Agriculture Departments jointly issued a folder of documents under the general title *Farming and Rural Enterprise*, which addressed the issues facing the agricultural industry and speculated on some of the possibilities for farm land and income diversification. Most controversial of all was the rather vaguely worded suggestion, given much media publicity, that restrictions on the industrial and residential development of agricultural land might be eased, primarily through a diminution of the Ministry of Agriculture's role in the planning process.

A draft DoE circular, 'Development involving agricultural land', suggested that proposals involving the development of agricultural land should only be referred to MAFF if they involved the loss of Grades 1 and 2 land of 20 ha or more. For land outside of Areas of Outstanding Natural Beauty, National Parks and Green Belts, the presumption that farming had first claim on the countryside was to be replaced by a view according equal importance to the economic and environmental aspects of development. However, other areas of 'good countryside' not covered by existing designations were to be protected from development by local authorities. 'Good countryside' was, however, not clearly defined, but clearly implied that not all rural land was worthy of protection. The final version of the circular (DoE 16/87), made a number of concessions to conservation opinion. MAFF was now also to be consulted about development on Grade 3a land, and attention was directed towards the prospects for the re-use of urban land. Most importantly, instead of safeguards for 'good countryside', there was a promise to 'protect the countryside for its own sake' rather than simply for its agricultural value.

None the less the real message from these moves was that non-agricultural economic activities were to be treated far more sympathetically in the countryside. An easing of planning control provided one plank for this policy. The other was the provision of grant aid in England through MAFF's FDGS and also through grants from the Rural Development Commission. FDGS grants were allowable for capital expenditure incurred in connection with the establishment or expansion of businesses ancillary to the farm, including horse stabling, farm shops, food-processing, catering, bunkhouses and camping barns. Tourist accommodation was dropped from the scheme in January 1991 as its rate of return had been discovered to be lower than other types of business – averaging 10 per cent, compared to 20 per cent for recreation enterprises, 26 per cent for pick-your-own enterprises, 31 per cent for craft shops, and 34 per cent for horse-based activities. The maximum grant available was £35,000 at a rate of 25 per cent, or 31.25 per cent for young farmers. In addition, grants for feasibility studies were available for the employment of consultants at a rate of 50 per cent on expenditure up to £6,000, and the work of marketing agents could be eligible for grant aid of 40 per cent, 30 per cent and 20 per cent over a three-year period. Between 1 January 1988 and 1 October 1992, 3,200 applications were received in England and Wales, of which 2,443 were approved and 1,396 completed, at a total cost to the public purse of £10.4 million (Edwards *et al.* 1994).

The press release announcing the removal of accommodation from the scheme heralded it as a move to 'concentrate grant payments on enterprises yielding the highest rate of return to farmers'. What the press release failed to point out was that the majority of grant schemes in the first two years were for tourist accommodation, and that the move might also be expected to lead to a reduction in public expenditure. In Devon, for example, the provision of accommodation accounted for 66.8 per cent of all grant moneys on the scheme in the first two years (Table 6.7).

Removing accommodation from the list clearly removed the type of diversification with the widest appeal to farmers, serving to narrow the opportunities for diversification to particularly well-placed farmers. By the same token it has to be said that the provision of tourist accommodation probably does not require a grant to stimulate it. Indeed this was true for most investments carried out under the scheme. Both of MAFF's own sponsored policy evaluations revealed that the majority of farmers in the scheme would have diversified without its aid (Edwards *et al.* 1994; Ilbery and Bowler 1993):

the grant appears either to subsidise farmers in carrying out develop-ment schemes they would have completed anyway, or to encourage those on the verge of making an investment decision. This seems to be

Table 6.7 The farm diversification grant scheme in Devon 1988–1990

Type of enterprise	Number of schemes	Total expenditure £			Total
		1988	1989	1990	
Tourism and recreation					
Accommodation	151	23,397	218,971	186,557	428,925
Catering	5	0	904	4,450	5,354
Recreation/education	56	9,968	35,292	45,738	90,998
Horses	19	4,616	1,856	8,015	14,487
Other					
Food processing	17	28,329	31,250	6,682	66,261
Farm shop	12	374	2,357	5,672	8,403
Miscellaneous	8	0	0	27,245	27,245
Total	268	66,684	290,630	284,359	641,673

Source: MAFF (unpublished data)

confirmed by the fact that 43 per cent of adopters had used the FDGS
to expand existing diversified enterprises rather than to start new
ventures.

(Ilbery and Bowler 1993: 168)

Ilbery and Bowler produced evidence that the low take-up of the scheme
was unlikely to be improved upon without major modifications to the EC's
own guidelines. For example, the exclusion of occupiers deriving less than 50
per cent of income from agriculture, working less than 1,100 hours per year
on the holding, or who had not been in farming for a minimum of five years,
all served to bar many farmers from entry into the scheme. Such findings
should not have entirely surprised the Ministry, for evidence from earlier
research had clearly suggested that the provision of holiday accommodation
is a cyclical activity, based on the family development cycle (Bouquet 1985,
1987; Denman and Denman 1990; Winter 1987). For example, in the West
Country, 'despite the publicity given to diversification and farm tourism,
there does not appear to have been a major growth in involvement by farmers
in tourism in the region in the past 15 to 20 years' (Denman and Denman
1990: 30). Whilst 45 per cent of providers had become involved in the last ten
years and 36 per cent in the last five, 11 per cent of farmers had been involved
in the past and had given up. Linked to the family cycle is evidence from the
use made of earlier grants in the LFAs that farmers utilise grants for
accommodation with long-term residential requirements in mind as well as
the desire to generate an income from tourism (Winter 1987).

The demise of the FDGS at the end of 1992 was, officially, a response to
the success of the scheme. Its purposes had been fulfilled, but it is hard to
avoid the conclusion that the government had retreated from diversification,
somewhat politically embarrassed, and now placed far greater emphasis upon
the environmental aspects of agricultural policy, particularly the expanding
ESA programme. Indeed, the deregulationist thrust of the 1980s also received
something of a reversal in the 1990s as planning policies again became tighter
in the aftermath of the 1992 Planning Policy Guidance note, *The Countryside
and the Rural Economy*, which in the run-up to the 1992 election, opined
that:

> The guiding principle in the wider countryside is that development
> should benefit the rural economy and maintain or enhance the
> environment.
> Building in the open countryside, away from existing settlements ...
> should be strictly controlled.
>
> (Department of the Environment 1992)

The Conservatives were clearly keen not to make the same mistake with
preservation-minded shire Tory voters as it had with the launch of the
ALURE package in 1987. None the less it would be a mistake to argue that
nothing of the rural diversification policy initiative was left by the mid-1990s.
Certainly planning authorities had become more likely to grant consent for

new economic activities on agricultural land and, in particular, in farm buildings than had been the case a decade earlier. But diversification had not lived up to its promise as a major plank in the efforts to reform the CAP, although its importance is clearly far greater than suggested by the débâcle of the FDGS.

THE AGRICULTURAL POLICY COMMUNITY IN THE 1980s

Before turning to the further reform of the CAP in the early 1990s, it is important to draw together some of the strands of the underlying policy process theme of this book. What did these developments amount to for a hitherto closed, even corporatist, agricultural policy community? There is little doubt that the relative consensus that characterised the policy community for most of the post-war period was severely tested during the early and mid-1980s. Internally, the key actors – the NFU and MAFF – found that the depth of the crisis affecting agriculture rendered invalid traditional solutions and approaches. Inevitably, in such a context relations within the community were at times strained. Externally, the policy community was beset by an ever-increasing range of interests keen to be involved in the policy process, as acknowledged by MAFF in a letter in 1992 to researchers examining the changing agricultural policy-making 'universe':

> Not only are there more bodies wishing to be consulted, but there is a much greater awareness of the interaction of policy changes in the various sectors. To give an obvious example, changes in agricultural policy are no longer of interest simply to the NFU; conservation and countryside organisations are very aware of the implications of such changes for the countryside and the environment, consumer bodies of the implications for consumers, charities of the possible implications for developing countries, etc.
>
> (Jordan, Maloney and McLaughlin 1994: 506)

These changes suggest the need seriously to question earlier characterisations of a closed policy community:

> while there is still an expectation that the ministry will 'fight the agricultural corner', civil servants are now much more cautious in going in a producer-led direction. Interviewees in MAFF suggested that the department has lost any stomach it ever had to defend agriculture in an unreserved manner. These broad pressures lead us to query whether characterisations of the agricultural policy process which emphasise exclusion are still appropriate.... Our agricultural policy map is congested with detailed and overlapping sub-sectoral policy communities.
>
> (Jordan, Maloney and McLaughlin 1994: 506–507)

The existence of sub-sectoral policy communities is an important point. In many respects it was ever so. Agricultural policy making has never been dominated solely by a single core policy community. Many issues – tenancy, milk marketing, animal welfare, food – have warranted sub-sectoral communities where particular specialist groups can mount successful challenges to NFU dominance. However, prior to the 1980s, these groups tended to be relatively few in number, dealing with minor issues and, essentially, being subservient to the dominant MAFF–NFU alliance. The increased politicisation of agriculture, and of issues hitherto at the margins of policy concern, has served to increase the number of these groups. Sometimes they become looser and more akin to issue networks, arenas of contest, with the outcomes heavily influencing deliberations at the centre. In other instances, the centre has ceded power to sub-sectoral groups. This can happen for very different reasons as the following two examples show. A perceived need to reform the agricultural tenure system (which eventually resulted in the Agricultural Tenancies Act 1995), an issue from the late 1980s on, was very much an outcome of new free market thinking on the political right. The NFU, in its initial dogged defence of a highly regulated system, was perceived as clinging to an out-dated corporatist and regulated system. The formation of the Tenant Farmers' Association, committed to market reforms, signalled a fundamental shift in the balance of thinking within the sub-sectoral policy community on tenure. The NFU was effectively defeated because its monopoly of representation on the issue had been broken. In much the same way, although for very different reasons, policy debates on farm animal welfare issues now largely take place within the Farm Animal Welfare Council which includes farmers, hauliers, slaughterers and welfarists (Jordan, Maloney and McLaughlin 1994). Here the issue has become so politically sensitive that both the NFU and MAFF are content to see it removed from the central policy community in the hope that consensus views will eventually emerge.

Even mainstream agricultural policy issues are now subject to a much wider scrutiny. As is discussed further in chapter 9, the NFU, having recognised that in many instances total resistance is no longer plausible, has had, as a key element of its strategy, to defend its policy territory by offering to take on board fresh issues (Cox, Lowe and Winter 1986b). Indeed, part of its strategy has been to reinterpret history to show that, taking just one example, environmental issues have always been a major concern to farmers and that the NFU and MAFF are perfectly able to respond to the need for new policies. Thus, by the close of the 1980s, the agricultural policy community was altered but not entirely transformed. New issues had emerged and new actors were engaged in the policy process, but the NFU and MAFF remained key players commanding significant political resources.

THE 1992–1993 REFORM

The agricultural situation in the Community in the early 1990s seemed little better than that of a decade earlier, with cereal surpluses a particularly pressing problem. Both at the European level and nationally, expenditure remained tied to commodity support despite the widespread recognition that structural measures were required. The budget cost of the CAP grew by 20 per cent between 1990 and 1991. Moreover, by the early 1990s, it was increasingly clear that the eventual outcome of the Uruguay round of the GATT talks was likely to require fundamental reforms to the CAP.

Publicly, the UK government was as anxious as ever for reform of the CAP. But when a further round of concrete proposals was made in 1991, its reaction was far from appreciative. 'We start by saying that we oppose Mr MacSharry's proposals. We hate them. We condemn them' (House of Commons debate, 14 February 1991, col. 1019). In such tones John Gummer, the Minister of Agriculture, launched a savage attack on the European Commission's new proposals for reform of the CAP published on 1 February 1991 (Commission of the European Communities 1991). The antagonism towards Ray MacSharry, the EC's Agriculture Commissioner, was based on the strong belief that his sympathies lay firmly with the small farmers of Europe, particularly his Irish compatriots. Indeed, the UK's main objection to his proposals concerned the in-built bias towards small farmers in the proposals, which would have seriously damaged the interests of many UK producers. The Commission in recognising that 80 per cent of FEOGA support was going to 20 per cent of farmers had determined not only to seek ways of cutting the growing cost of the CAP, but also to redistribute the benefits. Thus the proposals suggested that support should be tapered, with farmers producing above a certain level of produce receiving only market prices beyond that point. The levels suggested would have served to limit greatly the amount of support flowing to the UK which in comparison to its European neighbours is dominated by large farms.

In July the Commission published a revised and more detailed set of proposals in its paper *The Development and Future of the Common Agricultural Policy* (Commission of the European Communities 1991). These proposals removed what the UK considered to be the worst features of the first set of proposals, although to many they remained far from satisfactory, in particular as they implied such a major degree of intervention and bureaucratic control. Debate continued and further concessions were granted until a settlement was reached in May 1992. By this time the UK had secured such major concessions that the threat to the UK's large-farm structure seemed to have been removed. For example, the July proposals for sheep premiums had suggested a limit on payment to the first 750 ewes in the LFAs and the first 350 ewes elsewhere. These figures were raised in the final settlement to the first 1,000 ewes in the uplands and 500 in the lowlands.

In broad terms, the reform did not call into question the fundamental principles of the CAP-price unity, Community preference and financial solidarity; but was based on a recognised need for change so as to play a positive part in the GATT agreement, to reduce the food surpluses created by intervention buying, and to limit the uncontrolled rise in budgetary expenditure (General Secretariat of the Council of the European Union 1993). The fundamental underlying feature of the reform lies in a shift from payments based on levels of production to direct payments. The main feature of the reform package which was eventually agreed and implemented in 1993 concerned cereals. Cereal prices were to be cut by 29 per cent by 1995/96. But the blow was to be softened considerably by the Arable Area Payment Scheme (AAPS), providing substantial compensation for the reduced cereal prices. The arable area payments are dependent upon compliance with a prescribed area of set-aside (initially set at 15 per cent of the arable area). In addition to these measures, attractive subsidies were to be made available for new non-cereal crops, some of which could be grown on set-aside land.

Price reductions were also imposed in other sectors. Beef prices were to be cut by 15 per cent over three years from 1993/94 (with ceilings on intervention purchases to be introduced progressively from 1993/94). By way of compensation a special beef premium was introduced be paid on the first 90 male animals at 10 months and 22 months (including a system of 'passports' with farmers required to register births of all male cattle with MAFF) and a suckler cow premium for cows (with no upper limit but based on the number of animals in 1992). In the case of both these premiums, a prerequisite is a reduction in stocking density, which has to be reduced in stages to 2.0 GLUs (Grazing Livestock Units) per hectare by 1996/97. Farmers wishing to claim at a lower stocking rate (1.4 GLU per hectare) were also entitled to claim an extensification premium.

Butter intervention prices were set to fall by 5 per cent by 1994/95. Other measures included a reduction in milk quotas (no cut in 1992/93 but an estimated need for a cut of *c.* 2 per cent in 1994/95) and upper limits on sheep premiums. Although the original proposals to discriminate in favour of small farmers had largely disappeared, there remained some special treatment for small arable and dairy producers. Accompanying measures included an agri-environment programme (see chapter 9), an afforestation programme (which resulted in no additional measures in UK) and an early retirement programme (which was not implemented in UK).

The influence of GATT

General agreement on Tariffs + Trade.

Although the GATT round was not concluded until after the 1992 CAP reform its influence was, and remains, fundamental to the CAP reform process. The two systems were required to intermesh, and much negotiation took place in order to achieve this. The Uruguay Round of the GATT sought to harmonise world trade by reducing trade barriers between nations or groups of nations. The objectives for the agricultural element of the negotiations were to reduce support prices, export subsidies and border controls applying to agricultural commodities. The Uruguay Round represented the first time that agricultural commodities had been a major element in the Agreement. A key element in the outcome was the agreement eventually struck by the USA and the EU. In negotiations concerning the agriculture sector in autumn 1992 the European Union's position was consistently that the May 1992 CAP Reform provided an adequate basis for any concessions that the EU could make on international agricultural trade arrangements.

Considerable negotiating difficulties arose between the US and the EU over a number of issues, including oil-seeds production, subsidised export limitation and improved access to EU agricultural markets. France, in particular, insisted that there should be no question of the EU making deeper cuts than those agreed as part of the 1992 CAP Reform.

In November 1992 bilateral talks between the US and EU produced the Blair House Agreement between the two blocks on a number of issues, the most notable being that of reducing EU and US subsidised export volumes by 21 per cent (relative to the average of exports in 1986–1990), not 24 per cent as originally proposed, and the reduction of the EU's direct export subsidies by 36 per cent over six years. The European Commission saw the deal as a consolidation of the CAP reform and expected to be able to take the EU–US agreement into the mainstream of the GATT negotiations. But a number of EU countries, especially France and to a lesser extent Denmark and Italy, opposed the deal, arguing that it went beyond the CAP reforms. Opposition was further fuelled by an analysis done by COPA which appeared to confirm that the Blair House Agreement did in fact go beyond the limits of the 1992 CAP reform.

As a result of the G7 summit in Tokyo in July 1993, a deadline for concluding the Uruguay Round was set for 15 December 1993. Agreement was reached and the provisions came into effect in July 1995 running until June 2001. They include:

- the conversion of all border protection measures to tariffs, and a reduction in these tariffs by an average of 36 per cent (minimum 15 per cent) between July 1st 1995 and July 1st 2000;
- an undertaking to reduce aggregate domestic market support for agriculture by 20 per cent over a six-year period with 1986–1988 as the base period;
- a reduction in expenditure on export subsidies of 36 per cent over six years and in the volume of subsidised exports of 21 per cent using 1986–1990 as the base period;
- harmonisation of food safety, and animal and plant health regulations.
(after Ockenden and Franklin 1995: 56–57)

In the short term, the GATT agreement presents few problems for the European Union and has been seen as a considerable negotiating success for the European Commission. However, in the longer term, the limits imposed on subsidised exports are likely to present an increasing challenge unless production levels within the EU are kept very firmly under control:

The results of these negotiating successes are that the commitments on tariff reductions and the aggregate measure of support should be honoured easily. Even assuming low world prices in 2000, the new, fixed tariffs would still give the EU market significant protection in most fields.... Overall the GATT agreement will be broadly delivered by the 1992 CAP reform in the implementation period.... [However] the true impact of the Uruguay round will not be felt until after this.... All assessments show that at some point between 1998 and 2003 the current price level and GATT obligations will collide. The current CAP reform can only be the first step on the path of adjusting to the GATT obligations.

(Ockenden and Franklin 1995: 58–59)

The CAP reform in Britain

But what of the impact of the CAP reform in Britain? Although the schemes are in theory voluntary (except for milk quotas), in reality compliance is virtually compulsory if incomes are to be maintained. And this is the case even in circumstances where the expected price cuts have not materialised as a result of the UK withdrawal from the European Monetary System in September 1992. This led, over the following three years, to successive increases in the value of sterling in relation to the ECU, and consequently to increases in the commodity prices received by UK farmers. Table 6.8 provides an illustrative balance sheet produced by Lloyds Bank soon after the 1992 reform was agreed and shows how participation in the AAPS was likely to prove beneficial for the overwhelming majority of arable farmers.

Table 6.8   Implications of the 1992 reform package for a farm of 100 hectares growing winter wheat

		£ Projected gross margin
1992 Pre-reform		47,600
1993 In AAPS		
85 hectares at reduced prices	33,320	
Area payments on 85 acres	10,285	
Set-aside compensation on 15 hectares	3,285	
	⇒	46,890
1993 Not participating in AAPS		
100 hectares at reduced prices		39,200

It should be noted that the figures take no account either of cost savings likely to be made in the reduced need for management of set-aside land or of possible earning from alternative crops on set-aside land. This combined with the sterling effect already mentioned has meant that in practice earnings for cereal farmers have been high in the first three years of the AAPS; indeed they rose by a staggering 44 per cent in the UK in 1993 reaching the highest level since 1986/87. The Ministry announced the reform settlement in the following terms:

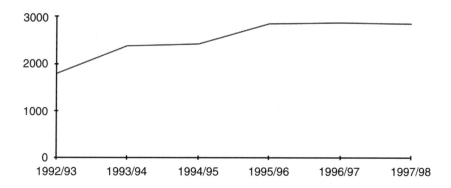

Figure 6.6 CAP market support expenditure in the UK (£ million)

Source: Cmnd 2803 (1995)

> support prices for the main agricultural commodities will be reduced and direct payments to farmers will increase. The result of these changes is that the cost to consumers will fall as more of the cost of agricultural support is borne by the EC budget. Overall, the increase in the budgetary costs of the CAP resulting from the reform measures will be more than offset by the reduced burden on consumers.
>
> (MAFF Press Release, November 1992)

This was putting a brave face on it, to say the least, for the budgetary consequences of the package have been grave. The Chancellor's autumn statement for 1992, based on Treasury estimates for CAP funding in the UK, anticipated an increase from £1,790 million to £2,730 by 1995/96. In the event the increases were marginally more than this, as shown in Figure 6.6.

CONCLUSIONS

The reform of the CAP certainly did not result in a sudden and sharp reduction in the budget, but it was hoped that in the long term a reduction in surpluses would result. But less than a year into the reform package it became clear that the crisis of the CAP was far from over. The successful completion of the Uruguay GATT round late in 1993 reopened a debate on the future of the CAP that remains unresolved at the time of writing, with some believing that the combined effect of the CAP reform and GATT is to place a permanent ceiling on production levels. Severe criticism of set-aside and the environmental implications of the package also raise serious questions over the future of the policy. These issues are returned to in chapter 9.

Doubts continue to be expressed about whether the CAP reform package will ultimately be successful in reducing surpluses and bringing European agriculture sufficiently into line with world markets (Ockenden and Franklin

1995). Others suggest that, even if it is successful in this respect, it will be at the cost of an unacceptably high degree of regulation and bureaucratic intervention. In view of the extraordinary difficulties encountered in reforming the CAP, it is not surprising that agricultural policy has recently been considered by some commentators as a serious candidate for subsidiarity and the renationalisation of policy (Grant 1995; Kjeldahl and Tracy 1994). But most commentators would agree that this is an unlikely scenario: 'for all the protests of the economic purists, a Community without a CAP is a practical impossibility' (Ockenden and Franklin 1995: 1), for three main reasons. First, despite its difficulties, the CAP is still seen, and defended, as the central achievement of the European Union. It is, after all, a genuinely *common* policy. To dismantle it at a time of deepening European integration would be deeply humiliating for the Commission and the Union as a whole. Second, it is by no means clear that the policy difficulties would be any easier if devolved to the national level. The GATT problems facing non-member countries, such as Sweden, are as acute as those within the Union. There are probably few, if any, member states which would relish the prospect of being solely responsible for such a problematic policy sector. The politics of this are important, for the UK government is not alone in having used the EU as a scapegoat for aspects of agricultural policy which have excited public disquiet. Third, the farm lobby, at national and European level, remains powerful despite the changes of the past few years. In this context, Grant (1995) argues for a modification of the traditional view of agricultural political strength residing in farming as such. Instead, he points to the benefits that the CAP has given to other sectors associated with farming. He sees the suppliers of agricultural inputs (fertiliser, agrochemical and machinery manufacturers for instance), and the banks (agricultural credit) as providing a powerful set of interests committed to the retention of the CAP.

Part III

RURAL ENVIRONMENTAL POLICY

7

THE ORIGINS OF ENVIRONMENTAL CONCERN

INTRODUCTION

This chapter traces the growing concern for environmental protection in Britain during the last century, paying particular attention to the emergence of environmental pressure groups and to the distinctive English preoccupation with the protection of landscapes and scenery. The cultural and ideological distinctiveness of English environmentalism are examined, along with mention of the emergence of more radical, and international, ecological concerns from the 1960s. The development of legislation, with the exception of brief mentions where important to the story, is not dealt with in this chapter as it provides the main focus for later chapters. This chapter, therefore, is concerned with ideas, and with the cultural as well as the economic underpinnings to the emergence of what today constitutes one of the most significant social movements of the late twentieth century.

Philip Lowe (1983) has identified four main overlapping phases in the development of nature conservation in Britain:

- The natural history and humanitarian period   1830–1890
- The preservationist period   1870–1940
- The scientific period   1910–1970
- The popular/political period   1960–present day

                                                           (Lowe 1983: 329)

To these four might now be added a fifth: the regulatory period from 1980. The increase in regulatory intervention, and the thinking behind it, is dealt with in greater detail in chapters 8, 9 and 10. Here our discussion is confined to the four phases identified by Lowe. As Lowe explains, 'the dominant ideas within each of these periods were not discrete, but each had a distinct value orientation, which was superseded by and subsumed in the next' (ibid.). Thus the main four sections of this chapter examine the development of ideas in each of these periods, with particular reference to the emergence of voluntary sector bodies representing the interests discussed.

THE NATURAL HISTORY AND HUMANITARIAN PERIOD

The roots of natural history can, of course, be traced back far earlier than the nineteenth century. Exploring the origins of natural history, David Allen writes that:

> There comes a point in the history of every pursuit at which its following becomes sufficient to entitle it to be termed a social activity.... It takes on a life over and above that of its individual adherents and through the pattern of its own development starts to influence, and sometimes even govern, the way in which they think and act. For natural history in Britain this point was reached some time in the course of the seventeenth century.
>
> (Allen 1976: 5)

Prior to the eighteenth century, most early natural history investigations were based on a strongly utilitarian and anthropocentric approach, in which creatures were classified or categorised according to their utility and virtue to humans rather than in any modern scientific manner (K. Thomas 1983). Thus, one of the earliest bodies devoted to the furtherance of the knowledge of plants was the Society of Apothecaries (Allen 1976). Medical properties provided a particularly important basis for classification, but moral characteristics were also invoked. A moral basis of classification was even reflected in law, which 'distinguished between animals "of a base nature", like ferrets, mastiffs and cats, and those which were "noble and generous", like falcons' (K. Thomas 1983: 58). By the eighteenth century, classifications were changing to reflect a growing interest in plants and animals in themselves, an interest rooted in the growth and development of science:

> it was indeed a revolution in the manner of understanding plants when, instead of describing their usefulness, abundance, size, smell and colour, attention was given exclusively to the disposition and form of the parts of the flower and seed....
>
> (Haudricourt 1973: 267)

Probably the first natural history society was the Temple Coffee House Botanic Club founded in 1689, which was closely linked to the Society of Apothecaries. In zoology the earliest body was the Society of Aurelians (lepidopterists), established sometime in the first forty years of the seventeenth century (Allen 1976). Much of the institutional vigour of these early years was lost in the middle of the century, not least because of the stagnation in British science generally at this time and the associated decadence of academic life at Oxford and Cambridge. But dedicated individuals and informal local associations continued to thrive, so much so that the *Critical Review* in 1763 could declare that 'Natural History is now, by a kind of national establishment, become the favourite study of the time' (quoted in Allen 1976: 45; see also Lowe 1976). The Reverend Gilbert White, whose *Natural History and Antiquities of Selborne* was published in 1788, is the best known and, in terms of his contribution to English literature, the most accomplished of many pioneers of natural history of this time.

By the close of the century much institutional progress was again being made. In 1788 the Linnean Society was formed, a society which flourishes to this day and whose *Transactions* represent the longest continuously published natural history periodical in the world. Collections and classifications of all forms of life, not to mention the emergence of interest in earlier forms of life through the study of fossils and geology, continued apace from the closing decades of the eighteenth century into the nineteenth century, by which time natural history had become one of the most popular and widespread pursuits amongst the middle classes, particularly the clergy.

Before considering the implications of all this for the birth of environmental concern as evidenced by pressure groups and by policy, it is important to place these developments in their wider context, in terms of culture and society. The emergence of natural history should not be associated only with the rise of science. On the contrary, it owes much to a broader set of developments and changes in culture and religion. Indeed it has been argued that the general decline in intellectual life that took place after about 1725, which includes a decline or at best a stasis, in natural science, was a direct consequence of the decline in Christianity's appeal to the educated at that time (Clark 1969).

The revival of religion in the nineteenth century was crucially important to the development of natural history, as was the reverence for nature reflected as much in Romantic art and literature (Bate 1991) as in the development of natural science as such. Romanticism and science were intricately and explicitly linked in some cases, as in the work of Goethe. Indeed, one of the articles of faith for those in the vanguard of the Romantic movement was the need to understand the natural world through artistic appreciation, a search for truth and beauty that represented a true fusion of art, religion and *natural* science. 'The purpose of art, according to Ruskin, is to reveal aspects of the universal "Beauty" or "Truth". The artist is one who, in Carlyle's words, "reads the open secret of the universe"' (R. Williams 1960: 135). Christianity was too deeply embedded in society for it to be supplanted by any new religion of nature. Instead religion itself adapted to the new thinking:

> Romanticism was well fitted to be a vehicle for religious thought.... This was the movement of taste that stressed, against the mechanism and classicism of the Enlightenment, the place of feeling and intuition in human perception, the importance of nature and history for human experience.... Its quintessence was what has been called 'natural supernaturalism', the ability to discern spiritual significance in the everyday world.
>
> (Bebbington 1989: 80–81; see also Abrams 1971)

Romanticism influenced all strands of Christianity including the Evangelical revival of the nineteenth century. Thus the popularity of natural history amongst the middle classes was based as much on the Evangelical revival as on the Romantic movement:

> Those transcendental moments of ecstasy which earlier Romantics had seen as the workings of the life-force were now reinterpreted in orthodox, theistic terms. Nature's magical enchantments still continued to be acknowledged ... and to be valued for the obvious bounties that they brought to human minds; but the uplift men received from them was envisaged no longer as sensuous and neutral, but as spiritual and prescriptive.
>
> (Allen 1976: 74)

Thus for Allen, Victorian natural history was 'in its whole essence an Evangelical creation'. The strands of this Evangelical–Romantic natural history were several. The study of the natural world was seen as a worthy pursuit for those seeking to laud and honour the Creator; the hard work required was fitting, for idleness was a vice; the natural world was given by God for mankind's use, so there was also a utilitarian streak involved. Ultimately, of course, the Darwinian controversy broke asunder much of the rapport between natural science and religion but this was not a serious issue until much later in the century.

A remarkable portrait of the correspondence of science and religion in one man is to be found in Edmund Gosse's study – one of the minor classics of Edwardian literature – of his father, the well-known Victorian geologist and naturalist Philip Gosse (Gosse 1949). The extreme devotion of Gosse senior to his scientific work of cataloguing and chronicling the natural world, except on the Sabbath, provides a compelling insight into the devotion to nature that provides one of the well-springs for modern environmental concerns. The emergence of professional ecology from its roots in natural history marks a continuity of tradition. At the same time Gosse's refusal to accept the implications of fossil evidence for his literal interpretation of the Genesis story is equally instructive about the way in which the relationship between religious belief and natural science has been transformed over the same period.

Middle-class involvement in natural history was not confined to the luminaries of the London-based Linnean Society. Every bit as important was the expression of the new movement in the burgeoning of local natural history clubs from the 1840s to the 1890s, by which time nearly all shire counties, and many districts as well, had associations or societies devoted to the pursuit of local natural history. The oldest of the survivors is the Ashmolean Natural History Society in Oxford dating from 1828 (Allen 1976). The Woolhope Society, founded in 1852, takes its name from a tiny village in Herefordshire; yet its *Transactions* had a national reputation. County-based associations were similarly at the forefront of the publication of local studies of natural history. The Devonshire Association for the Advancement of Science, Literature and the Arts founded in 1862 numbered among its nineteenth-century presidents such notables as Charles Kingsley and Sabine Baring-Gould. A survey undertaken in 1873 discovered 169 local scientific societies in Great Britain and Ireland, and by the close of the century it was estimated that 50,000 people belonged to local natural history societies (Allen 1976). Many more people followed developments in natural history through numerous natural history publications which outsold popular novels in the first half of the century (Burrow 1968).

The emergence of issues

Our discussion in this section has so far focused on the realm of ideas and beliefs as a foundation for the emergence of natural history as a social activity. This is not surprising insomuch as widespread concern for environmental degradation in rural areas is largely confined to the present century and especially the post-war period, but there were issues to excite the attention of campaigners during the last century. The first was a direct consequence of the growth in natural history. For all the reverence for nature felt by some, the popularisation of the pursuit also led to major excesses in the collection of plants and animals for natural history purposes. In some instances scientific collection rapidly escalated into commercial activity to supply the wants of middle-class consumers whose knowledge and interests in natural history were fashionable rather than informed. For example, the fashion of decorating women's hats with plumage led to massive destruction of sea-birds, with plumage hunters often cutting the wings off and flinging the victims into the sea. The passion for ferns as decorative houseplants led to a plundering of ferns on the western seaboard, particularly Cornwall, one professional boasting of sending five tons to London by rail on one occasion (Allen 1969; Sheail 1976b).

The topic that attracted most attention from embryonic environmentalists in the nineteenth century was cruelty. Keith Thomas (1983) has demon-strated how cruelty became more of an issue from the seventeenth century onwards. Initially the concern of moralists was primarily directed at the likely impact of cruelty to animals on those who themselves inflicted the cruelty. Cruel behaviour came to be seen as morally degrading to the perpetrators and fears were expressed that a capacity to be cruel to animals could easily turn into violence against fellow humans. This anthropocentric theology lies at the heart of the emerging concern for animal welfare in the eighteenth and into the early nineteenth centuries. Thus, the Society for the Suppression of Vice, dating from the eighteenth century, campaigned against bull baiting 'on the grounds that it corrupted those who indulged in it' (R.H. Thomas 1983: 65). In addition to concern over the individual moral degradation of those who might inflict cruelty, increasing attention was turned to the creatures themselves. Remarkable notions were developed to explain how all creatures were designed by God for human benefit:

> Savage beasts were necessary instruments of God's wrath, left among
> us 'to be our schoolmasters', thought James Pilkington, the Elizabethan

bishop; they fostered human courage and were useful training for war. Horse-flies, guessed the Virginian gentleman William Byrd in 1728, had been created so 'that men should exercise their wits and industry to guard themselves against them'.... The louse was indispensable, explained the Rev. William Kirby, because it provided a powerful incentive to habits of cleanliness.

(K. Thomas 1983: 19–20)

Strange though these views appear to us, they represented a considerable advance on disregard for creatures as being below the dignity or concern of man, and a concern merely for the moral and spiritual well-being of perpetrators of cruelty. If God had placed creatures in the world for a purpose then respect should be accorded to them. Thus, increasingly, the language of moral and theological argument was couched in terms of man's duty before God. According to the philosopher David Hartley, writing of the animal kingdom in 1748, 'we seem to be in the place of God to them and we are obliged by the same tenure to be their guardians and benefactors' (K. Thomas 1983: 155).

With the emergence of Romanticism came a further transition from guardianship to the declaration of oneness with animals. Keith Thomas (1983) provides many examples of changing attitudes. Coleridge addressed an ass as 'brother'. Robert Burns was 'truly sorry man's dominion/Has broken nature's social union', and William Blake asked the fly:

> Am not I
> A fly like thee
> Or art not thou
> A man like me?

The types of action considered cruel also changed from outrageous examples such as bull- and bear-baiting to the cruelty prevalent in more 'accepted' practices such as slaughtering animals for meat, slaughter for fur or plumage and in hunting and shooting. In 1808, the Society for Preventing Wanton Cruelty to Animals was formed in Liverpool, but rapidly sank into obscurity (E.S. Turner 1964). It was followed in 1824 by the Society for the Prevention of Cruelty to Animals, which received its Royal Charter in 1840. One of its founding members was William Wilberforce, the Evangelical churchman and anti-slavery activist. For much of its first thirty years the RSPCA concerned itself primarily with policing laws, the first of which was passed in 1822, to safeguard the welfare of domestic animals, including a ban on bull- and bear-baiting and cock-fighting. In 1835 its inspectors were given powers to prosecute those organising such activities (Sheail 1976b). In the 1860s the RSPCA extended its activities to the protection of wild birds and their eggs; to campaigning against vivisection; and to seeking to place limits on some of the excesses of hunting and shooting (Sheail 1976b).

The excesses of shooting were an increasing cause of concern during the nineteenth century, as technical advances vastly improved the deadly efficiency of guns (Allen 1976). To take but one example, Osgood Mackenzie boasted from his estate in the Highlands that, in 1868, he shot:

> 1,314 grouse, 33 black game, 49 partridges, 110 golden plover, 35 wild duck, 53 snipe, 91 blue rock pigeons, and 184 hares. He had no compunction in shooting 'anything that moved'. Mackenzie recorded that he was shooting snipe one day on the Isle of Ewe when he started 'a thrush which had a broad white ring round its throat, just like that of a ring ouzel. I promptly shot it', and he described how 'on another day on the same island we kept putting up nearly as many short-eared owls as grouse and snipe.... I shot five.'
>
> (Sheail 1976b: 3, quoting from Mackenzie 1980)

The RSPCA's links with the sport-loving nobility (its first president was the Earl of Carnarvon) and with royalty, whilst giving it considerable influence, also limited its sphere of activity. Thus, it did not oppose hunting as such until 1976 and shooting remains a problematic issue to the Society.

But the bulk of the RSPCA's members, and its origins, were far from patrician:

> The triumph of the new attitude was closely linked to the growth of towns and the emergence of an industrial order in which animals became increasingly marginal to the process of production. This industrial order first emerged in England; as a result, it was there that concern for animals was most widely expressed.... Such feelings were not just urban. They were those of the professional middle classes, unsympathetic to the warlike traditions of the aristocracy. For hunting was notoriously a military exercise and a training ground for cavalry.
>
> (K. Thomas 1983: 181, 183)

Thus was the conflict between patrician society and the new industrial commercial classes, already discussed in chapter 4, played out over the issue of animal cruelty too. Following the early successes of the RSPCA a number of other bodies were founded to further the cause of protection from cruelty and exploitation. These included the East Riding Association for the Protection of Sea Birds founded in 1868, the Association for the Protection of British Birds (1870), the Selborne Society for the Protection of Birds, Plants and Pleasant Places (1885) and the Society for the Protection of Birds (1889, Royal Charter 1904). The emphasis upon individual birds and bird species remains a powerful element in contemporary conservation. The RSPB has, in recent times, made determined efforts to shift part of its attention to the protection of habitats and to broad principles of ecological management but, as with the RSPCA, much of its attention remains focused on issues of cruelty and protection.

THE PRESERVATIONIST PERIOD

The main focus of this second period was the move from natural history and cruelty to a much greater concern for the protection of landscapes, and the beginnings of a concern for amenity and public access. In the same way that Keith Thomas identified the worries over animal welfare as arising just as the practical importance of animals in many people's lives declined, so too preservationism can be linked to the declining importance of agriculture and rural society. Thus, in contrast to the natural history period, the preservationist period is much more explicitly concerned with issues of change. These were seen as social – for example the decline of the country estate – as well as environmental. The main purpose of this section is to show how concern for the physical environment is deeply rooted in the place of landscape in English culture and traditions. Britain's early industrialisation and the incorporation of some landowners into structures of power and influence alongside industrial and financial interests gave rise to a strong vein of nostalgia for 'rural roots', a respect for rural and provincial life within the dominant culture of England's ruling class, and corresponding positive representations of English rural landscapes in art and literature.

The most compelling ideas were those associated, however loosely, with Romanticism, but while this provides some of the intellectual roots of preservationism, a new and very different impetus was given by Darwin's *The Origin of Species*, first published in 1859. The direct influence of Darwinian thought on the practice of erstwhile natural historians is dealt with in the next section on the rise of ecological science. But the Darwinian controversy also influenced those outside the realms of practising scientists, in its vision of an uncaring and changing nature and 'whole species formed and then blindly squandered' (Lowe 1983: 336; Fleming 1961; Houghton 1957). For many, Darwinism dealt a savage blow to the optimism of the Enlightenment. The idea of the survival of the fittest, whilst it could be pressed into service to support an atomistic and individualist commercial society, could also point to the folly of assuming that civilisation and progress were automatically beneficial to all. On the contrary, as the century advanced it became increasingly clear that industrialism and technological advance left in its wake many losers, socially, economically and environmentally. The increasing awareness of the loss of individual species and landscapes led to a growing concern to protect and to preserve which went far beyond the concerns of the early naturalists.

A strong current in the new thinking was an explicit antagonism to urban-industrial society, which originated in the last century and has continued into the present. What started as a strongly moralistic critique of industrialism, of accumulation, and of the owners and managers of manufacturing industry had, by the end of the century, become a full-blown idealisation of the countryside, a ruralism which appeared to grow in inverse proportion to the importance of the rural economy itself (R. Williams 1973). This idea was taken by Martin Wiener (1981) and elevated to a comprehensive explanation for the economic and industrial ills of late twentieth-century Britain, whose roots are not, after all, in its indifferent workforce, but in the self-doubts of a Janus-faced bourgeoisie looking uncertainly at its own economic base in industrial and mercantile Britain and, with far greater reverence, at the English countryside.

It is not part of our purpose to examine whether cultural trends can, in fact, be linked to economic decline in this way, save to note that a number of commentators have expressed considerable reservations on this matter. For example Daunton (1989) has cautioned that aesthetic preferences may tell us very little about business dynamism. And, in a vigorous assault, Rubinstein (1993) attacks Wiener on almost every count in an iconoclastic assertion of the significance of commercial values in Britain's economic and social development. However, while Rubinstein undoubtedly dents Wiener's reputation as a student of economic history, he pushes his case rather too far in suggesting that ruralism is unimportant in British culture. For the evidence assembled by Wiener tells us much about society and politics in terms of patterns of consumption and taste, and, crucially for us, the emphasis accorded to rural and environmental concerns. Wiener suggests that the rural nostalgia which he identifies could appeal to both radical and conservative elements within British political culture, and that in both cases industrialism was culturally domesticated. As this occurred, a vision contrary to that of urban industrialism – of the true England as a traditional and rustic way of life – was increasingly promoted. This growing antipathy to the industrial spirit reflected, according to Wiener, the absorption of the urban bourgeoisie into the upper reaches of British society and its genteel value system – a value system which disdained trade and industry, which stressed civilised enjoyment rather than the accumulation of wealth, and which preferred social stability to enterprise. Echoing Raymond Williams, Wiener explains how, as the practical importance of rural England diminished, 'the more easily could it come to stand simply for an alternative and complementary set of values, a psychic balance wheel' (Wiener 1981: 49).

Historically, his argument is that England's particular path towards industrial capitalism had been incomplete, both culturally and socially, whatever the crude economic data might indicate. England's political revolution of the seventeenth century was too early, the industrial revolution of the nineteenth too shallow:

If society was transformed with a minimum of violence, the extent of the transformation was more limited than it first appeared to be. New economic forces did not tear the social fabric. Old values and patterns of behaviour lived on within the new, whose character was thus profoundly modified. The end result of the nineteenth century transformation of Britain was indeed a peaceful accommodation, but one that entrenched premodern elements within the new society, and gave legitimacy to antimodern sentiments.

(Wiener 1981: 7)

In practice this meant a two-fold *rapprochement* between the old landowning pre-industrial ruling class and the new financiers, professionals and industrialists. On the one hand, some landowners and aristocrats themselves diversified into non-agricultural economic activities such as mineral working and manufacturing. Thus a proportion of the industrial-commercial bourgeoisie had deep roots in rural landholding. On the other hand, while Britain was one of the earliest industrialised and urbanised countries, it was also one of the first in which rural living became a significant attraction for those whose wealth came not from the land but from manufacture and commerce (Beckett 1986; Perkin 1969). The urban rich invested in rural land not for economic purposes but for status and recreation. The extent to which land purchases of this nature occurred has been disputed by Rubinstein (1981a, 1981b, 1992), but Daunton (1989) has cited new evidence in support of the thesis that the English landowning class was a remarkably open one, with new owners of land emerging from the ranks of the industrial and commercial bourgeoisie until late in the nineteenth century (see also Anderson 1987). And F.M.L. Thompson (1990; 1992), in open dispute with Rubinstein, concludes that roughly 60 per cent of business magnates worth more than half a million pounds at the time of death (between 1873–1875) acquired landed estates.

The teachings of the public schools (Berghoff 1990) and the Church of England, and above all the lure of the country house and the landed estate, meant that 'at the moment of its triumph, the entrepreneurial class turned its energies to reshaping itself in the image of the class it was supplanting' (Wiener 1981: 14):

> The London merchant banker Baron von Schroder bought a country house in Cheshire about 1868, became a magistrate in 1876, and was high sheriff and returning officer at the time of the first county council elections in 1889. He was a well known follower of the Cheshire hounds. In Cheshire, as elsewhere, it was increasingly difficult to distinguish between the habits of a banker, like 'Fitz' Brocklehurst, who insisted on spending three months in every year shooting in Scotland, and those of the aristocracy.
>
> (ibid.: 13)

Linked to the appeal of rural living was a hearty distaste among many of the intelligentsia for industrial and commercial life:

> the immediate and home effect of the manufacturing system, carried on as it now is upon the great scale, is to produce physical and moral evil, in proportion to the wealth which it creates.
>
> (Southey 1829: 197)

> England is not yet a commercial country in the sense in which that epithet is used for her: and let us still hope that she will not soon become so.... Merchants as such are not the first among us; though it perhaps be open to a merchant to become one of them. Buying and

selling is good and necessary; it is very necessary, and may, possibly, be very good; but it cannot be the noblest work of man; and let us hope that it may not in our time be esteemed the noblest work of an Englishman.

(Trollope 1980: 12)

Social commentators, such as John Stuart Mill, John Ruskin, Matthew Arnold and William Morris, could be equally scathing:

Ruskin's vision of the good life was just what 'progress' was destroying: order, tranquillity, and harmony. The true satisfactions of human life had not changed: 'To watch the corn grow, and the blossoms set; to draw hard breath over the ploughshare or spade; to read, to think, to have, to hope, to pray.... The world's prosperity or adversity depends upon our knowing and teaching these few things: but upon iron, or glass, or electricity, or steam, in no wise.' ... His hope for the future was that 'England may cast all thoughts of possessive wealth back to the barbaric nations among whom they first arose' and turn instead to the cultivation of noble human beings.

(Wiener 1981: 39)

Arguably one of the most important points of Wiener's analysis is the claim that ruralism appealed to all strands of political thought. To some, there was the appeal of the country house and the social order associated with that; to others a new idealised democratic order based around the cottage. As one socialist militant, a disciple of William Morris, put it, one could lie back in the country and be enveloped by the 'permanence of the loveliness of England' and know the 'transience of modern civilisation' (Glasier 1921). In terms of the development of environmentalism, preservationist thought is consequently deep-rooted not as part of a radical alternative to capitalism but as a conservative reaction to urban-industrialism, equally strong in both socialist and conservative thought. A weakness of Wiener's case, though, is his underestimation of the ability of politicians to manipulate symbols whilst, at the same time, pursuing policies completely at odds with their rhetoric, as with the 'countryman' Baldwin's support for the national grid (W. Grant, personal communication, 1994).

Thus the policy outcomes and prescriptions of preservationism do not easily correspond with conventional left–right politics. On the one hand, we find an early radical preservationism. The movement to safeguard what was left of the commons following the enclosures of the previous two centuries led to the formation of the Commons Preservation Society (CPS) in 1865 and the Open Spaces Society in 1893 (later merging as the Commons, Open Spaces and Footpaths Preservation Society). A leading figure in the CPS was the radical Liberal MP George Shaw-Lefèvre who, as First Commissioner for Works in the Liberal administrations of 1881–1883 and 1892–1893, sought to improve the public access to royal parks and metropolitan open spaces (Shoard 1987). The CPS fought its campaign primarily through the law courts, although direct action to remove illegal fences was resorted to on occasions (Eversley 1910; Hoskins and Stamp 1963; Shoard 1987; W.H. Williams 1965). The democratic and populist thrust of this movement is undeniable and it provides a direct antecedent to contemporary movements to increase public access to the countryside. Similarly the first secretary to the Society for the Protection of Ancient Buildings (SPAB), when it was founded in 1877, was William Morris. His SPAB activities whilst pre-dating his socialist activities coincided with his increasing democratic radicalism and anti-militarism (MacCarthy 1994).

By contrast, perhaps the earliest national preservation body, the Society for Checking the Abuses of Public Advertising founded in 1863 (Matless 1990a), showed a typically conservative concern with the appearance of the countryside. So too, the National Trust's concern for the preservation of grand properties seems more deeply rooted in patrician sentiment, or at least in an attempt to preserve the homes and gardens of patricians after their demise. However, the Trust's origins were less clear-cut than might be suggested by its current preoccupation with managing its estates along traditional lines. The National Trust for Places of Historic Interest and Natural Beauty, to give it its full title, was formed in 1895, a response in part to some of the difficulties experienced by the CPS whose status prevented it from acquiring land in its own right. The CPS solicitor, Robert Hunter, suggested that a separate property holding company be formed (Fedden 1968; Wright 1985). Hunter was joined in his campaign by the social reformer Octavia Hill and by Canon Hardwicke Rawsley, a friend of Ruskin and an active member of the Lake District Defence Society, formed in 1876. Like the CPS, the Lake District Defence Society faced legal difficulties in acquiring land in the Lakes. The genius of the National Trust lay in its appeal to landowners to donate land and property for the nation. The Trust set itself up, not merely as a campaigning group pressing for the rights of ordinary people to enjoy rural property, but also as a holder of property and an arbiter of taste. Its breakthrough came in 1907 when legislation gave a privileged status to the Trust allowing it to hold property 'inalienably'. Without denying the undoubted subsequent achievements of the Trust in the conservation and recreation fields, its focus on property-holding certainly served to deradicalise the preservationist movement and secure its incorporation within a conservative British establishment:

> the inalienability of the Trust's property can be regarded (and also staged) as a vindication of property relations: a spectacular enlistment of the historically defined categories 'natural beauty' and 'historic interest' which demonstrates how private property simply *is* in the public interest.
>
> (Wright 1985: 52)

A potentially more radical initiative was the emergent town and country planning movement, initially drawing its main impetus from Ebenezer Howard's vision of a urban–rural utopia through garden cities. The Garden

City Association (which became the Garden Cities and Planning Association in 1909 and the Town and Country Planning Association in 1941) was formed in 1899 and drew strong support from a number of long-standing campaigners for land nationalisation (Hardy 1991). Their campaign bore early fruit in the establishment of Letchworth and Welwyn garden cities in 1903 and 1919 respectively. Notwithstanding the movement's roots in social, even socialist, concerns, in the longer term the establishment of a comprehensive land-use planning system became a central component in inherently conservative preservationist policies.

A more obviously preservationist campaign was launched in 1926 with the formation of the Council for the Preservation of Rural England (CPRE), followed in 1927 by the Association for the Preservation of Rural Scotland and in 1928 the Council for the Preservation of Rural Wales. The influential planner and architect Patrick Abercrombie, active in the garden city and planning movement, was the CPRE's first secretary and the inspiration behind the new organisation (on Abercrombie see Matless 1993a):

> we speak eloquently of the obligation that is on us to preserve and save from destruction the ancient monuments of this land ... but we are apt to forget that the greatest historical monument that we possess, the most essential thing which *is* England, is the Countryside, the Market Town, the Village, the Hedgerow Trees, the Lanes, the Copses, the Streams and the Farmsteads.
>
> (Abercrombie 1926: 6)

The CPRE became a highly successful campaigning group, with a considerable degree of responsibility for steering planning away from decentralisation (the garden city idea) and towards the containment of urbanisation. It was also instrumental in mounting a campaign for national parks (see Sandbach 1978; Sheail 1975). Lord Bledisloe, Parliamentary Secretary at the Ministry of Agriculture for the Conservative government of the day, made a private visit to Yellowstone National Park in the USA in 1925 and, in 1928, wrote to the Prime Minister, Baldwin, advocating the conversion of the Forest of Dean to a National Park, even offering part of his own estate for the purpose (Sheail 1976b). The CPRE in 1929, by which time the Conservatives were out of office, wrote to the Labour Prime Minister, Ramsay MacDonald, urging him to establish an inquiry into the need for parks. The appointed committee was chaired by Christopher Addison, an influential Secretary in MAF. The Addison Committee reported in 1931, advocating the establishment of parks:

- to safeguard areas of exceptional natural interest
- to improve means of access
- to protect flora and fauna.

It was proposed that the parks should be under a single authority in England and Wales and another authority in Scotland. The report recommended two types of park: national reserves of scenery and wildlife (for example the Lakes, Snowdonia, South Downs) and regional reserves for outdoor recreation (for example the Peak District). They were to be managed through a combination of land acquisition and planning controls. Progress towards establishing a national parks authority was delayed until after the passage of the first Town and Country Planning Bill (Sheail 1976b). But in 1931 the Labour government fell and, in the financial crises that followed, national parks were lost sight of. However the CPRE continued its campaign in the mid- and late 1930s and seemed on the point of success at the outset of war.

The campaigns to preserve landscapes and protect the countryside were therefore an important feature of the inter-war scene. Luckin (1990), for example, has chronicled the opposition to electricity pylons during the construction of the national grid between 1927 and 1934, highlighting the role of influential individuals such as Patrick Abercrombie and the historian G.M. Trevelyan, about whom Luckin writes:

> the overriding impression, as with other preservationists who were deeply concerned about the pylon and super-station issues, is that he was as preoccupied with the collapse of patrician social order as with any explicitly technological threat to an unspoilt rural terrain.
>
> (Luckin 1990: 159)

Luckin distances himself from Wiener in suggesting that the rural preservationism of the inter-war years was not based on the seductiveness of the rural life as such. After all only a minority of Britons could afford to buy into that way of life. Instead Luckin identifies two strands emerging in a highly conservative preservationism: nature mysticism dating back to early nineteenth-century romanticism and 'a barely suppressed antagonism towards the urban working class and urbanism in general' (ibid.: 167). Matless (1990a, 1990b, 1993b) argues for a third strand, that of preservationist-planning:

> Writers like Hoskins have been very successful in aligning ideas of conservation and landscape against modernity and 'progress'. But it was not always so. Inter- and post-war writers such as Stamp, Abercrombie and Thomas Sharp, all active in the cause of landscape preservation, did not turn from the modern world in disgust.... the planner-preservationists strove to revive an eighteenth-century spirit of design in assertively modern and planned form.
>
> (Matless 1993b: 191–192)

Thus, as with much of the preservationism of the late nineteenth century, so the inter-war preservationist movement was not quite as conservative and backward looking as it is sometimes made out to be. The public access movement in the inter-war years never appeared conservative at all. Writers such as Lowerson (1980) and Shoard (1987) have documented the emergence of a working-class movement to improve access, especially to the unenclosed grouse moors of northern England, a movement which inherited the mantle of earlier advocates of land reform (see chapter 4). The Ramblers' Association, founded in 1935, became the chief protagonist in this debate.

Despite the extent of campaigning during the inter-war years, solid achievements were relatively few. However there were some exceptions. A small number of vigorous local authorities had implemented a modest degree of planning control under the facilitating mechanisms offered by the Town Planning Acts of 1919 and 1925. Surrey went further in sponsoring its own Surrey County Council Act 1932, which allowed it to protect a number of key sites from development and to improve public access (Sheail 1976b).

Another example of non-legislative progress occurred in the Lake District. In the 1930s the Friends of the Lake District and the CPRE obtained 13,000 signatures on a petition against further afforestation of the Lakes. Despite the rejection of parliamentary calls for a select committee on the work of the Forestry Commission, the Commission was embarrassed enough by the strength of public opinion to invite the CPRE to join a joint informal committee. The committee forged a voluntary agreement effectively banning planting from a central zone of 300 square miles of the Lake District.

THE SCIENTIFIC PERIOD

Scientific work on the natural environment during much of the first half of the last century was largely concerned with the continuing task of classification, revealing the variety of nature. Thereafter, the field was increasingly replaced by the laboratory as attention turned from observation to examination. Biologists such as T.H. Huxley pioneered a new kind of natural history with a 'strong bias towards anatomy and morphology' (Allen 1976: 181), a bias so strong that field studies and the efforts of amateur naturalists were neglected and even derided.

The clash between the old and the new biologies was ultimately debilitating to both and arguably delayed the emergence of ecological science. There was little such acrimony between agricultural science and chemistry and, ultimately, some of the work of the last century more relevant to the development of environmental science occurred in this area. Soil chemists seeking to find ways of optimising fertiliser applications obtained valuable new evidence, and made methodological advances on the loss of nutrients from the soil (Wilmot 1993), work of use a century later as new concerns over agricultural pollution emerged.

It was during the inter-war years that it first became apparent that agriculture might have a negative polluting effect. The concern was not yet with on-farm pollution but with the consequences for watercourses of the discharges from newly established factories for the processing of sugar-beet and dairy products. In 1921 MAF appointed an advisory Standing Committee on River Pollution and in 1927 a Water Pollution Research Board (Sheail 1993).

But the emergence of ecology had less to do with immediate concerns of this sort and more to do with the resolution of the tensions between field and laboratory work. The word *ecology*, popularised in America in the late nineteenth century (Bate 1991; Clarke 1973), was first used in 1866 by the German zoologist Ernst Haeckel who defined it in 1870 as:

> the body of knowledge concerning the economy of nature – the investigation of the total relations of the animal both to its inorganic and to its organic environment; ... ecology is the study of all those complex interrelations referred to by Darwin as the conditions of the struggle for existence.
>
> (Translated in McIntosh 1985: 7–8)

In Britain one of the roots of ecology lay in the quasi-religious doctrine of Vitalism.

> Refusing to accept the bleak and pessimistic view of those who saw only the harsh northern aspect of Evolution – the terrible inevitability of the Struggle for Existence, the remorselessness of a Nature 'red in tooth and claw' – the Vitalists built instead a positive and optimistic creed upon its other, warm, south-facing side.... they put their faith in a quality of 'insurgence' inherent in every particle of life, which, once harnessed, enabled an organism to surmount all the constraints and pressures that opposed it. This led on to the conviction that all life is deserving of reverence.
>
> (Allen 1976: 200–201)

There were roots also, of course, amongst a very small number of the best of the amateur naturalists. As early as 1819, Nathaniel Winch was attempting to relate his findings on the flora of Northumberland to edaphic conditions (Sheail 1987). It is in Britain that the first ecological society was formed in 1913: the British Ecological Society's first president was Arthur Tansley, a lecturer in Botany at Cambridge, whose *Types of British Vegetation* was published in 1911 (see also Tansley 1939a) and who, in the 1930s, first developed the conception of the ecosystem (Tansley 1935).

However, without major government or university backing, progress in plant ecology was slow and in the inter-war years Tansley discouraged students from specialising in a subject that was so poorly resourced (Tansley 1939b; Lowe 1976). At University College London, coastal ecology was developed under Francis Oliver (Tansley's original mentor) and in Cambridge the Bureau of Animal Population was started by Charles Elton in 1932. Animal ecology received some government funding in the inter-war period because of its significance for pest control. But essentially the inter-war years were years of struggle. Tansley, like many of his generation, was deeply affected by the bleak sense of carnage of the 1914–1918 war (P. Lowe, personal communication, 1994) and he resigned his Cambridge post in 1922, subsequently spending some months studying psychology with Freud in Vienna before returning again to ecological studies (Sheail 1987). His interests in ecology and psychology were not completely divorced, as both were informed by an interest in how communal order is established and in the relationships between individuals and groups (Lowe 1994).

Alongside the development of ecology, and very much linked to it, was the emergence of a campaign for the promotion of nature reserves, primarily for scientific purposes. The Society for the Promotion of Nature Reserves was established in 1912 (and became the Society for the Promotion of Nature Conservation in 1977 and the Royal Society for Nature Conservation in 1981). In much the same way that ecology was slow to develop outside a relatively small world of devotees, so too the SPNR struggled to make an impact in the inter-war years:

> Whereas issues such as national parks and rights of access to open country were built up by the CPRE and Ramblers' Association as popular causes, the case for nature reserves received little publicity and generated no widespread interest. Instead it remained an esoteric matter, viewed even by naturalists as a costly and impractical expedient only to be contemplated as a last resort when a unique spot was threatened by an improving farmer or speculative builder, and certainly no substitute for protective wildlife legislation.
>
> (Lowe and Goyder 1983: 153)

It was during the 1939–1945 war that ecologists took their place alongside others in planning for the future. The British Ecological Society established a committee in 1943 under Tansley's chairmanship, to investigate the need for nature reserves and nature conservation, which led to the formation of the Nature Conservancy in 1949, an agency which was very much science-led. Its duties were to conduct and sponsor ecological research, to give advice and information on nature conservation and to acquire and manage nature reserves.

It was not only ecology which came to the fore in the period of post-war optimism. The belief in the rational approach of applied science was also reflected in the expectations for a comprehensive land-use planning system. Thus, Dudley Stamp, one of the doyens of wartime and post-war geography, argued for a scientific approach to the use of land resources with a belief that central and local planning could achieve this. It would be a mistake to see the enthusiasm for planning as not carrying with it a strong element of rural nostalgia and agricultural fundamentalism, reflecting its preservationist roots. However, in other respects it was far from preservationist in tone. Throughout the 1950s, 1960s and into the 1970s scientific planning continued to find favour. Economic planning was adopted as a central tenet of the politics of managerialism by Harold Wilson, a tenet that Edward Heath was loath to challenge. This mood was reflected in thinking about rural environmental issues. For example, Nan Fairbrother in *New Lives, New Landscapes*, published in 1970, accepts the changes of modern farming and urges landscape planning to accommodate them through the cultivation of a new aesthetic:

But the new open farmland, if we cease to look at it nostalgically, has its own distinctive beauties, its very openness being one.... In large-scale arable farming we are conscious too of the land, the earth itself. We can see the shape of the ground as we never can in small hedged fields....

And the new ways may suit other new uses than farming – as the landscape of travel for instance. Travelling man is seeing man and our contact with the landscape is almost entirely visual. We see it as a changing view, a panorama unrolling along our route like a Chinese scroll, and from our fast-moving cars we see a completely new countryside. The details have gone; flowers which charm us at a strolling three miles an hour are invisible at sixty, trees in hedges flash by as distractions, and winding lanes are inconveniences not delights.... The best landscape for fast travel therefore is spacious, a bold and uncluttered composition of wide views and clearly-defined effects.... Our new farmland and new fast roads can make fine landscape combinations in the same large scale and simple functional style, and in its different way swift motorway travel brings us as vividly close to the countryside as walking.

(Fairbrother 1972: 236; 239–240)

The 'Countryside in 1970' conferences held during the 1960s reflected a similar concern to find ways of accommodating the changes wrought by science and technology through processes of rational planning. As a response to mounting environmental concern this approach was ultimately to be seen as flawed, as discussed in the next section, but some of the organisational responses to which it gave rise have certainly stood the test of time. For example, the 'Countryside in 1970' conferences contributed to two very concrete outcomes. First, a commitment was made by local authorities to broaden the scope of their activities to include a more active consideration of the rural, and particularly the agricultural, environment, in part a response to Section 11 of the Countryside Act 1968, which required all public authorities to 'have regard to the desirability of conserving the natural beauty and amenity of the countryside'. Second, the origins of the Farming and Wildlife Advisory Group (FWAG) can be traced, in part, to the conferences and their stress on the establishment of new conservation and landscape features (for more on FWAG see chapter 9).

THE POPULAR/POLITICAL PERIOD

In this section we explore the rapid rise of popular environmentalism during the last three decades. Space is limited and the treatment is necessarily somewhat cursory. Many words have been written on this topic in recent years – probably the best general accounts remain the pioneering studies by Timothy O'Riordan (1976) and Philip Lowe and Jane Goyder (1983), although the works of, *inter alia*, Dobson (1990), Pepper (1986), Rudig

(1990) and Yearley (1991) also warrant close attention.

As we have already seen, many of the pressure groups at the centre of contemporary environmental campaigning were founded in the last century or the first half of the present century. However, for many of these organisations growth was slow until the 1980s, as shown in Figure 7.1. For example, membership of the RSPB during the immediate post-war period, by which time the organisation might be said to have come of age in terms of legislative advance, remained static between 1945 and 1960 at around 8,000 (probably nearer 12,000 individuals if family memberships are taken into account). A slow rise commenced in the 1960s, accelerating rapidly in the 1970s and 1980s so that by 1995 its membership stood at 750,000 with an additional 140,000 children as members of the Young Ornithologists Club. Less spectacular, but none the less solid growth has been achieved by similarly mainstream organisations such as the Royal Society for Nature Conservation whose membership now stands at a quarter of a million.

The membership expansion of the established groups represents one aspect of the growth in popular appeal of environmental issues. The second is the development of new, and often more radical, national or international groups. Friends of the Earth was formed in 1970 (see Lowe and Goyder 1983) and Greenpeace in 1971 (see Eyerman and Jamison 1989; Yearley 1991). The growth of Greenpeace in the 1980s was dramatic so that by 1993 it had 410,000 members in the UK. The politicisation proper of the environment is perhaps best symbolised by the formation of the first political

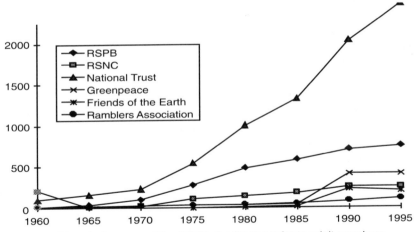

Note: In the case of the RSPB, the figures refer to adult members only; also for the RSPB, the pre-1990 figures involve an estimation of the number of *individual adult members*, because their membership figures prior to this date recorded membership rather than members with many individuals subsumed within family membership.

Figure 7.1 Membership of environmental groups (thousands)

Source: Information supplied by environmental groups

party solely devoted to environmental issues. The People Party was formed in 1973 and relaunched as the Ecology Party in 1974 (renamed the Green Party in 1985). The established political parties responded with the formation of groups specifically dealing with environmental issues. The Socialist Environment and Resources Association (SERA) dates from 1973 and the Liberal Ecology Group and Conservative Ecology Group were both founded in 1977.

A third aspect of this period was the expansion of local amenity societies from the 1950s through to the 1980s. These varied enormously in scope and focus. Some were little more than narrowly focused preservation groups anxious only to preserve the visual amenities (positional goods) of already privileged shire county communities. Others sought radical improvements to blighted urban environments with radical implications for lifestyles covering such issues as transport, recycling, and energy use.

In seeking to steer a course through the labyrinth of organisations covering contemporary environmental concerns, Lowe and Goyder (1983) divided the groups into four main sectors of concern – resources, conservation, recreation and amenity. They further established the range of contacts of each groups with other organisations so as to identify a core group of organisations at the centre of an extensive network of campaigning organisations. As Figure 7.2 shows, this central group encompasses CPRE, RSPB, FoE, the Civic Trust, the National Trust and CoEnCo (the Council for Environmental Conservation). The approach is very useful for establishing the focus of a particular group in terms of its likely engagement in a particular policy sector. It is somewhat less helpful in indicating some of the fundamental differences between groups in terms of contrasting views on economy and society as a whole. The fact that Greenpeace and the Conservative Ecology Group occupy the same segment of the chart is a case in point.

A distinction, much in vogue in recent years, has been drawn between groups or individuals espousing either deep or shallow ecology, a distinction credited to the Norwegian, Arne Naess (1973). Shallow ecology involves both a recognition that environmental problems exist and a commitment to seeking means to tackle them. There is, however, no underlying critique of the social and economic orders that have generated the problem in the first place.

> For Naess this is having it both ways – trying to promote environmentally friendly policies, while carrying on with the whole growth-oriented, car-driven, polluting jamboree. Naess and others writing from a deep ecology, or dark green perspective argue that such tidying up, light green, cosmetic approaches do not address the main issues.
>
> (Young 1992: 14)

By contrast, deep ecologists argue for a fundamental recasting of society and economy, the key phrases being decentralisation, 'small is beautiful',

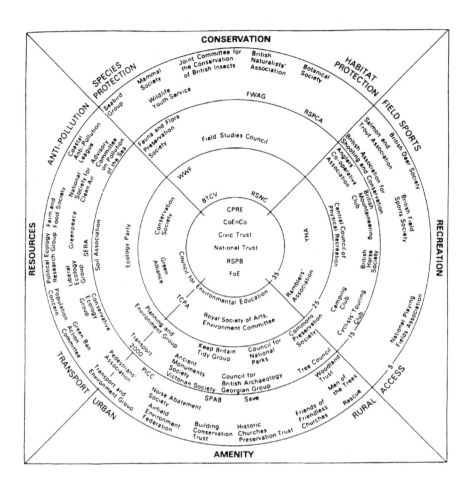

Figure 7.2 The structure of the environmental lobby

Source: Lowe and Goyder (1983)

sustainability, even post-industrialism and post-materialism and zero growth. Deep ecology can be traced back to the late 1960s and early 1970s. Events such as the Torrey Canyon oil spillage off the coast of Cornwall in 1967 and the impact of Rachel Carson's *Silent Spring* (1962) in highlighting the impact of pesticides, served to underline the failures of post-war incremental environmental legislation. The capitalist-industrial system seemed increasingly to be caught up with seemingly intractable commodity and fuel constraints. In retrospect, the oil crisis and escalating world commodity prices of the early 1970s can be seen to have been almost solely a consequence of complex geopolitical factors but, at the time, many environmentalists expected an escalation of the problems consequent upon a strictly finite resource base.

Key publications in the emergence of deep ecology include the apocalyptic *Blueprint for Survival* (Goldsmith 1972; see also Goldsmith 1971) and *Limits to Growth* (Meadows *et al.* 1983, first published 1972), both reports producing much anxiety. Rather more creative optimism was generated by the vision of a new de-centralised and harmonious society in E.F. Schumacher's *Small is Beautiful*. Of equal importance to these seminal works were the month-by-month articles in magazines such as *The Ecologist* and *Resurgence*. Later expositions of deep ecology include Ekins (1986), Fox (1984), Naess (1989) and Porritt (1984), with some of the sociological and political implications discussed in Redclift and Benton (1994). Deep ecologists profoundly call into question the enlightenment faith in economic growth, the application of ever more sophisticated technologies, growing mobility, and an increasingly bureaucratic and managerialist society. This is a revolutionary faith which

> seeks nothing less than a non-violent revolution to overthrow our whole polluting, plundering and materialistic industrial society and, in its place, to create a new economic and social order which will allow human beings to live in harmony with the planet. In those terms, the Green movement lays claim to being the most radical and important political and cultural force since the birth of socialism.
>
> (Porritt and Winner 1988: 9)

Andrew Dobson (1990) in commenting on this passage goes so far as to suggest that Green politics represents an even more profound critique of the capitalist order than did the emergence of socialism.

All this might seem a far cry from the debates on the greening of agricultural policy that lie at the core of much recent policy initiative and provide a major focus for this book. But it is as well to be reminded of the radicalism of the ecology movement and consequently the caution that reformist zeal is unlikely to satisfy all actors in the environmental field. None the less policy reforms are now very much at the centre of environmental politics. This places individuals and groups espousing a deep ecology in an ambivalent and difficult position. There are parallels with the dilemma that faced revolutionary socialists in Britain and Europe in the late nineteenth and early twentieth centuries as parliamentary democracy opened up opportunities for legislative advance, but arguably at the risk of sacrificing revolutionary principles (see the discussion in MacCarthy 1994). Friends of the Earth is perhaps the best example of an organisation that has confronted these problems. Rooted in radical deep ecology, but committed to a localism and to a campaigning style that has depended upon rational and forceful scientific argument, FoE has found itself increasingly brought to the centre stage of policy debates especially where its technical expertise commands the respect of government or other pressure groups (although the Green movement as a whole has a deeply ambivalent relationship with modern science: Yearley 1992).

To take just one example: FoE, in principle at least, espouses the necessity of a ubiquitous system of small-scale, organic farming and gardening. However this does not stop it from entering the mainstream debate on the reform of the Common Agricultural Policy. It is scarcely surprising that Jonathan Porritt, a former director of FoE and former leader of the Green Party (though now no longer a member of the party), has been gently chided for an ambivalence between reformism and dark Green ecology (Dobson 1990).

CONCLUSIONS

Whatever the ambivalence or even contradictions between deep and shallow ecology, there can be little doubt that the popular and political period of environmentalism has had a profound and lasting impact upon the way in which environmental issues are addressed in mainstream political discourse and decision making. Some of the key phrases from deep ecology, such as sustainability, have entered the common political vocabulary. Most of the radical campaigning organisations are, to a greater or lesser extent, embroiled in practical policy debates. By the same token traditional organisations, such as the CPRE and even the National Trust, have found themselves debating a much broader set of issues than would have been the case in the past. The CPRE is an increasingly radical voice on issues such as energy, transport and agriculture. And the National Trust has conducted an exhaustive internal countryside policy review involving grassroots employees and external 'experts' in order to investigate the impact of new thinking on its ways of holding and managing large tracts of rural land. Whilst it is too soon to talk of a convergence between the interests of different groups within the extraordinarily diverse spectrum of environmental organisations, a combination of increasing memberships and national and international politicisation of the issue has given rise to a growing recognition of a common set of problems.

An important point about the phases of environmental development outlined in this chapter is not only that they overlap, but also that their effects are cumulative. The fact that the preservationist period had its origins and heyday in the last decade of the nineteenth century and the first forty years of the twentieth should not detract from its continuing salience for contemporary analysis. Indeed, many would argue that the heritage explosion of the last decade or more is a direct result of such preservationism.

> Today's city dwellers' main contact with the countryside is scenic or sportive; the landscape is as superficial a splendour as Madame Tussaud's, as exotic as the Elgin Marbles. We domesticate this alien presence by paying it homage as heritage.
>
> (Lowenthal 1991: 222)

That such sentiment can be mixed, on occasions, with a brand of radical environmentalism only serves to emphasise the essential fluidity of the environmental movement. And alongside the popularity of moves to preserve cherished landscapes or eliminate harmful pollution there is, of course, a continuing emphasis on scientific endeavour to buttress political debate and policy initiative.

8

PROTECTING LANDSCAPES, HABITATS AND WILDLIFE

INTRODUCTION

This chapter considers the development of rural environment protection policies for landscapes, habitats and wildlife in the post-war period. Although the implications for agriculture of many of these developments have been profound, this chapter does not deal with a range of specific and relatively recent measures taken by the UK agriculture departments themselves, or with agricultural pollution, as these topics are dealt with in the next two chapters. Instead I examine first the development of policies for protecting landscapes and wildlife up to the early 1970s. This is followed by a section which examines the intensifying countryside debate in the 1970s culminating in the passage of the Wildlife and Countryside Bill, which reached the statute book in 1981. The Act and subsequent developments are considered. Finally attention is turned to the growing role of European environmental policy and, in particular, its impact on habitat protection measures in the UK.

Throughout, the chapter attempts to discuss the development of countryside legislation in the light of the changing nature of the environmental movement, and the view that the environmental movement has slowly but surely mustered sufficient political strength to be able to challenge the dominant agricultural and landowning interests in the policy process. As a particular instance of this, the Wildlife and Countryside Act 1981 is considered not so much for its legislative importance but as an example of a political coming of age for environmental groups.

THE POST-WAR SETTLEMENT

National parks

The progress towards establishing a framework for environmental conservation during the 1939–1945 war and the immediate post-war years has been discussed extensively by a number of authors (Blunden and Curry 1990; Cherry 1975; MacEwen and MacEwen 1982). A key development was the Scott Committee's *Report ... on Land Utilisation in Rural Areas* of 1942

(Cmnd 6378), which considered the impact of urban and industrial changes on the countryside, stressing that national parks were long overdue. The report was based on two key assumptions: first, that the planning system was the most appropriate mechanism for securing the kind of countryside management required to provide for recreation and conservation objectives; and second, that there was no contradiction between these objectives and a prosperous farming industry. The Scott Report was

> a passionate reaction against the disastrous depression years and the devastation that a depressed agriculture had had on farming, the rural communities and the landscape itself.... It firmly declared that 'there is no antagonism between use and beauty', a doctrine repeated *ad nauseam* by the farmers' spokesmen ever since. Herein lay the fatal contradiction of the Scott report, for it believed in having the best of all worlds – traditional mixed farming, rural living standards raised to urban levels and the 'traditional' landscape.
>
> (MacEwen and MacEwen 1982: 10)

But in a dissenting minority report, Professor S.R. Dennison denied the link between a prosperous agriculture and the preservation of a traditional countryside. Anticipating the post-war trends of agricultural specialisation and efficiency he argued, with considerable prescience, that farmers could be paid to conserve landscape as 'landscape gardeners and not as agriculturalists'. But it was the Scott Report's assumptions that held sway in the Dower Report on *National Parks in England and Wales* published in April 1945 (Cmnd 6628), just three months before the Labour election victory. Dower defined a national park as:

> an extensive area of beautiful and relatively wild country in which, for the nation's benefit and by appropriate national decision and action, (a) the characteristic landscape beauty is strictly preserved, (b) access and facilities for open-air enjoyment are amply provided, (c) wild life and buildings and places of architectural and historic interest are suitably protected, while (d) established farming use is effectively maintained.
>
> (Cmnd 6628, 1945: para 4)

It is worth contrasting this definition with the more robust one adopted by the International Union for the Conservation of Nature and Natural Resources in 1969 (IUCN 1975) which includes the following elements:

- a natural landscape unaltered by humans
- the elimination of occupation and exploitation by the state
- visiting only under special conditions.

Consequently the English and Welsh National Parks (there are none in Scotland or Northern Ireland), which are peopled and used for many economic activities, are not included on the IUCN list. Dower insisted that national parks, in as populous and long-occupied a country as Britain, must provide for both recreation opportunity and for the conservation of wildlife and landscape. Moreover, the continuation of economic activity would also have to be permitted:

> Dower reconciled the preservation of the landscape with the main-tenance of effective agriculture by stating that with attention to detail in agricultural practice and a limited extension of development control (e.g. to encompass farm buildings), there should be no irreconcilable contradictions between national park purposes and the pursuit of agriculture. Agriculture is of course not the only intrusive land-use in national parks. Mining, hydro-electric power, reservoirs, utilisation for defence purposes and latterly motorways are all possibilities. Dower recommended that these should be permitted only upon clear proof of requirement in the national interest and that no satisfactory alternative site could be found.
>
> (M.V. Williams 1985: 360)

Labour's Minister of Town and Country, Lewis Silkin, was a convinced advocate of national parks and of the cause of ramblers. Within days of taking office in 1945 he had appointed Sir Arthur Hobhouse to chair a Committee on National Parks which, with its two sub-committees (Footpaths and Access and Wildlife and Conservation), reported in 1947 (Cmnd 7121), the precursor to the National Parks and Access to the Countryside Act 1949, which covered England and Wales. The Hobhouse Committee added little to Scott and Dower except in so much as it spelled out in rather more detail the way in which parks would be established and also how some land might be acquired (a facility scarcely ever used as the money was never made available) to supplement the main controls to be vested in planning authorities. Hobhouse's counterpart in Scotland, the Ramsay Committee (Cmnds 6631, 1945; and 7235, 1947) recommended land acquisition, by compulsion if necessary. However,

> National parks in Scotland foundered on the resistance of the landed interests to the radical proposals of the Ramsay report. Five areas proposed by the Ramsay report as national parks in Scotland lingered on as 'National Park Direction Areas', within which proposals for development were notified to the Secretary of State for Scotland, the title 'national park' serving only as a buoy to mark the wreck and to recall the ideas of the 1940s. A direction by the Secretary of State finally extinguished National Park Direction Areas in 1980.
>
> (MacEwen and MacEwen 1982: 12)

The 1949 Act led to the designation of ten parks by the National Parks Commission between 1951 and 1957, all within the English and Welsh uplands (with the exception of part of the Pembrokeshire Coast National Park) as shown in Table 8.1. The Norfolk and Suffolk Broads Act 1988 enabled the establishment of the Broads Authority, and a new national park in all but name.

Dower's suggestion that the Cornish coast should be one of the national

Table 8.1 The national parks of England and Wales

	Date of designation	Areas sq km 1990	Population 1981
Peak District	April 1951	1,404	37,400
Lake District	May 1951	2,292	40,000
Snowdonia	October 1951	2,171	23,800
Dartmoor	October 1951	945	29,100
Pembrokeshire Coast	February 1952	583	23,000
North York Moors	November 1952	1,432	27,000
Yorkshire Dales	October 1954	1,760	18,600
Exmoor	October 1954	686	10,000
Northumberland	April 1956	1,031	2,200
Brecon Beacons	April 1957	1,344	32,200
Norfolk and Suffolk Broads	March 1988	288	5,500

Source: Edwards 1991

parks was rejected as were Hobhouse's proposals to include the Broads and the South Downs. In recognition of the importance of other areas (not only those which failed to gain national park status but also a list of 52 'conservation areas', mostly located in the lowlands, suggested by Hobhouse) Part 6 of the 1949 Act provided for the designation of Areas of Outstanding Natural Beauty (AONBs):

> In terms of landscape quality, the only distinction made by the Act is that their natural beauty must be 'outstanding', whereas that of national parks need not be! They were given no nature conservation or recreational functions, although recreational use could be inferred from the power to provide wardens.... Despite the proximity of many of them to centres of population and recreational use as intensive in some AONBs as in national parks, they received no special funds.
>
> (MacEwen and MacEwen 1982: 16)

Not that the finances and administration of national parks were initially much better:

> The low priority given to national parks in 1950 was evident immediately in the appointments to the National Parks Commission. The chairman (part-time of course) for the first four years was Sir Patrick Duff, a retired civil servant, who had no previous interest or commitment in the field. Few of the early commissioners and its tiny staff had any understanding of the national park concept or knowledge of the English landscape, let alone the catalogue of essential qualities prescribed by Dower and Hobhouse. The staff was headed by Harold Abrahams, an Olympic athlete whose consuming passions were athletic journalism and broadcasting.
>
> (ibid.: 21)

On the ground the weaknesses were compounded by the insistence by county councils that they should retain considerable controls. Having only recently been granted powers to administer the new town and country planning system, the councils were loath to cede responsibilities to new authorities within their territory. Moreover, the farmers and landowners on many county councils in national park areas were anxious to ensure that the parks did not take initiatives that might damage their own interests. Only in the Peak and the Lakes, the first two to be designated, were fully independent park authorities established. In Wales, proposals that Snowdonia should be constituted in the same manner ran into fierce opposition from Welsh farming interests, fearful of committees appointed by an English government department. Consequently, in the remaining national parks, complex and cumbersome administrative arrangements were entered into, involving multiple committees. Only in Northumberland and Dartmoor, located within single-county authorities, were arrangements somewhat simpler. Meanwhile, the finances available for practical management and the provision of recreational facilities remained pitiful.

The Countryside Act 1968 represented the first attempt to improve matters. The National Parks Commission was replaced by the Countryside Commission which was given a grant-awarding facility. The Commission has proved a persistent advocate for the strengthening of the national park system, but its interests are broad; and much of its early years was taken up with establishing country parks as 'honey pots' close to urban centres (Curry 1985; 1994), arguably to relieve some of the pressure on national parks rather than funding recreational and conservation developments within the national parks themselves.

But the 1968 Act left the park authorities in limbo on this and many other matters. The MacEwens tell a remarkable story of how the Conservative government of 1970 to 1974 refused to address a wider set of issues and rode rough-shod over normal protocol. In March 1971, John (now Sir John) Cripps, chairman of the Commission since 1970, informed the government of the Commission's proposal to undertake a review of national parks 'with special reference to national considerations which should affect national park policies' (Cripps 1979):

He received no reply, but two months later Peter Walker, the Secretary of State for the Environment, descended by helicopter on the shores of Windermere to address the National Parks Conference. He gave Mr Cripps five minutes notice of his announcement that the Department of the Environment and Countryside Commission would make a joint study of national park policies, through a committee chaired by Lord Sandford, his Under Secretary. The Commission tried again in December 1971 to persuade the Department to take a wider look at countryside problems, suggesting the setting up of a Royal Commission on the countryside modelled on the Scott Committee of 1941–2. It specifically drew attention to the danger that not only recreational pressures but also developments in farming and forestry might be combining to destroy the 'natural beauty of the countryside'. The Commission's suggestion was ignored.

(MacEwen and MacEwen 1982: 26)

Denied the opportunity to deal with the vexed issue of park administration, the Sandford Committee took a strong consensus and management approach, stressing the benefits of co-operation and voluntarism. In response, the government accepted many of the recommendations. Crucially it agreed that priority should be given to conservation where there was an irreconcilable conflict with recreation, but it refused to allow NPAs to exercise compulsory purchase powers. New administrative arrangements were inaugurated and after 1974 national parks certainly achieved more than in earlier decades. However, the dispute about their underlying purpose and the role of local populations has never been far from the surface. In 1989 a further committee was appointed to consider matters. Some of the recommendations of the Edwards Panel (Edwards 1991; see also Phillips 1993; Shaw 1991) are (at the time of writing) before Parliament as part of the 1994/95 Environment Bill. In particular new independent authorities, though still with strong links to local authorities, are proposed; and greater weight is to be given to conservation and to parks as 'tranquil places for quiet enjoyment'.

Nature conservation and the great divide

In the war and immediate post-war period, nature conservation appears, at first sight, to have been something of an afterthought in the generation of countryside policies. John Dower's report was supplemented by six memoranda produced by the Nature Reserves Investigation Committee, which operated in the slipstream of Dower's investigations. And the Hobhouse Committee appointed a sub-committee, the Wild Life Conservation Special Committee (Cmnd 7122, 1947), under the chairmanship of Sir Julian Huxley. The fact that the work on nature conservation seemed to be relatively less high-profile had less to do with its perceived relative importance than with the extent to which agreement had already been forged in this area. This was in stark contrast to the uncertainties surrounding national parks:

A particularly telling episode occurred back in 1941 when the Conference on Nature Preservation in Post-war Reconstruction set up a drafting committee to compile its first memorandum to government.... There was agreement that the principle of establishing national nature reserves should be incorporated in any national planning scheme, that an official body, representing scientific interests, should draw up detailed proposals, and that the management of reserves should be placed in the hands of persons with expertise and experience of wildlife management, but there was considerable controversy as to how far detailed reference should be made to the question of selecting and administering national parks.

(Sheail 1988: 3)

Thus it was that Part 3 of the 1949 Act has often been seen as its most coherent and successful part. It provided for the establishment of National Nature Reserves (NNRs) and Sites of Special Scientific Interest (SSSIs) by a newly established Nature Conservancy to be responsible for nature conservation, defined as 'the flora and fauna of Great Britain'. According to the Huxley Report, NNRs should:

> preserve and maintain ... places which can be regarded as reservoirs for the main types of community and kinds of plants and animals represented in this country, both common and rare, typical and unusual, as well as places which contain physical features of special or outstanding interest.... Considered as a single system, the reserves should comprise as large a sample as possible of all the many different groups of living organisms, indigenous or established in this country as part of its natural flora and fauna.
>
> (Cmnd 7122, quoted in D. Evans 1992: 76)

It was expected that the majority of these natural laboratories, which eventually numbered more than 150, would be owned and/or occupied by the NC itself. By contrast, the SSSIs were expected to remain in private ownership. Like NNRs they were selected as *representative* sites covering areas of high-quality natural or semi-natural flora or fauna, or containing rare or endangered species or with key geological characteristics. Under the 1949 Act, local authorities were empowered to designate local nature reserves, whether or not they had SSSI status and to introduce by-laws, where necessary, to protect them. In the first ten-year period after the 1949 Act, just 5 were designated and by 1974 there were only 36; thereafter the number grew rapidly, rising to over 120 in Britain by the mid-1980s, but their management has been heavily criticised (Tyldesley 1986).

In addition to its site responsibilities, the NC was to have a research function (this was lost in 1973 when the NCC replaced the NC, with most of the research effort taken over by a new Institute of Terrestrial Ecology). The National Parks Commission and subsequently the Countryside Commission were also given responsibility for flora and fauna within a wider definition of 'natural beauty':

> In practice, as the Nature Conservancy took nature for its province, 'natural beauty' has come more and more to be equated with the appearance of the countryside.... the 1949 Act departs from one of Dower's central tenets that national parks and nature conservation should be in the same hands.
>
> (MacEwen and MacEwen 1982: 16)

This so-called great divide between, on the one hand, the interests of nature conservation and, on the other, those of landscape protection, has been a characteristic and, many would argue, a debilitating feature of the British arrangements.

The divide was intellectually flawed in the sense that integrated management with multi-purpose objectives would have been more appropriate, especially given the context of a relatively small national land mass already subject to diverse pressures. But more than that the divide fuelled the development of two distinctive cultures of conservation, within the Nature Conservancy and the Countryside Commission, which, whilst they may have promoted a healthy debate on objectives, also undermined the united front needed when facing cynical, or at best financially prudent, central government departments or the impenetrability of the agricultural policy community. In short, the divide proved enervating, more so for national parks than for nature conservation, for from the outset the Nature Conservancy was far better endowed with resources than were the national parks.

In the light of comments in the previous chapter on the relative weakness of science in relation to preservation in the founding ideas of environmentalism in Britain, this suggests an apparent reversal of the fortunes of the two camps. There are a number of reasons for this. The first, and probably the most important, has to do with the strength of the farming and landowning lobby. Landowners' interests appeared to be more threatened by the demands of recreationists and those seeking to preserve landscapes than by a relatively small number of scientists concerned with the safeguarding of a number of small key sites, many of them not in agricultural use in any case. This perception was almost certainly reinforced by the apparent relative strengths and stridency of the two lobbies in the inter-war years. The movement for national parks was a well publicised and orchestrated campaign which threatened to have an impact on large tracts of farmed and forested land. The NFU and influential individual landowners did their best through well-established contacts to minimise the danger. It was during the debates on these issues during the 1930s and 1940s that the NFU learnt the importance of presenting a unified front with MAF to non-agricultural sectors of government. The arguments presented were used on many subsequent occasions to justify the privileges of agriculture and depended on the following key elements:

- an unfettered agriculture was crucially important to guaranteeing the nation's food supply;
- farmers were already the natural custodians of a much-loved countryside and no special measures were needed to promote its protection and enjoyment;
- agriculture was best managed and regulated by farmers themselves in concert with their own Ministry.

The success of the NFU (and to some extent of the CLA) in ensuring that recreational issues were marginalised was, of course, primarily based on their position of strength within the emergent agricultural corporatism and on the strength of food production interests within central government. By contrast, the natural scientists owed their political strength less to corporatism and more to their operation within an elitist system. A small number of well-placed individuals were able to make representations at a high level of government to safeguard their own interests and distance themselves from the national park cause. Sheail (1987, 1988) has demonstrated how A.G. Tansley, C.S. Elton and E.M. Nicholson from the world of science exerted a high degree of influence in pressing for a particular set of arrangements to sustain ecological science:

> Ecologists secured not only unprecedented opportunities for developing their subject, but, by being the driving force in this new enterprise, they were able to play a large part in removing any stigma of antiquarianism from protecting wildlife. The National Nature Reserves, chosen primarily on scientific criteria, and schedules of Sites of Special Scientific Interest, could be portrayed as key components of a new and rational use of land and natural resources.
>
> (Sheail 1988: 4)

If the scientists were careful not to alienate landowning interests, they had also been careful to avoid becoming embroiled in the complex and contentious debates over local and national government which had bedevilled the national park issue.

In addition to these political factors, it also has to be conceded that part of the story of the 'great divide' is to do with misplaced confidence in the planning system. It was assumed that both national parks and SSSIs would be adequately protected by planning mechanisms. The proponents of national parks, in particular, displayed a belief in planning which served to deflect their attentions away from some of the weaker aspects of the legislation. Development control was seen as the key to pursuing national park objectives. Large tracts of land would be protected from unsuitable development in this way and for many, in particular park residents, the work of park authorities in planning decisions became their key public role – often a highly contentious and controversial role as parks sought to balance the interests of local populations with national objectives.

By contrast, the role of planning in the pursuit of nature conservation objectives was much more modest. Under the 1949 Act the NC was obliged to notify Local Planning Authorities (LPAs) of the sites it had designated as SSSIs, and LPAs were required to consult the NC when any development was proposed (Town and Country Planning General Development Order 1950, article 9). The duty to notify was extended to river authorities in England and Wales under Section 102 of the Water Resources Act 1963. But landowners need not be notified, reflecting the assumption that agriculture and other traditional rural uses could be relied upon to safeguard a site's interest. The key emphasis on active management, protection and research was largely confined to National Nature Reserves, which became the flagship of the NC. Occupying relatively small areas of land, usually in public ownership, the NNRs presented few problems for the broad interests of landowners.

With the environmental movement weak, fragmented and small, the NC was certainly loath to challenge the farming community by extending its concerns to the management of SSSIs (it had quite enough to do just designating them: N.W. Moore 1987). Moreover, steps to ensure good management were virtually impossible in the absence of a system whereby either land-occupiers could be notified that they had an SSSI on their land or the NC be notified of land-management operations. In both 1963 and 1968, Private Members' Bills to compel farmers to notify the NC of proposed operations on SSSIs were lost. Section 16 of the 1949 Act had empowered the NC to enter into management agreements with owners of NNRs, as an alternative to outright purchase or long lease. The 1949 Act provided a similar mechanism for national parks to negotiate access agreements. In both cases the opportunity was used but rarely. The opportunity to negotiate management agreements was only extended to SSSIs under Section 15 of the Countryside Act 1968, but as the Nature Conservancy 'still had no formal relationship with owners and occupiers of SSSIs' (Withrington and Jones 1992: 91), the Council had little opportunity to offer such an agreement on an SSSI before discovering that farmers had carried out damaging operations, the farmers themselves being in ignorance of the fact that their land had been designated. Section 17 of the 1949 Act gave powers of compulsory purchase to the NC, powers which it has rarely exercised.

THE FARMING AND CONSERVATION DEBATE
INTENSIFIES – THE 1970s

Following the 1968 Countryside Act, the 1970s was a decade of deepening debate on countryside issues, all of which can be seen, with the benefit of hindsight, as providing the build-up to the passage of the Wildlife and Countryside Bill between 1980 and 1981. The reason for the new attention to countryside issues was the increasing realisation that agriculture, with its exemptions from most aspects of planning control, could not be relied upon to protect the countryside either for the interests of recreationists and landscape preservationists or for nature conservationists and ecologists (the latter had been very much brought into the wider debate following revelations about the extent of pesticide pollution – see chapter 10, p. 258). Evidence to back up the concern began to emerge slowly during the 1970s. For example, in 1976 the Nature Conservancy Council issued a report which showed that approximately 4 per cent of SSSIs were severely damaged each year.

In an attempt to reconcile the interests of farmers and conservationists, the Farming and Wildlife Advisory Group (FWAG) had been born (see chapter 9, pp. 240–249), following a joint initiative of a number of leading agriculturalists and conservationists in the late 1960s. It gathered pace in the 1970s with the formation of county FWAGs (Cox, Lowe and Winter 1990c). The concern was reflected within the leadership of the NFU and CLA, which hitherto had eschewed much involvement in such debates. As revealed in Figure 8.1, the decade saw a dramatic increase in the extent of coverage given to conservation issues in the farming press, in this case *British Farmer*, the house journal of the NFU published roughly fortnightly during this period.

The varying nature of the coverage demonstrates vividly the delicate path the NFU had to take on this issue as it sought to appeal to its own members, agricultural policy makers and to an increasingly sceptical environmental audience. For the benefit of its own members, many articles provided evidence of the Union's tough stance with hard-hitting attacks on the so-called conservation *cranks* and *extremists*. For example, increased coverage in 1977 is almost entirely down to the Exmoor moorland controversy (see next section) with most articles attacking the conservation case. Other pieces, though, were directed towards educating members either with informative articles on practical conservation or, by way of political education, with the increasing emphasis given to demonstrating how much farmers are already doing, and always have done, for conservation. For other members of the agricultural policy community and for environmentalists, there was the need to impress upon them just how much farmers 'cared' and also could be relied upon to introduce appropriate measures voluntarily. In 1977, the NFU went further, issuing a joint statement with the CLA entitled *Caring for the Countryside*, to press the message home. It represented the beginning of a high-profile and sustained publicity campaign to promote farmers and landowners as the natural custodians and stewards of the countryside.

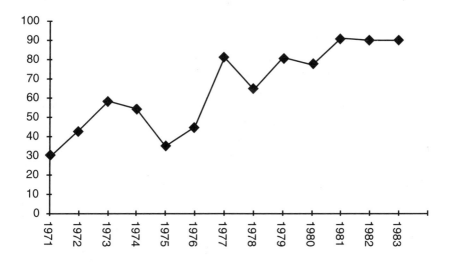

Figure 8.1 Percentage of issues of *British Farmer* with coverage of environmental matters

The moorland issue

The issue that most tested the stewardship claim in the 1970s was that of the reclamation of heather moorland for agricultural grassland, particularly on Exmoor but also on the North York Moors and elsewhere. Pressure on upland habitats increased during the 1970s, particularly with regard to Exmoor, where the heather moorlands were especially vulnerable to agricultural improvement due to the relatively low terrain, mild climate and lack of surface rocks. Exmoor became something of a *cause célèbre* in the 1970s with the Park Committee showing considerable reluctance to oppose proposals to reclaim moorland. The passage of the 1968 Act had been seized on as an opportunity by the CPRE, the Exmoor Society, the Devon and Somerset County Councils and the Exmoor National Park Committees, to promote amendments which would have given national park authorities the power to control moorland change (Lowe *et al.* 1986). A voluntary notification agreement forged in the aftermath of the Countryside Act 1968, whereby the NFU and CLA agreed that their members would inform the park authority of plans to reclaim moorland, resulted in not a single management agreement nor any occasion when the Ministry of Agriculture either advised a farmer not to proceed with a reclamation scheme or withheld grant aid (Lowe *et al.* 1986).

Eventually exasperation with the pro-agricultural stance of the park committee led to combined pressure upon the government from Somerset County Council and the Countryside Commission. This resulted in the appointment of a committee chaired by Lord Porchester whose report recommended that:

- the notification system should be statutory
- the National Park should be empowered to make conservation orders binding in perpetuity, based on a one-off compensation payment
- the Ministry should withhold grants where conservation objectives were paramount.

(Porchester 1977)

The unanimous acceptance of the report by the park committee probably owed more to the threat of disbandment hanging over it, than to any radical change of heart. Indeed, at the very same meeting agreement was given to improve forty hectares of land identified by Porchester as worthy of rigorous protection (Lowe *et al.* 1986).

Most of the Porchester recommendations were taken up by the Labour government and contained within the terms of its 1978 Countryside Bill. Notwithstanding Lord Porchester's credentials as a Conservative and a landowner, the Bill was opposed by the Conservatives. When Parliament was dissolved for the May 1979 general election, the Opposition was unwilling to allow the legislation through 'on the nod'. Instead it agreed to promote its own bill, should it be successful in the election.

Thus it was that within days of the Conservatives assuming office in

1979, officials of the CLA and the NFU had met with civil servants from MAFF and the Rural Directorate of the DoE to consider plans for drafting the new legislation. The government's own statutory advisers on countryside conservation, the NCC and the Countryside Commission, were not party to these discussions (Cox and Lowe 1983a, 1983b). In many ways the passage of the Bill through parliament and the controversies that followed, exemplified the continuing political strength of the farming and landowning lobby but also the coming of age of the conservation movement in Britain.

THE WILDLIFE AND COUNTRYSIDE ACT 1981

The full story of the Bill's passage through parliament has been well documented and the following account draws heavily on these sources (Cox and Lowe 1983a and 1983b; Lowe *et al.* 1986), allowing some very significant lessons to be drawn for our understanding of the policy process and the relative strengths and weaknesses of the actors involved. During the period from August to October 1979, the DoE issued six consultation papers covering the following key issues:

- bulls and public footpaths
- review of definitive maps of public rights of way
- species protection
- conservation of habitat
- countryside provisions
- moorland conservation in national parks.

The first two need not concern us here and the third was broadly acceptable, due to the input from the RSPB, the only environmental pressure group to have been consulted prior to the issue of the consultation papers (Lowe *et al.* 1986). On the protection of birds, the RSPB was, and largely remains, more influential than the government's own statutory adviser, the NCC. The other three consultation papers set out the government's approach which, reflecting the NFU and CLA inputs, essentially revolved around the voluntary principle.

The only elements of compulsion proposed were:

reserve powers to require landowners or tenants to give up to 12 months' notice of any intention to convert moor or heath to agricultural land in specified parts of national parks (this became the basis for Section 42 of the Act); or of any intention to undertake operations which could be detrimental to the scientific interest of selected SSSIs (the basis of Section 29 conservation orders). . . . Ministers assured the NFU and the CLA that there was no intention to use the reserve powers for national parks (a similar reserve power in the 1968 Act providing for six months' notice had only been used twice on small parcels of land in Exmoor) and that only a few especially important

SSSIs would be given this extra safeguard (a maximum of about 40 was suggested out of a total of some 3,500 SSSIs).

(Lowe *et al.* 1986: 135)

The proposals represented a considerable logistical triumph for the NFU and the CLA, showing just how secure the agricultural policy community remained. Through MAFF, the NFU/CLA had gained direct access to those drafting the legislation in the DoE, a demonstration of how a close relationship with one government department could easily secure good relations with another. The environmental pressure groups and even the agencies remained seemingly isolated and outside the policy process, reduced to rather crude parliamentary lobbying tactics. Not surprisingly in this context, the environmentalists' response was hostile but, initially, ineffectual due to lack of co-ordination. The RSPB placed most emphasis upon problems of enforcement and resources. The NCC concentrated on the need for marine nature reserves and control on the imports of non-native flora/fauna, thus effectively distancing itself from the mainstream countryside concerns. The RSNC urged greater protection for SSSIs, whilst the CPRE pressed for moorland conservation orders. Friends of the Earth, meanwhile, did not even introduce its own ideas until January 1981, two months after the Bill was introduced in Parliament. Matters were not helped by the demise of the Council for Nature, for twenty-two years a forum for co-ordination, disbanded at this time as a result of internal rivalries. In haste, Lord Melchett, a Labour peer and keen conservationist, convened a replacement umbrella organisation, Wildlife Link.

If the government's initial timetable, a period of five to seven weeks for consultation, had been adhered to, a very weak Bill would probably have been passed. But on 22 January 1980, a postponement of the Bill was announced because of pressure on parliamentary time. This allowed Wildlife Link time to co-ordinate more effectively those who found key provisions in the Bill too weak. It urged, in particular, first that there should be prior notification of agricultural changes on all SSSIs, not just the forty so-called 'super-SSSIs', and second that protection orders on SSSIs should be available as a last resort.

The NCC, however, took a different view:

Ever since calls for improved safeguards for SSSIs were first made in the early 1960s it had been markedly less enthusiastic than voluntary conservation groups in seeking controls which might overstretch its staff and resources and draw it into confrontation with farmers and landowners. Thus ... it broadly accepted the government's proposals and expressed itself fully satisfied once assurances had been given that the criteria for special protection would be broadened and the NCC consulted before the selection of any site. This accommodation set the NCC on a collision course with the voluntary conservation groups.

(Lowe *et al.* 1986: 136)

The collision was averted only after a direct appeal by Lord Melchett addressing a council meeting of the NCC in November 1980, which resulted in the NCC pressing for an appropriate amendment after the Bill had started its course through Parliament. Lord Melchett was aided in his task by changes in the rules applying to general agricultural grants which would allow farmers to apply for grants retrospectively, thus removing the possibility of averting damaging operations even from ADAS. Such a change prompted both the government and the NCC to reconsider the prior notification issue in national parks and SSSIs.

The parliamentary phase, mostly fought out in the House of Lords, lasted from November 1980 until October 1981, the Bill eventually receiving Royal Assent on 30 October. No less than 2,300 amendments were tabled, and in its efforts to stem the effects of many of these the NFU produced thirteen parliamentary briefing papers, maintained contacts with 150 peers and 350 MPs, and employed three full-time officers on the Bill:

> The NFU ... was the only group with the resources to be present at all stages of the bill's passage through both houses, and its diligence and thoroughness were greatly admired and envied by conservation lobbyists.
>
> (Cox and Lowe 1983a: 62–63)

By contrast, neither the CPRE, the CNP, the RSPB nor the RSNC were able to devote a full-time officer to the Bill.

During the parliamentary phase there were both gains and losses as far as conservationists were concerned. The gains included protection for limestone pavements and provisions for marine nature reserves, as well as a requirement for owners of SSSIs to notify the NCC of intentions to carry out potentially damaging operations. In this the conservationist cause was helped by new evidence of the level of destruction of SSSIs, evidence which flew in the face of the government's continued and highly vocal commitment to the voluntary principle. In 1976 the NCC had found that approximately 4 per cent of SSSIs were being severely damaged each year. A new survey, the results of which were not available until February 1981, by which time the Bill was at committee stage in the House of Lords, revealed a massive increase in damage to SSSIs, now running at the rate of 13 per cent per annum.

However, the list of losses included the rejection of moorland conservation powers and the emasculation in committee of an amendment successfully tabled by Lord Sandford which would have given MAFF the duty to further conservation in national parks and SSSIs. Moreover, in response to both the Sandford amendment and the tougher clauses on SSSIs, a new principle was inserted into the legislation, to the effect that compensation payments would be offered when agricultural grants were refused on conservation grounds. The NFU and the CLA, which had throughout emphasised the resource implications of retaining goodwill, welcomed the new measure as providing the necessary financial safeguards to farmers adversely affected by conservation objectives. In so doing, the Union was turning its back on opposition to conservation in principle or as being inevitably inimical to the interests of farmers. Following years of successfully defending the right to production grants and subsidies, the Union was now asserting a new right to environmental payments that reflected profits and grants forgone when conventional agricultural practices were not to be permitted, a principle that became firmly established in the decade that followed (Cox, Lowe and Winter 1988). Robin Grove-White, Director of the CPRE, summarised the objections of environmentalists to the new clause in a letter to *The Times*:

> It gives legal expression to the surprising notion that a farmer has a right to grant aid from the tax-payer: if he is denied it in the wider public interest, he *must* be compensated for the resulting entirely hypothetical 'losses'.... The Bill requires compensation to farmers to be paid ... from the meagre budgets of the conservation agencies.
>
> (quoted in Lowe *et al.* 1986: 147)

In terms of parliamentary time and the amount of public debate, the passage and immediate aftermath of the Wildlife and Countryside Act provides the most important legislative episode in rural policy in the post-war period, excepting only the legislative programme of the 1945–1951 Labour administration. It marks a watershed in post-war agri-environmental policies. Prior to the Bill, a tight and narrowly defined agricultural policy community had been open to relatively few outside influences and interests. Responses from the mainstream actors within the agricultural policy community had been to resist change through the assertion of an agricultural exceptionalism (i.e., that agriculture is a special case exempt from many of the rules operating in the wider society) and this had characterised agricultural policy deliberation for some years.

However, for all the successes scored by the NFU in the passage of the Act, it represented a defeat for agricultural exceptionalism. Success had only come because agricultural policy actors were prepared to engage in a new set of relations with a wider group of interests and government agencies. And for the environmental groups, the passage of the Bill represented something of a baptism of fire. They engaged head on with the might of the farming lobby and emerged scathed but experienced. After 1981 most of the mainline environmental groups made it their business to understand agriculture and to confront many of the assumptions of the post-war settlement.

Implementation of the Act

But what of the Act itself? Part 2 of the Wildlife and Countryside Act 1981 provided for five main innovations in site protection. It removed the formal distinction between NNRs and SSSIs. All NNRs now have to be notified as SSSIs and the basic legal framework for protection applies equally (as set out for SSSIs), to both designations. Second, Section 28 requires the NCC to notify owners and occupiers of SSSIs as well as LPAs. The notification includes a statement of the special interest, a boundary map and a list of Potentially Damaging Operations which could damage the flora, fauna, geological or physiographic features for which the site was selected (Withrington and Jones 1992). Third, Section 28 also makes it an offence for an owner or occupier to carry out any operation specified in the site notification unless formal consultation procedures with NCC have been fully adhered to or unless planning permission has been granted. Fourth, where Management Agreements are offered, following a process of formal consultation, the level of payment should reflect the profits forgone by the farmer or landowner. And finally, the Secretary of State is empowered to impose a Nature Conservation (Section 29) Order to protect endangered sites by extending the period for negotiation from four to sixteen months (thereafter compulsory purchase orders could be made, although this has only happened in one instance).

The most pressing problem that faced the NCC in 1981 was the need to (re-)notify all SSSIs as speedily as was reasonably possible. This was a time-consuming and demanding procedure and progress in the period immediately following the Act was necessarily slow, as indicated in Table 8.2. However by 1994 the process was virtually complete, with 5,600 SSSIs designated and notified in the UK, covering 8 per cent of the land surface. The notification procedure, under the 1981 Act, is illustrated in Figure 8.2, which shows how a three-month loophole existed after notification during which time it was not an offence to damage the site. During this period the NCC could serve a Nature Conservation Order, but even these took three weeks to prepare. Court orders could be applied in certain circumstances but could hardly be relied upon. Subsequently, this loophole was closed by the Wildlife and Countryside Amendment Act 1985.

The importance of adhering rigorously and strictly to the correct

Table 8.2 SSSI (re-)notification progress by NCC in the UK

	Total hectares notified under the 1981 Act	*% of total SSSI area i.e. total of areas notified under 1981 Act + 1949 Act*
1982	831	0
1983	18,487	1
1984	229,823	17
1985	415,465	29
1986	690,158	48
1987	1,021,958	67
1988	1,190,183	78
1989	1,414,335	86
1990	1,618,641	94
1991	1,721,502	97

Source: NCC annual reports

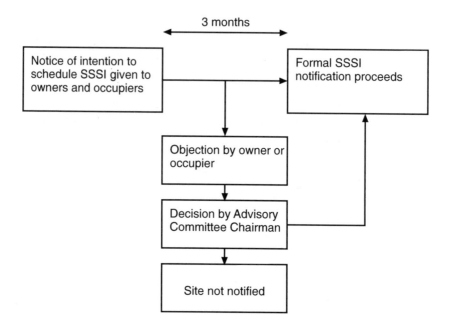

Figure 8.2 Notification procedure for new SSSI

Source: Adams 1984: 30

procedures was demonstrated in a court case in 1985 when the NCC brought an action against a farmer in Wales for ploughing an SSSI. The case was dismissed with costs awarded against the NCC. The magistrates issued a written ruling, clearly aimed at advising the NCC how to avoid further embarrassment, in which they stated that:

> the original notification of the SSSI under section 28 was invalid, because the NCC had invited the owner to 'comment' on the proposed notification rather than make 'representations or objections'. Another point raised by the defence was that the NCC had not served notification on all the owners, as the farmer's wife was a part-owner – a fact not known to the NCC.
>
> (Withrington and Jones 1992: 99)

In much the same way that notification was a complex and time-consuming business, so too was the negotiation of management agreements. The procedure for this under the terms of the 1981 Act is set out in Figure 8.3. Once again the procedures proved less than satisfactory in practice. The original period for the NCC to act upon notice of intention to carry out a Potentially Damaging Operation (PDO) was just three months; it was only raised to four months by the 1985 amendment act. Three months had proved too short a period for the NCC in many instances. In practice the situation is often now resolved through the agreement of a provisional or interim

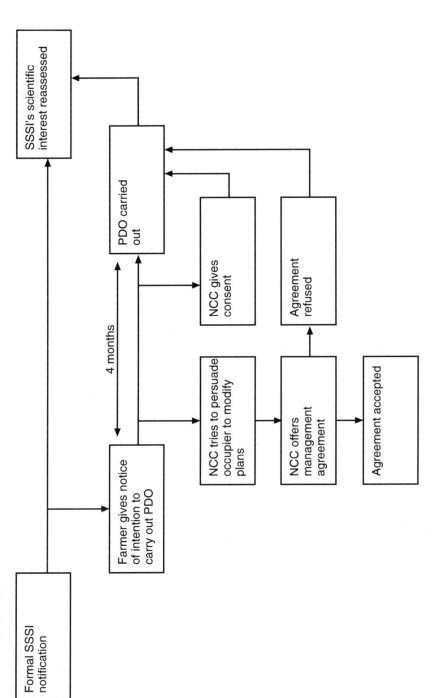

Figure 8.3 Response to notice of intention to carry out a Potentially Damaging Operation (PDO)

Source: Adams 1984: 32

Table 8.3 NCC management agreements in Great Britain

Year	Number of agreements	Hectares	Annual cost £	Cost as a % of NCC's expenditure
1975	10	236	not known	not known
1979/80	43	2,228	25,000	0.3
1981/82	90	3,032	60,550	0.6
1983/84	139	4,648	206,605	1.6
1985/86	360	16,690	1,967,715	9.1
1987/88	1,053	38,447	4,555,339	13.5
1989/90	1,759	86,919	6,852,226	16.7
1990/91	2,032	98,545	7,238,932	15.7

Source: NCC annual reports

short-term arrangement allowing payments to be made to the occupier with a financial reconciliation at a later date when full agreement has been reached. Table 8.3 shows the extent and development of management agreement coverage.

In the aftermath of the 1981 Act, attention was by no means confined to the procedural difficulties outlined above. The period from 1981 to the mid-1980s represented one of the most sustained periods of intense media and public interest in countryside protection in the post-war period. In 1984 both the *Sunday Times* and the *Observer* ran series attacking the loopholes in the 1981 Act, including expression of concern over the vast majority of the farmed countryside not covered by any designation, and therefore not subject to any protection by Section 2 of the Act. The CPRE launched a Campaign for the Countryside at the same time as initiating talks with the CLA and the NFU to seek common ground (Lowe and Winter 1988). In June 1984 the NCC published a report detailing 'the overwhelmingly adverse impact of modern agriculture on wildlife and its habitat in Britain'. And in Parliament two select committees produced reports with devastating criticisms, not so much of the operation of the 1981 Act although failings were highlighted, but of MAFF and its continued failure to integrate fully policies for agriculture and the environment. The House of Commons Environment Committee considered the operation and effectiveness of the 1981 Act itself:

> Our underlying concern is that, even if the changes we recommend are made to the Act and its administration, the wider agricultural structure will fuel the 'engine of destruction'. This is the responsibility of Government, not farmers.... MAFF must reappraise its attitudes.
> (House of Commons 1985a: para 73)

And the House of Lords Select Committee on the European Communities criticised MAFF's approach to a proposed new Structures Regulation from the European Commission:

The Committee consider that the draft Regulation is too closely production-orientated, despite its gestures in other directions. MAFF, by their narrow interpretation of the few innovative features it contains, reinforce this backward-looking tendency.

(House of Lords 1984a: para 134)

We return to the impact of some of these developments for the agricultural policy community in chapter 9.

How well are SSSIs protected?

A number of problems remain for the protection of SSSIs. One major difficulty lies with enforcement. The case of the Welsh farmer referred to above is not unique, as demonstrated in Table 8.4, which also shows that even where prosecutions have been successful the fines have been modest. Another problem of enforcement lies with the issuing of Section 29 orders which apply only to SSSIs of 'special importance' and 'national importance', or where 'compliance with an international obligation' is required (no orders have yet been in this category: Withrington and Jones 1992). The ruling against the NCC in a public inquiry in Essex in 1985 means that not all SSSIs have this level of protection because not all are considered to be of national importance (for details see Withrington and Jones 1992; also Adams 1986). The NCC continues to press for an amendment to Section 29 so that all SSSIs are covered, but failed in its best chance to date – the passage of the Environmental Protection Act 1990.

Between 1982 and 1989 a quarter of SSSIs suffered some form of damage (RSNC 1989). Thus the incidence of damage to SSSIs through agricultural operations has by no means been eliminated (Figure 8.4; see also Adams 1991), and the damage or complete destruction of a small number of SSSIs each year as a result of other forms of development remains a serious problem. In England 3.5 per cent of the area of SSSIs was damaged in some way in 1992/93, but only 0.001 per cent was actually destroyed (NAO 1994).

The protection offered by the planning system by no means guarantees the safety of an SSSI and planning committees or Department of the Environment inspectors may decide that other compelling factors, such as the need for jobs, should override SSSI protection when development is proposed. Legally, it is perfectly legitimate to destroy SSSIs if planning permissions for development are granted or if private acts of Parliament are passed.

There is, of course, also an issue concerning the long-term protection offered to SSSIs by the management agreement system. Agreements are for between three and twenty-one years, with most being for the maximum period. Those forged soon after the 1981 Act are now more than half-way through their term. Questions remain about what should happen after this period. It is clear to many of those involved in the complexities of site-by-site negotiations that the system is too cumbersome and costly. The cost of

Table 8.4 Prosecutions brought by the NCC under Section 28 of the Wildlife and Countryside Act 1981, 1981–1990

Date	Name of SSSI	County	Nature of operation(s)	Outcome
31 October 1984	Ulverscroft Valley	Leicestershire	Spreading lime	Guilty verdict. Fine £200 with £50 costs
12 March 1985	Gwenydd Yr Afon Fach	Dyfed	Ploughing 8 ha	Case dismissed. NCC paid £250 costs
4 April 1985	Broadstone Meadow	Hereford and Worcester	Ploughing 1.5 ha	Case dismissed. NCC paid £1,437 costs
6 October 1987	Chesil Beach	Dorset	Extraction of sand and shingle	Guilty verdict. Fine £1,500 with £1,000 costs
4 January 1988	Hurn Common	Dorset	Ploughing 7 ha of heathland	Guilty verdict. Fine £800 with £800 costs
4 March 1988	Westhay Moor	Somerset	Application of herbicide	Guilty verdict. Fine £250 with £500 costs. Unsuccessful appeal to Crown Court, 18 July 1988
5 May 1988	Tealham and Tadham Moors	Somerset	Application of herbicide	Case dismissed
17 October 1988	East Devon Pebblebed Heaths	Devon	Grazing and stock feeding	Guilty verdict. Fine £300 with £100 costs
30 September 1988	Swithland Woods	Leicestershire	Extraction of slate, construction of paths, use of vehicles	Guilty verdict. Fine £1,000 with £1,262 costs
9 January 1990	Alverstone Marshes	Isle of Wight	Drainage operations	Guilty verdict. Fine £6,000 with £2,326 costs. Conviction quashed by High Court and costs awarded against NCC.
11 April 1990	Cors Llyferin	Gwynedd	Extension of golf course	Guilty verdict. Fine £800 with £700 costs
3 September 1990	Cwm Gwynllyn	Powys	Drainage operations	Guilty verdict. Conditional discharge with £50 costs

Source: Withrington and Jones 1992: 100

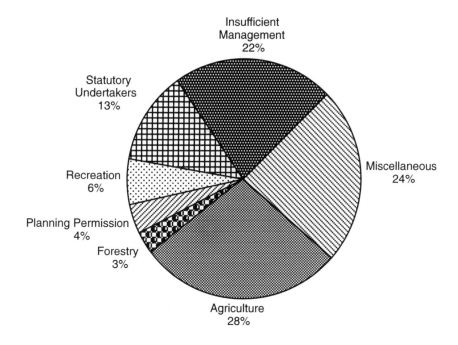

Figure 8.4 Causes of loss and damage to SSSIs 1992–1993

Source: English Nature in NAO 1994

management agreements rose from £0.3 million in 1983/84 to £7 million in 1993/94 (constant 1992/93 prices: NAO 1994). Extraordinarily, these payments cover a mere 7 per cent (1991/92) of the total land area of SSSIs. Whilst many of the remaining areas are in public ownership there is a substantial number of sites (65 per cent of SSSI land is in private ownership) which might potentially come under the management agreement process. English Nature has come under pressure to revert from compensatory payments to enhancement payments and to simplify its procedures, and in 1992 introduced the Wildlife Enhancement Scheme (WES). This has been introduced to cover all SSSIs sharing similar conditions within a locality and where the individually negotiated agreements with occupiers have produced very similar payments. The WES introduces a set of standardised and simplified payments to encourage positive management on SSSIs. Administration costs for WES agreements amount to 10 per cent of total cost compared to 24 per cent in standard SSSI management agreements (NAO 1994).

Three areas were chosen for piloting the WES – the Culm Grasslands of Devon and Cornwall, the Pevensey Levels in East Sussex, and the Coversand Heath and Peatlands of North Humberside and Yorkshire. In addition to the simplified payment formula, the scheme differs from the standard SSSI management agreement in so much as English Nature has informed all land-occupiers and invited them to participate. This is in stark contrast to the formal situation in other SSSIs where the occupier has to threaten to perform a damaging operation in order to trigger the offer of a management agreement, which has inevitably resulted in accusation of bluff and counter-bluff in some of the negotiations. It seems likely that the WES formula alongside other standardised packages, such as ESAs and Countryside Stewardship, will become the norm for environmental land management schemes in the future. Another approach is to encourage land purchase: county wildlife trusts have received grants from the NCC to buy SSSIs. In one case, a £100,000 grant to the Essex Wildlife Trust has led to expected savings in management agreement payments of nearly £600,000 (NAO 1994).

An opportunity to remedy some of the perceived weaknesses of the 1981 and 1985 Acts was provided with the passage through Parliament of the Environmental Protection Act 1990, although many suggestions made by the NCC and others were rejected (Ball 1991). One innovation was a clause allowing management agreements to be negotiated with owners and occupiers of land adjacent to SSSIs, a development that could help where existing SSSIs are adversely affected by activities, such as drainage, on near-by land. But an amendment to extend this to the wider countryside was rejected 'on the grounds that the Council's money was better targeted on the protection of key sites' (Ball 1991: 82). However there has long been scientific recognition that it is inappropriate to consider isolated sites without reference to conditions in the remainder of the countryside (Diamond 1975) and there is growing pressure to recognise that 'the survival of the wildlife interest of the SSSIs themselves depends to an extent on the nature of the landscape of the wider countryside' (Adams et al. 1992: 247).

Arguably, the most significant part of the Environmental Protection Act 1990 is Part VII which deals with the reorganisation of the conservation authorities. Together with the Natural Heritage (Scotland) Act 1991 this led to the devolution of responsibilities in Scotland and Wales where the Countryside Commission and the NCC were merged to form Scottish Natural Heritage (SNH) and the Countryside Council for Wales (CCW). In England the two bodies remained separate, but a merger of the Countryside Commission and the NCC (English Nature), was widely rumoured in 1992/93. Instead, in October 1994, the Environment Minister, John Gummer, called for a closer working relationship between the two agencies. Much concern was expressed over the way in which co-ordination of responsibilities and the establishment of UK priorities would be handled, and to this end a new Joint Nature Conservation Committee was established, although its remit is confined to conservation and science despite the fact that the two new merged authorities in Scotland and Wales have wider countryside, recreation and landscape functions.

A particular concern for many environmental groups was the fear that the changes heralded a dilution of conservation standards and objectives, especially in Scotland. As the director and a policy officer of the CPRE, firm opponents of the changes, explain:

it must be acknowledged that a prime motivation for the reforms was the unpopularity of the NCC in Scotland. During the 1980s tension had grown between well-meaning, but possibly less than tactful, NCC officials and Scottish landowners over certain instances of SSSI designation or proposed controls over potentially damaging operations. . . . Nicholas Ridley (Secretary of State for the Environment) and Malcolm Rifkind (Scottish Secretary) were persuaded that an independent and Scottish-based NCC rather than a UK organisation based in Peterborough, England were likely to be less antagonistic to the Scottish interest, and were more likely to carry those interests with them in discharging their functions.

(Reynolds and Sheate 1992: 74)

The suspicion that the devolved approach might, in the case of Scotland at least, lead to a softening in the pursuit of conservation objectives came to a head during the passage of the Natural Heritage (Scotland) Act 1991. During the third reading in the House of Lords, Scottish peers took the opportunity to press successfully for an amendment that would require SNH to review all SSSI designations during a five-year period, notwithstanding the exhaustive process of renotification following the 1981 Act. Moreover, the amendment contained a clause that would allow any objections to designation from owners and occupiers to be put to an inquiry to be conducted by the Secretary of State:

[This] is certain to reinforce the worst fears that conservationists harbour about the break-up of the NCC. The scientific criteria governing SSSIs are called into question by political factors, and increased influence is given to landowners and to the Scottish Department, whose conservation credentials have never appeared strong.

(Ball 1991: 84)

To conclude this section, the Wildlife and Countryside Act did much to resolve, albeit in a manner which deeply offended some, the specific issues of concern regarding habitat loss through agricultural development in SSSIs. And as the 1980s progressed, further key developments took place which tackled even more explicitly the agricultural issue and these are explored further in the following two chapters.

THE IMPACT OF EUROPEAN ENVIRONMENTAL POLICY

The main aim of this section is to consider the development of European environmental policies and to give some indication of the extent to which the UK arrangements for habitat protection are consistent with EU policy principles. As Liefferink et al. (1993) have explained, environmental policy in the EC is the outcome of a complex interplay of forces:

The game is played in Brussels, but at the national level too: the positions taken by representatives of member state governments or interest groups are themselves usually products of lengthy domestic processes. This leads not only to clashes of material interests that are often quite sharp, but also to clashes of different social, political and legal cultures.... the obverse problem to the clash of national cultures in EC decision making is the difficulties of assimilation that occur when European rules have to be implemented through domestic administrative structures.

(Liefferink, Lowe and Mol 1993: 7)

The focus in this chapter is on habitat and landscape protection, although some of the background information is also relevant to both chapters 10 and 11. In sharp contrast to agriculture, which was accorded a special legal status by the Treaty of Rome, the Treaty made no provisions for common environmental policies. Thus it was that when attempts were made to forge common environmental policies for the members of the EC, legal objections were put by certain states, including the UK.

Three main legal justifications were utilised by the Commission from the early 1970s to justify its interest in this area:

- Article 2 of the Treaty of Rome states that *the Community shall ... promote throughout the Community a harmonious development of economic activities, a continuous and balanced expansion, ... an accelerated raising of the standard of living.*
- The Preamble to the Treaty speaks of the Community's need for *the constant improvement of the living and working conditions of their peoples.*
- The need to promote and ensure the harmonisation of trade, as discrepancies in national environmental laws could be used as a means of trade discrimination.

The first two Treaty elements allowed the Commission to argue that the importance attached to environmental concerns by the public could be translated into environmental provisions based on balanced expansion and raised standards of living and living conditions.

The reason for emphasising this legal background is that it explains, in part, a certain anthropocentrism in the initial development of policy. EC policies tended to focus on the prevention of pollution incidents and environmental disasters, and on resource issues rather than the protection of the environment on aesthetic or moral grounds. The initial impetus for policy development was provided by the UN Conference on the Human Environment held in Stockholm in June 1972. This was followed by a Commission paper discussed by the heads of government at the Paris Summit in October 1972. The Summit Communiqué called for an EC environmental programme and acknowledged that economic expansion was 'not an end in itself' but should be linked to 'improvement in living and working conditions of life of the citizens of the EC'. Within a year the objectives of the first five-year Environmental Programme had been agreed.

Its objectives were ambitious, if somewhat lacking in specific and attainable targets:

- to abolish the effects of pollution;
- to manage a balanced ecology;
- to improve working conditions and the quality of life;
- to combat the effects of urbanisation;
- to co-operate with states beyond the EC on environmental problems.

Perhaps of greater importance than these objectives, were the two underlying principles set out by the EC, principles that have stood the test of time – *prevention at source* and the *polluter pays principle*, both of which are dealt with in greater detail in chapter 10.

There followed environmental programmes at five-yearly intervals, and a wide range of directives, until in 1987 the Single European Act provided a much clearer legal basis for environmental policy, whereby the EC members committed themselves to the following objectives:

- to preserve, protect and improve the quality of the environment;
- to contribute towards protecting human health;
- to ensure a prudent and rational utilisation of natural resources.

The Act expanded and considerably broadened principles for environmental policy:

1 The principle of prevention is better than cure.
2 Environmental effects should be taken into account at the earliest possible stage in decision making, and should be a component of all EC policies.
3 Exploitation of nature or natural resources which causes significant damage to the ecological balance must be avoided. The natural environment can only absorb pollution to a limited extent. It is an asset which may be used, but not abused.
4 Scientific knowledge should be improved to enable action to be taken.
5 The polluter pays principle: the cost of preventing and eliminating nuisances must be borne by the polluter, although some exceptions are allowed.
6 Activities carried out by one Member State should not cause deterioration of the environment in another.
7 The effects of environmental policy in the Member States must take account of the interests of the developing countries.
8 The Community and the Member States should act together in international organisations and in promoting international and worldwide environmental policy.
9 The protection of the environment is a matter for everyone. Education is therefore necessary.
10 The principle of the appropriate level. In each category of pollution, it is necessary to establish the level for action (local, regional, national,

Community, international) best suited to the type of pollution and to the geographical zone to be protected.

11 National environmental policies must be co-ordinated within the Community, without hampering progress at the national level. This is to be achieved by the implementation of the action programme and of the 'environment information agreement'.

In the light of such a major development in thinking, it is hardly surprising that the fourth Programme (1987–1992) was both more ambitious and more specific than the previous programmes, including, *inter alia*, the aim of ensuring the effective implementation of existing legislation, such as the Birds Directive; the development of a Habitats Directive; the compilation of a comprehensive list of protected areas designated within member states; and ensuring the provision of finance for the protection of sites of Community importance.

Clearly the implications of EC policy for mainstream wildlife conservation were becoming greater and it is worth considering some of the more important developments of this nature. The European Community Directive on the Conservation of Wild Birds was enacted in 1979 (79/409/EEC) and placed a general duty on member states to:

maintain the population of all 'species of naturally occurring birds in the wild state' in the European territory at a level which corresponds in particular to ecological, scientific and cultural requirements, while taking account of economic and recreational requirements.

Member states are required to take measures for the creation of protected areas; the upkeep and management of habitat in accordance with ecological needs (including, where necessary, outside protected areas); the re-establishment of biotopes previously destroyed and the creation of new ones; restrictions on the sale of live birds; the protection of migratory birds; and special habitat conservation measures for so-called Annex 1 species leading to the establishment of Special Protection Areas (SPAs).

The number of Annex 1 species was raised from 74 to 144 in 1985. There was also a prohibition on hunting and shooting of most Annex 1 species. In the same year as the passing of the Birds Directive, the Berne Convention on the Conservation of the Wildlife of Europe's Natural Environment was approved by the Council of Europe. The Convention stressed the importance of habitat protection and was ratified by the UK in 1983. Also in 1979, the Bonn Convention on the Conservation of Migratory Species of Wild Animals was approved under the UN Environment Programme, whereby EC members and twenty other states agreed on respect and protection for migratory species. To these measures must be added the earlier 1971 Convention on Wetlands of International Importance (Cmnd 6465), which established an international list of Ramsar sites, committing the contracting parties to 'formulate and implement their [land use] planning so as to promote the conservation of the wetlands in the List, and as far as possible the wise use of wetlands in their territory'.

The implications of these various international and EU obligations are significant, with the UK somewhat slow to comply with every aspect of the obligations. By 1990, the UK had designated just 44 of a potential 218 Ramsar sites and 40 of a potential 154 SPAs. It should be noted that whilst all Ramsar sites and SPAs will also be SSSIs, not all SSSIs will be Ramsar sites or SPAs and that some Ramsar sites and SPAs cover more than one SSSI. Following the passage of the Wildlife and Countryside Act (as much concerned with obligations under the Bern, Bonn and Ramsar Conventions as with the Directive itself) the UK government notified the EC of its SPAs, protected through NNR or SSSI designation.

In 1992 a further directive of considerable importance was passed, the Habitats Directive, or to give it its full title, the Directive on the Conservation of Natural Habitats and of Wild Fauna and Flora. Under this legislation, the EU is to establish a European Ecological Network, known as Natura 2000, of Special Areas of Conservation (SACs) which will include SPAs. An SAC is defined as a:

> site of Community importance designated by the Member States through a statutory, administrative and/or contractual act where the necessary conservation measures are applied for the maintenance or restoration ... of the natural habitats and/or populations. ...

In exceptional circumstances the EU will designate sites not designated under domestic legislation by member states and may also provide EU finance in some circumstances. The Habitats Directive also seeks a strengthening of Environmental Impact Assessment.

It is far from clear that the UK, for all its long history of legislative protection for bird species and rather more recently habitats, is in a position to comply fully with either the Birds Directive or the Habitats Directive. Indeed it has been strenuously criticised for its lack-lustre approach, which relies primarily on a restatement of the value of existing legislative provision (Phillips and Hatton 1994). Problems include an inadequate networking of sites as birds may use some sites and habitats for very short periods of time. Outside SSSIs, damage of habitats for species listed under Conventions or the Birds Directive can occur quite legitimately, with the sole exception of bat habitats. There are also inadequate mechanisms for notifying SSSIs below mean low-water mark. But, perhaps above all, there is continuing uncertainty regarding the level of protection for sites of international importance within the town and country planning system. In an effort to provide reassurances, a DoE circular on Nature Conservation in 1987 urged careful consideration of all planning applications within SPAs and suggested that planning permission should be granted only where disturbances to birds or habitats will not be significant or where 'any such disturbance or damage is outweighed by economic or recreational requirements' (Department of the Environment 1987: para. 26). The problem with such an approach is that it offers no guidelines on how such an evaluation is to be carried out, particularly the crucial question of what economic or recreational benefit might be considered great enough to outweigh the value of the protected habitats and birds. Moreover, two years after the circular the UK government's position was severely dented by a ruling in the European Court. In a case brought by the EC against the Federal Republic of Germany (supported by the UK), the Court allowed coastal protection work in the Leybucht SPA as *exceptional* (that is, there was a competing superior general interest, in this case flood protection). But, at the same time, the Court made it clear that *general* economic and recreational interests do not allow for removal or destruction of SPA land.

Consequently, the DoE produced a consultation paper on Planning Policy Guidance on Nature Conservation in February 1992 which suggested a further modest strengthening of planning controls in SSSIs, but mainly for NNRs, SPAs, SACs and Ramsar sites. The paper avoided stating a presumption against development in SSSIs even within such super-SSSIs, but in its final form issued in October 1994, it affirms that developments affecting SSSIs will be 'subject to special scrutiny'.

A major pillar in the government's defence of its refusal to strengthen further the planning system with regard to SSSIs is its implementation of another EC directive, the Environmental Assessment Directive (85/337). It is claimed that Environmental Assessment (EA) provides a set of procedures and tools for ensuring compliance with the terms of other directives such as the Birds Directive. This is the first directive 'concerned directly with the introduction of anticipatory decision-making procedures for a wide range of land-use related projects' (Sheate and Macrory 1989: 68). Section 7 of the directive defines EA as:

> a technique for the systematic compilation of expert quantitative analysis and qualitative assessment of a project's environmental effects, and the presentation of results in a way which enables the importance of the predicted effects, and the scope for modifying or mitigating them, to be properly evaluated by the relevant decision-making body before a planning application decision is rendered.

The directive identifies two classes of project – *Schedule 1* comprising major types of development (e.g. oil refineries or power stations) where the requirement for EA is mandatory, and *Schedule 2* where EA might be necessary depending on the scale and exact nature of the development proposed. Decisions on the categorisation of projects rest with Local Planning Authorities, not with any of the environmental protection agencies although they may be consulted. Agriculture is a Schedule 2 operation:

> Concerns about the environmental effects of agricultural intensification were reflected in early drafts of the Directive, and received high profile during its political passage. Yet by the time the Directive was finally agreed by the Council of Ministers, the process of inter-governmental negotiation and compromise had relegated agriculture to

one of the non-mandatory classes of project in Annex II of the Directive.

(Sheate and Macrory 1989: 70)

In the UK, the directive was implemented as a Statutory Instrument in 1988 – the Town and Country Planning (Assessment of Environmental Effects) Regulations. Agriculture operations subject to EA include water-management schemes, poultry and pig rearing, fish farming and reclamation of land from the sea. It is up to individual countries to set thresholds for such developments and in the UK the thresholds for EA affecting agriculture are new pig-rearing buildings housing more than 400 sows or 5,000 fattening pigs and new poultry buildings for more than 100,000 broilers or 50,000 hens (Selman 1992). The thresholds may be lower in designated or sensitive areas. 'However, the Secretary of State will not always deem that Schedule 2 projects affecting a sensitive area will require an EA, nor operate an automatic presumption against development within a "designated" area' (ibid.: 148). In much the same way as the UK government has suggested that the planning system is the main form of effective protection for special sites, so too it has suggested that EA will often not be needed because of the strength of the UK planning system. According to the DoE (Circular 15/88):

> the existing development control system is, in principle, already capable of ensuring that the Directive's objectives are met in most instances.... the tenor of the DoE ... is that an EA should be incorporated into existing planning control laws with as little additional red tape as possible.

(Mertz 1989: 484)

Consequently, agriculture is but marginally affected by EA and the system appears, contrary to initial expectations, to do relatively little to assist in the protection of significant habitats and landscapes. Indeed, the system appears to be inherently weak in so much as the need for EA appears to depend on the likely significance of the environmental impact, 'while one of the principal reasons for producing the EA is to determine the significance of the environmental impact' (ibid.: 489).

CONCLUSIONS

This chapter has illustrated the gathering pace of reform in the area of rural environmental protection, broadly defined. To an increasing extent these policies have begun to impinge directly on farmers. Thus the key agricultural policy actors (chiefly the NFU and MAFF) have played an important role in the policy process. As a consequence of their historically strong position, they have been able to exert pressure to resist some forms of policy development and to influence the details of other policies. The NFU and the MAFF increasingly sought opportunities to assert their centrality in the policy process, even to attempt to re-establish a closed policy community, through themselves becoming the promoters of environmental policy for agriculture.

By the mid-1980s, the NFU and the CLA had developed a coherent position on nature conservation. This was based around the assertion of new property rights, payment for the provision of environmental goods, and the desire for controls over farmers to be based, as far as possible, on the voluntary principle and self-regulation and, in the last resort, on MAFF.

For the conservationists, an increased engagement in the political process had brought its victories, but also an increasing awareness of some of the problems facing the environmental movement. The plethora of designations and government and other agencies concerned with conservation had led to piecemeal provision and the complete absence of a national land-use strategy (Winter 1990a, 1990b). And the emphasis on designation had tended to lessen the protection of undesignated areas, which prompted a number of authors (e.g. Mabey 1980; King and Clifford 1985) to make a passionate plea for the protection of the wider countryside local features, even if they were of no national scientific importance.

9

THE GREENING OF
AGRICULTURAL POLICY

INTRODUCTION

At the outset of the 1980s, environmental concerns were of relatively minor importance within mainstream agricultural policy. By the early 1990s the environment had become, at least in terms of policy rhetoric, one of the main policy planks on which the agriculture departments in the UK took their stance. Environmental policies increasingly impinged upon agriculture with a consequent opening of the policy community. This chapter examines the extent to which agricultural policy has 'greened' during the 1980s and early 1990s. The discussion is inevitably wide-ranging, looking at a range of policy initiatives. It is not enough to consider only mainstream agricultural commodity policies and politics. For there are many other instances of greening: initiatives to encourage environmentally friendly farming and new uses for agricultural land, the issues of conservation advice for farmers, and agricultural research. The final section provides an examination of the agri-environment package of the 1992 reform of the CAP. A specific issue, pollution, is too large a topic to deal with alongside these other case studies and is left to the next chapter.

THE NEW POLICY AGENDA

Evidence of the opening of the policy process requires evidence of increased involvement of other interests, in this case environmental pressure groups seeking to exert a direct influence upon agricultural policy making. Whilst there may be considerable debate amongst political scientists over the influence of select committees on the content of government policy, there can be little doubt that the list of expert witnesses and published evidence casts considerable light on shifts in policy debate and political agendas. For example, by the mid-1980s parliamentary select committees and environmental pressure groups had both latched on to the extent of environmental problems in agriculture and the significance of CAP reforms to the environment. Several inquiries were carried out which were either critical of the excesses of the CAP or of MAFF for its failure to reorientate itself in the light of environmental problems. Increasingly the two issues were linked and environmental pressure groups were drawn into the debate. For example, the House of Lords Select Committee on the European Communities, when it considered the CAP in 1980, drew evidence solely from farming and food organisations or individuals. But, when their lordships turned to the same topic in 1984–1985, the group of witnesses had expanded to include several environmental groups, such as CPRE, the Countryside Commission and the Institute for European Environmental Policy, although oral evidence continued to be the preserve of the actors in the agricultural policy community.

It may be open to debate as to whether groups such as the CPRE can really claim to have achieved insider status within the agricultural policy community but, certainly by the mid-1980s, the CPRE had achieved as high a profile as a lobbyist on farming issues as on other issues more traditionally within its purview, such as planning and landscape protection. By the end of the 1980s and early 1990s, it was sponsoring outside consultants to undertake a series of rigorous analyses of the drawbacks of the CAP and make firm recommendations for policy reform (Baldock and Beaufoy 1992; T.N. Jenkins 1990; Lawrence Gould Consultants Ltd 1989). Some of this work was undertaken jointly with the Worldwide Fund for Nature which, despite its public reputation for overseas concerns, had in fact been one of the first lobby groups to sponsor work of this kind on mainstream agricultural policy and its implications in Britain (Potter 1983).

Of particular importance to the CPRE's stance was its acceptance that agriculture should continue to be supported financially, but in ways which would encourage environmental protection rather than by only increasing still further agricultural production. The CPRE is by no means a radical organisation, either in terms of green or leftist tendencies or for any commitment to the radical liberalism of market Conservatism. It is, moreover, firmly based in the shire counties with strong links with the farming community. Its critique of agricultural policies, therefore, firmly rests upon adapting traditional interventionist policies to support farming. As far as the House of Lords Select Committee on the European Communities was concerned, the CPRE certainly seemed to be pushing at an open door. Whether their lordships' stance reflected a dissatisfaction with market Conservatism is open to conjecture, but certainly the 1984 Report on Agriculture and the Environment roundly condemned the Ministry of Agriculture for its narrow interpretation of certain EC policies and for its lukewarm commitment to conservation policies.

At first, the Ministry was reluctant to assume new responsibilities. Indeed for a time in the early 1980s it had to be encouraged and cajoled by the NFU to pick up the gauntlet of a new and expanding policy agenda. In a sense this appears to be the reverse of what might have been anticipated, given the account of agricultural policy developments described so far. The NFU, representing a beleaguered industry, might have been expected to have clung to the certainties of the past and avoided new policy experiments. By contrast, it would appear to have been in the interests of MAFF, as a government department, to act in a proprietary or even an imperial manner. In the case of MAFF, though, there had been concern for some time that its future was under threat. Thus, in the late 1970s, MAFF's jostling with the Department of the Environment over land drainage drew the following observation:

> There was considerable sensitivity within MAFF that if land drainage was lost, the rot would set in and other functions would be lost to other departments, for example ... health and safety, and education and training.... In theory a number of MAFF's functions could quite logically be transferred to other departments, thereby undermining the rationale for MAFF's existence.
>
> (Richardson *et al.* 1978: 54)

Thus MAFF's reluctance to press for new duties might be construed as a defensive posture in the light of threats to hive off other activities. Additional concern within the Ministry on the implications of the series of Rayner reviews of civil service departments only served to reinforce the uncertainty within the Ministry. The National Farmers' Union, fearing the loss of its special relationship with a central government department with a seat at Cabinet, publicly pressed the Ministry to take a more positive role.

This had less to do with farmers' commitment to environmental matters than with the NFU's determination to avoid ceding controls over agriculture to the DoE or any other conservation agency. If regulations were required then the NFU was determined that MAFF should be the agency to control matters, hopefully through the encouragement of self-regulation by the industry itself (Cox *et al.* 1986b). Another, and linked response was the farming lobby's assertion of new property rights, in terms of their emphasis upon the need for compensation payments if they were to suffer environmental restrictions.

However, as late as 1985 the government resisted attempts to insert changes in the Wildlife and Countryside Amendment Act to give the Ministry a statutory responsibility to further conservation, a much stronger injunction than its existing duty to take conservation considerations into account in carrying out its duties. Yet only one year later the Ministry was given just such a responsibility to promote conservation under the terms of the Agriculture Act 1986. This, of course, was a bill sponsored and promoted by the Ministry itself. The change of mood reflected a growing confidence within the Ministry that an integration of environmental concerns within its remit could serve to counter threats both to itself and to agricultural support policies within the EEC. In its efforts to limit the cost of agricultural support, particularly the horrendous cost of storing surplus produce, the UK government had turned to environmental conservation as a politically acceptable policy for the support of agriculture. The process had, in fact, begun as early as the autumn of 1984 when the UK successfully moved amendments to Article 19 of the new EC Structures Regulation, which had emerged from the ten-year review of the structures policies in 1983–1984.

These amendments were designed to enable agricultural departments to designate areas where '*the maintenance or adoption of particular agricultural methods is likely to facilitate the conservation, enhancement or protection of the nature conservation, amenity or archaeological and historic interest of an area*' (EC Regulation 797/85), and to give financial incentives to encourage appropriate farming practices in what came to be known as Environmentally Sensitive Areas (ESAs).

Once under way, the greening of the Ministry, at least as far as policy rhetoric was concerned, proceeded rapidly. By the early 1990s the picture had changed dramatically, with virtually every aspect of agricultural policy recast to reflect, however modestly, environmental concerns. This does not necessarily mean, as we shall see later in the chapter, that the changes have satisfied the desires of environmentalists – far from it – but the change in the terms of the debate has been dramatic. In terms of spending, the Ministry rapidly overtook the conservation agencies in the amount of its expenditure devoted to conservation payments to non-public landowners and occupiers, as shown in Figure 9.1. It should be noted that the position of the Forestry Commission is somewhat misleading, as the figure includes all afforestation spending, including upland afforestation.

In addition to environmental matters, the Ministry has also increasingly embraced a wider role with regard to issues such as food safety and animal welfare. Thus its 1993 departmental report identified five main policy areas which reflect a programme much broader than the support of the farming economy *per se*:

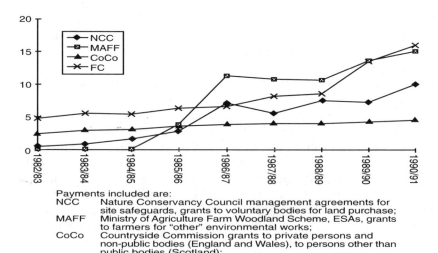

Payments included are:
NCC   Nature Conservancy Council management agreements for site safeguards, grants to voluntary bodies for land purchase;
MAFF   Ministry of Agriculture Farm Woodland Scheme, ESAs, grants to farmers for "other" environmental works;
CoCo   Countryside Commission grants to private persons and non-public bodies (England and Wales), to persons other than public bodies (Scotland);
FC   Forestry Authority grants to private woodland owners.

Figure 9.1 Conservation payments to non-public landowners and occupiers

Source: Hodge *et al.* 1994: 205

To protect the public
- by promoting food safety
- by taking action against diseases transmissible to man
- by planning to safeguard essential supplies in an emergency.

To alleviate flooding and coastal erosion
- by assisting and encouraging the building of flood defences and the provision of flood warning.

To enhance the rural and marine environment
- by encouraging action to reduce water pollution
- by encouraging positive environmental measures
- by making the countryside more attractive
- by protecting the rural economy, particularly in less favoured areas.

To promote a fair and competitive economy
- by seeking a CAP which enables UK producers to compete fairly with those elsewhere, and gives better value for money
- by creating the conditions in which efficient agriculture, fisheries and food industries can flourish
- by taking action against animal and plant diseases and pests
- by adopting measures to conserve fish stocks
- by providing chargeable services.

To protect farm animals
- by encouraging high welfare standards.

(Cmnd 2203, 1993: 1–2)

Of course, the mere statement of aims in this way tells us very little about the distribution of resources within the Ministry, and one of the aims of the remainder of this chapter is to examine the extent to which the policy rhetoric has been matched by a genuine shift of resources and a reorientation of policy priorities.

ENVIRONMENTALLY SENSITIVE FARMING AND NEW USES FOR LAND

The close link between the surplus crisis of the CAP and environmental concern is symbolised by the ESA policy which has become very much the flagship of the agriculture department's engagement with environmental issues, so much so that, by the close of 1994, 15 per cent of the agricultural area of the UK was covered by designated ESAs (although that does not mean that all farmland within ESAs is entered into the scheme). The budget for ESAs is expected to rise to £43 million in 1995–1996, making it easily the single most important aspect of agri-environmental policy in the panoply of policies promoted by the agriculture departments and countryside agencies.

ESAs have been heralded as an integral element in the new-look agricultural policy, reflecting the need to cut surplus production while still maintaining farm incomes. At the same time they reflect a widely held view that neither designations of special sites nor conservation advice alone can prompt the whole-farm management required to preserve an environmentally beneficial style of farming. Thus,

> ESAs signal a recognition that specific tracts or pieces of countryside can often only be effectively conserved by maintaining the traditional systems and styles of farming which lie behind them. According to the enabling legislation, ESAs are to be designated by agriculture departments of the EEC member states where 'the maintenance or adoption of particular agricultural methods is likely to facilitate [the] conservation, enhancement or protection of the nature conservation, amenity or archaeological and historic interest of an area'. This indicates an important shift towards husbandry as well as simply investment and project-based solutions to conflicts between agriculture and the environment.... ESAs mean that farm support moneys are being used for the first time to maintain what might justifiably be called 'environmentally sensitive farmers'.

> (Potter 1988: 301)

The criteria used by MAFF for the designation of ESAs reflect the recognition that particular farming regimes may be essential for environmental protection:

- They should be of national environmental significance.
- Their conservation must depend on the adoption, maintenance or extension of a particular form of farming practice.
- They should be areas where encouragement of traditional farming practices would help to prevent damage to the environment.
- They should comprise a discrete and coherent unit of environmental interest.

The distinction between landscape protection and nature conservation, so characteristic of the British approach to countryside management, is blurred by describing the new areas as *of national environmental significance*.

Following enabling legislation (the Agriculture Act 1986), the first eight ESAs were designated by MAFF in 1987 (Broads, Pennine Dales, Somerset Levels, South Downs (East), West Penwith, Cambrian Mountains, Breadalbane and Loch Lomond). These were selected from a list of fourteen prepared by the Countryside Commission, the NCC and English Heritage. More ESAs were designated in subsequent years, including extension of the scheme to Northern Ireland. The location of the UK ESAs is shown in Figure 9.2.

ESAs cut across existing conservation and landscape designations, so that nearly all include some SSSIs within their boundaries and many are within AONBs. Some of them are partly or wholly within national parks. ESA farmers have the option to enter into agreements whereby certain practices

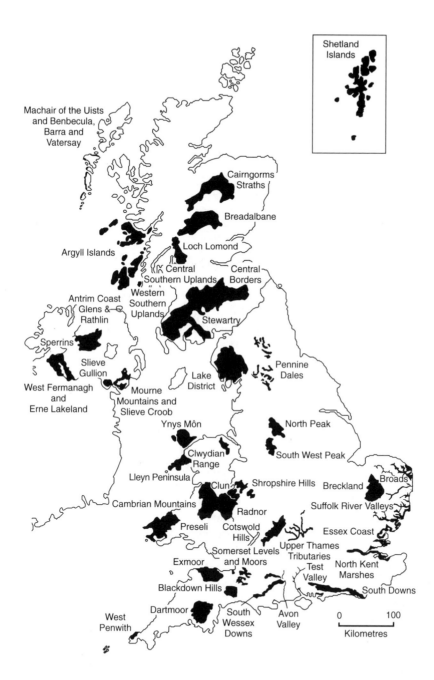

Figure 9.2 UK ESAs, 1994 (Crown copyright)

are undertaken or forgone in return for annual payments on an acreage basis. The take-up of the scheme has, in most cases, been high, for although the payments are not great the scheme is flexible and in many cases farmers are happy to receive payments for continuing to farm in a traditional manner. Although production levels cannot be directly controlled, limits on fertiliser applications or the extent of arable land taken into cultivation effectively reduce possible increases in production. As in other initiatives the voluntary principle is central; this can be seen as a problem as the scheme is less likely to be adopted by farmers within an ESA with higher existing or potential levels of production. Most ESA agreements include:

- restrictions on fertiliser use and stock densities;
- prohibitions on use of herbicides and pesticides and the installation of new drainage or fencing;
- maintenance of landscape features such as hedges, ditches, woods, walls and barns.

Whilst day-to-day farm management procedures are not subject to scrutiny, it would be a mistake to suggest, as have a few commentators, that the designation makes scant difference to farmers. Indeed, the fact that some schemes have been more successful than others would suggest this to be the case. For example while take-up in the Broads has covered 89 per cent of eligible land, in the Test Valley it has only been 27 per cent (Colman 1994). The Dartmoor ESA, designated in 1994, has encountered strong resistance from many farmers and take-up has been slow.

Colman (1994; Colman and Lee 1988) has argued that too little considera- tion has been given to whether other methods of environmental protection might not have been more cost-effective than ESAs in some instances. For example, in the Broads a combination of site purchase and management agreement would have been equally effective and cheaper than the ESA policy (Colman 1994). There is considerable variation between what is expected of farmers in different ESAs. This is reflected in the use of the payments made. In some cases, much of the money flows directly into farmers' pockets. For example, it is estimated that in the Suffolk River Valleys ESA as much as 61 per cent of the moneys paid out between 1988 and 1992 boosted farm incomes directly (Colman 1994). Such figures help to explain the enthusiastic support for ESAs from the NFU. This is indeed a successful translation of environmental protection into property rights. By contrast, though, the South Downs ESA imposes many management duties on farmers which cost money, so that only 30 per cent of payments received is used to bolster incomes (Colman 1994). The situation is complex – not surprisingly, given the beguiling simplicity of a single scheme (notwithstanding that prescriptions and payments vary from one ESA to another) adopted in such widely contrasting situations:

It is perhaps understandable, given the sort of estimates presented above, that a criticism which arose of many of the ESA management prescriptions was that they demanded too little change and action from farmers, that there were few costs, and that farmers were being paid to do what they would have done in any case. This was not true of all ESAs and all tiers of prescription, but it certainly was the case that the farming restrictions imposed in many of the lower tier contracts appear to have been designed to demand the minimum of change.

(Colman 1994: 241)

Another concern with ESAs is the so-called *halo effect*, a concern which also applies to set-aside. It is suggested that a natural response of farmers with an ESA agreement, is to seek means of intensifying production elsewhere. Whilst the evidence for this is far from conclusive in many evaluations of ESAs (Whitby 1994), it has certainly emerged strongly in one study. In the Cambrian Mountain ESA, it was found that the majority of farmers who had reduced their stocking of rough grazing land in compliance with the scheme had compensated by intensifying grazing elsewhere and/or renting additional land: *only 17 per cent of farmers with semi-natural rough grazing had to reduce their overall stocking rate as a result of joining the scheme* (Hughes 1994: 141).

In addition to the obvious advantages to farmers of an administratively relatively straightforward scheme with tolerably generous payments, ESAs also provide an example of a policy initiative acceptable to the farming lobby for political reasons. ESAs are in the style of the permissive corporatism of the post-war era, whereby the partnership between state and producers is based on markets and incentives rather than on controls and regulations (Baldock *et al.* 1990; Cox, Lowe and Winter 1988):

where control has been necessary within agriculture, the farming and landowning community has been anxious both that it take a permissive form and that the control remain within the industry. In terms of such long-cherished priorities the attractions of the ESA arrangements are clear enough.... Hence, whilst SSSI designation and its associated mechanism of the management agreement is an often protracted and highly bureaucratic procedure, entailing elements of compulsion and administration by a conservation agency and including the possibility of criminal liability, the ESA scheme is voluntary and administered by MAFF through the advisory service with which farmers are familiar and at ease.

(Whitby and Lowe 1994: 19–20)

Countryside stewardship

If the ESA initiative represents the flagship of agri-environmental schemes, there are several others in the fleet. One such is the Countryside Stewardship Scheme (CSS). One of the problems with ESAs identified by the Countryside Commission in its own deliberations during the formative stages of the programme was the somewhat sweeping and generalist approach adopted for

Table 9.1 Take-up of Countryside Stewardship 1991–1995

	Hectares
Chalk and limestone grassland	
• management and improvement of calcareous grassland	15,232
• regeneration of chalk and limestone grassland on cultivated land	2,467
Coastal land	
• management and improvement of salt marsh	3,328
• management and improvement of cliff tops, sand dunes and coastal grazing marsh	3,614
• re-creation of coastal vegetation on cultivated land	1,032
• restoration/regeneration of reed beds/fen/carr	28
Historic landscapes	
• management and protection of historic landscapes	7,737
• restoration of cultivated land to pasture in historic landscapes	1,410
• restoration and management of old orchards	526
Lowland heath	
• management and improvement of existing heath	9,888
• re-creation of heathland	540
Old meadows and pastures	
• management of old meadow and pasture	1,812
Uplands	
• regeneration of suppressed heather	12,191
• regeneration of heather moor on agriculturally improved land	652
• restoration/management of hay meadows and pastures	11,032
• restoration/regeneration of reed beds/fen/carr	24
Waterside landscapes	
• restoration and management of waterside land	17,273
• re-creation of waterside landscapes on cultivated land	2,674
• restoration/regeneration of reed beds/fen/carr	289
Hedgerow landscapes	
• pastures conserved and improved as part of hedgerow restoration programme	188
TOTAL AREA	91,937
Access	
• open access	13,517
• linear access (length)	397 km
Hedgerow restoration	
• hedgerow restoration (length)	3,031 km
• hedgebank restoration (length)	130 km
Traditional walls	
• wall restoration (length)	217 km
Field margins	
• grass margins created (length)	1,205 km

Source: MAFF and DoE 1995: 16–17

ESAs. They were designed to be 'user-friendly' with relatively straightforward management prescriptions for the purposes of whole-farm planning. The Commission felt strongly that there was a need for another scheme, similar to the ESA scheme in that it identified farming as a key management tool, but different insomuch as the sites would be smaller and more coherent with the management prescriptions correspondingly more detailed and demanding. The CSS (Tir Cymen in Wales is broadly similar) was launched by the Commission in 1991 for a five-year experimental period. Funded by the DoE, it was administered by the Commission in partnership with English Nature and English Heritage. In 1994 it was announced that MAFF would be assuming responsibility for its implementation. At the same time it was announced that another scheme, the Hedgerow Incentive Scheme, would be merged with the stewardship scheme. It is forecast that when the pilot ends in March 1996, 5,200 schemes will be in place at a total cost of £11.7 million per annum (MAFF and DoE 1995).

The CSS targets distinct categories of landscape and the take-up for each is shown in Table 9.1. A complex and wide-ranging menu was devised to meet all possible requirements and to ensure that management could secure precisely defined environmental targets (as shown in Table 9.2).

Farm forestry and alternative land uses

In addition to the ESA policy, attention was also focused in the mid-1980s on new uses for land, in particular forestry. The relatively small proportion of the UK under woodland coupled with the post-war decline in the extent and quality of ancient woodland in the lowlands began to excite much concern at this time (Rackham 1980; Marren 1990; Peterken 1981; Watkins 1990). Meanwhile in the uplands, coniferous afforestation was increasingly being lambasted for its impact on the number of sites of value to birds of prey as well as its impact on the openness and wide vistas of the familiar moorland landscape, as will be discussed in chapter 11.

Thus at the same time as the Forestry Commission came under pressure to review its upland policies (MacEwen and MacEwen 1982) it also undertook a review of its broadleaved woodland policy resulting in the first grant scheme designed specifically to encourage the establishment and management of deciduous woodland (Watkins 1986). The Broadleaved Woodland Scheme has now been merged with other grant schemes to form the Woodland Grant Scheme, but grants for deciduous planting remain set at a higher level. Although the impetus behind this move, which departed from established policy and silvicultural norms, was undoubtedly prompted by environmental considerations, it rapidly became apparent that the government also saw lowland forestry as having a potential contribution to make to utilising land that might otherwise be producing agricultural surpluses. This led, in 1987, to a Farm Woodland Scheme from the Ministry itself, this time offering annual management payments to farmers prepared to divert land from agricultural production to woodland. In the case of both the Forestry Commission and MAFF schemes, it is important to recognise that environmental objectives formed only one element in the thinking behind the schemes. Moreover the schemes were promulgated by a government committed to free enterprise and reluctant to impose onerous conditions upon

Table 9.2 Summary of annual payments in the Countryside Stewardship Scheme

Annual payments	Payments
Chalk and limestone grassland	
conservation	£50/ha
creation	£210/ha
Lowland heath	
base payment	£20/ha
quality improvement measures	£30/ha
recreation on improved land	£250/ha
Waterside landscapes	
conservation	£70/ha
creation or conservation	£225/ha
Coastal land	
conservation of salt marsh	£20/ha
cons. of coastal grazing marsh	£70/ha
cons. of coastal veg. on cliff tops and sand dunes	£50/ha
establishment of coastal veg. on improved land	£225/ha
Uplands	
regeneration of heather on enclosed moorland	£15/ha
plus £50 for first five years	
regeneration of heather on agriculturally improved land	£50/ha
plus £50 for first five years	
restoration and management of:	
– hay meadow	£80/ha
– additional payment of small hay meadow of 5ha or less	£50/ha
– in-bye pasture	£50/ha
– intake/allotment rough pasture	£20/ha
Historic landscapes	
conservation of existing features	£70/ha
restoration of land to permanent pasture or other vegetation	£225/ha
restoration of old orchards	£250/ha
Old meadow and pasture	
conservation and management	£70/ha

Capital payments		Payments
Scrub management		
scattered scrub under 25%		£100/ha
scrub between 25–75%		£250/ha
scrub more than 75%		£500/ha
Pond restoration	first 100m²	£2/m²
	thereafter	£0.50/m²
Pond creation	first 100m²	£3/m²
	thereafter	£0.50/m²
Creation of scrapes		£1.25/m²
Sluice for water-level control		
	soil bund	£40
	timber	£140
	brick, stone or concrete	£400
Stone wall repair		£7.50/m
	restoration	£15
	step-over-stile	£20
	step-through-stile	£30
Top-wiring stone wall		£0.60/m
Stone-faced hedge banks	repair	£10/m
	restoration	£25/m
Earth banks restoration		£3/m
Tree and shrub planting		
	whips and transplants	£0.65/plant
	maiden fruit tree in old orchard	£3/tree
	standard fruit tree in old orchard	£7/tree
	standard parkland/hedgerow tree	£6/tree
Tree guards	rabbit guards	£0.20/guard
	tube	£0.50/tube
	parkland	£30/guard
	orchard	£1.50/guard
Coppicing bankside trees		£12.50/tree
Pollarding		£17.50/tree
Tree surgery/major pruning		£40/tree
Frameworking of old fruit trees		£30/tree
Fruit tree pruning and restoration		£8/tree

Access

– on land available for public access	£50/ha
– for creation of new permissive paths, bridleways and paths for the disabled	
agreement plus paths	£100/
bridleways	£0.10/m
paths for disabled	£0.20/m
– on land used by educational establishment	£25/ha

Supplements

– for initial restoration or re-creation works on chalk and limestone grassland, waterside landscapes, coastal land or pasture and meadows	£40 first-year payment
– for initial work needed to conserve historical features	£40 for up to five years
– for special heather regeneration works	£50 for first five years

Hedge planting	£1.75/m
Hedge laying	£2/m
Hedge coppicing	£1.50/m
plus supplement for laying or coppicing hedges over 1.5m wide at base	£0.50/m
supplement for wire and post removal to aid laying or coppicing	£0.50/m
Bracken control mechanical	£50/ha
chemical	£85/ha
Clearance of eyesores	£120
Field gate	£125
Bridle gate	£100
Kissing gate	£130
Kissing gate for disabled	£200
Stile	£30
Ladder stile	£40
River gate	£125
Footbridge	£125
Culvert	£40/m
Ditch, dyke, rhyne restoration	£2/m
Post for sign or waymark	£4
Sign with map of agreement land	£20
Interpretation board	£80
Bench	£30
Car parks/hardstanding	£5/m²
Construction of path for the disabled	£7.50/m
Advice/application preparation	£100
Management plan preparation	£300

The following payments are available only for work that is essential to achieve good env. management

Fencing post and wire	£0.80/m
sheep fencing	£1.20/m
rabbit or sheep netting	£0.40/m
(For historic/deer parks only):	
Deer fencing	£3.50/m
Continuous iron-rail fencing	£6/m
Water supply	£0.40/m
trough	£25

businesses in receipt of government aid. Thus under both woodland schemes considerable freedom was given to recipients of grant to determine their own aims and objectives on a continuum from fully commercial hardwood planting to amenity planting. No attempt was made to determine the appropriate balance between different types of planting at national, regional or farm level. Therefore a scheme publicised by government as an example of its commitment to conservation could, in practice, result in monocultures

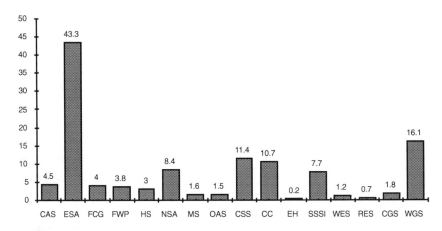

MAFF	CAS	Countryside Access Scheme
	ESA	Environmentally Sensitive Area
	FCG	Farm and Conservation Grant Scheme (conservation grants)
	FWP	Farm Woodland Premium Scheme
	HS	Habitat Scheme
	NSA	Nitrate Sensitive Area
	MS	Moorland Scheme
	OAS	Organic Aid Scheme
Countryside Commission	CSS	Countryside Stewardship Scheme
	CC	Other Grants
English Heritage	EH	Management Agreements
English Nature	SSSI	SSSI Management
	WES	Wildlife Enhancement Scheme
	RES	Reserves Enhancement Scheme
	CGS	Conservation Grants Scheme
Forestry Commission	WGS	Woodland Grant Scheme

Figure 9.3 Agri-environmental schemes in England: forecast payments in 1995/96 (£ million)

Source: MAFF and DoE 1995

of fast-growing species such as southern beech or sycamore which some ecologists consider to be as detrimental to woodland ecology as coniferous species. Indeed, coniferous planting is also acceptable. It has to be said that in both cases the Forestry Commission is obliged to consult local authorities and, in certain instances, other conservation agencies, on the desirability of the proposed planting schemes. This may result in modifications to some schemes according to environmental criteria but as each scheme is considered on its merits and no attempt is made to co-ordinate responses between authorities, no sense of overall environmental planning emerges or, indeed, is intended. There is a continuing risk that recreation and environment objectives may be sacrificed to commercial criteria.

Before leaving this section, it is worth remembering that there are several other schemes which have contributed to the greening of agriculture. A full list and the costs of the programmes are given in Figure 9.3.

CONSERVATION ADVICE AND TRAINING FOR FARMERS

One of the pillars of the post-war agricultural settlement was the importance attached to a free advisory service available to farmers. However in 1987 this came to an abrupt end. Under the terms of the Agriculture Act 1986, ADAS was empowered to charge for its provision of advice and the government made it very clear that its future would depend on adopting an aggressively commercial strategy. The government's reasons for adopting such a course were straightforward enough. It wished to cut public expenditure and agricultural advice seemed an excellent candidate at a time of surplus production, when it was clearly no longer in the public interest for there to be a state-sponsored advisory service showing farmers how to increase production.

But matters were not quite so simple. Another section of the 1986 Act dealt with the establishment of ESAs, plans for the Farm Diversification Grant scheme were already well advanced, and agricultural pollution was causing increasing concern. Farmers were facing pressures to make profound adjustments to their businesses, adjustments in large measure necessitated or promoted by government actions. On that count it seemed a profoundly inappropriate time to cut back on the modest expenditure incurred in funding ADAS. Under growing pressure from the environmental lobby, particularly the CPRE, during the passage of the Bill, the Minister gave promises that free advisory provision for 'public good' advice, taken to include diversification, conservation and pollution, would be retained after the implementation of the legislation.

ADAS has offered free conservation advice but the amount available has been restricted by financial considerations and there have been accusations that some farmers have been turned away (for example in unpublished evidence to the National Audit Office provided by CPRE in 1990). In 1992 ADAS became a 'next steps' agency, possibly a precursor to full-scale privatisation, and consequently faced even greater pressures to commercialise its activities. Since 1992 the amount of free conservation advice on offer has been determined and paid for by the Ministry which sub-contracts the work to ADAS. In addition the Ministry part-funds the Farming and Wildlife Advisory Group (FWAG) for an additional element of free conservation advice. During the debates in 1986, the Ministry reiterated its commitment to FWAG, a routine response to environmentalist criticisms since the passage of the Wildlife and Countryside Act 1981. One of the consequences of the wider debate engendered by the passage of the 1981 Bill was an increased awareness of the importance of the wider farmed countryside. The main policy response to this concern for areas outside those designated for special protection was the provision of conservation advice and small-scale grants for farmers and landowners throughout the countryside. The most obvious vehicle for advice was FWAG.

Formed in 1970, FWAG had initially been a national forum for leading farmers, agriculturalists and conservationists to exchange views. Through the promotion of farming and conservation 'exercises' on farms, the group sought to encourage farming practices that would accommodate conservation interests. County groups were formed in the mid- to late 1970s comprising interested farmers, MAFF officials and representatives from the voluntary and public conservation groups. A number of counties began to offer advice to farmers and from 1984 many extended this activity by taking on full-time specialist officers, with grant aid from the Countryside Commission, the Farming and Wildlife Trust (a national charity) and local subscriptions (Cox, Lowe and Winter 1990c). Nearly all counties in England and Scotland now have a FWAG adviser available to advise farmers on all aspects of conservation and to direct farmers to other specialist sources of advice or grant aid where necessary. The distribution of advice remains uneven because some counties are so much larger than others – as seen in Figures 9.4 and 9.5, which show FWAG representation in the UK in April 1994. Despite the rhetoric surrounding advisory provision, with frequent government avowals of its importance, the resources are clearly stretched and universal free conservation advice provision is a distant goal. The situation is more serious in some parts of the country than others. FWAG's reliance on effective voluntary local organisation and, indeed, a continuing element of local funding has meant that, inevitably, some counties are better served than others. Thus Berkshire and Oxfordshire FWAG now employs four advisers; whereas larger counties such as Hereford and Worcester have to be content with one adviser for the entire county. In Wales and Northern Ireland FWAG barely exists and the same applies in a small minority of English counties. FWAG is in receipt of significant sums of money nationally from the DoE and MAFF to support part of the costs of its advisory provision (£370,000 p.a. in 1993/4), but the deployment of those funds continues to depend on a policy of shared funding at the local level.

FWAG is a blend of public and private initiative with leading figures in the farming community heavily involved. It is thus 'acceptable' to many farmers

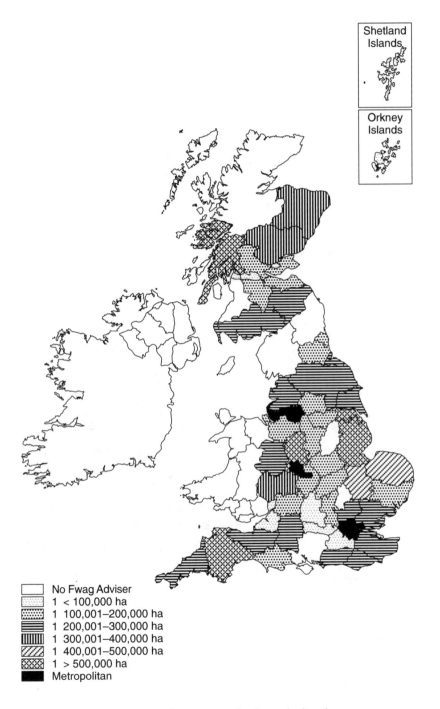

Shetland
Islands

Orkney
Islands

	No Fwag Adviser
	1 < 100,000 ha
	1 100,001–200,000 ha
	1 200,001–300,000 ha
	1 300,001–400,000 ha
	1 400,001–500,000 ha
	1 > 500,000 ha
	Metropolitan

Figure 9.4 FWAG representation by agricultural area

Source: Winter 1995

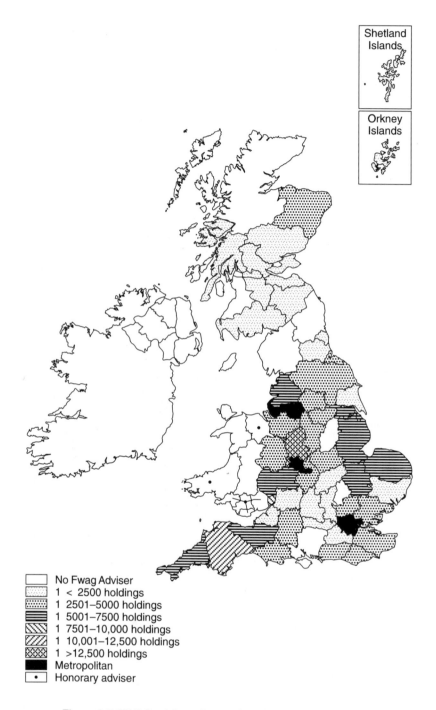

Shetland
Islands

Orkney
Islands

☐ No Fwag Adviser
▦ 1 < 2500 holdings
▥ 1 2501–5000 holdings
▤ 1 5001–7500 holdings
▨ 1 7501–10,000 holdings
▧ 1 10,001–12,500 holdings
▩ 1 >12,500 holdings
■ Metropolitan
• Honorary adviser

Figure 9.5 FWAG advisers by number of holdings per county

Source: Winter 1995

in a way that official conservation agencies might not be. It is based firmly on the principles of voluntarism and compromise. Farmers are encouraged to seek its advice and much emphasis is placed on the agricultural knowledge and sympathies of those giving advice. Consequently a compromise between the needs of modern agriculture and conservation is stressed, which may in practice lead to purely cosmetic changes. FWAG's capacity to provide advice is very small with one adviser covering an entire county, and it has tended to concentrate on wildlife habitats and new features with relatively less regard for pollution and landscape. Still less has it played a role with regard to recreation and access issues. Consequently its relationship with the Countryside Commission (with its twin concerns for landscape and recreation), whose financial assistance was crucial to FWAG's survival in the late 1970s and early 1980s, was somewhat strained at times.

Some would argue that the public–private mix has been unwieldy and insufficiently responsive to national agricultural and environmental policy initiatives. Indeed, although FWAG has scrupulously avoided political debate, its very existence in some measure trespasses on the territory of other public agencies with a conservation remit. This has been the case especially since MAFF assumed greater conservation responsibilities under the Agriculture Act 1986. For all the show of co-operation between ADAS and FWAG, there is a rivalry not far from the surface.

Provision of advice

In 1993/94, FWAG provided 34.8 person-years of advice to the farming community in England, a 40 per cent increase from 1990. This contrasts with 13 staff years of free conservation advisory provision from ADAS (1992/93), although it should be noted that ADAS devoted a further 56 staff years to free pollution advice. In terms of farm visits this translates into 1,400 annual visits by ADAS and 3,500 visits by FWAG. Thus the norm of 100 farm visits for an adviser each year established in the mid-1980s still holds true (Cox, Lowe and Winter 1990c).

There are approximately 120,000 agricultural holdings in England so between them ADAS and FWAG would appear to cover 4 per cent of all holdings each year. Perhaps one half of these represent return visits so that the rate of new contacts per year is perhaps between 1 per cent and 2 per cent of all holdings, with maybe an additional 5–10 per cent of farmers having received advice as a consequence of occupying designated land, particularly within ESAs. Others will have received advice from local and other miscellaneous agencies, although almost invariably this is rather narrowly circumscribed advice according to the objectives of the agency concerned. For example, local authorities have achieved a fairly deep penetration of the farming community in some areas, but almost entirely as a result of landscape enhancement schemes, especially new tree planting. Much of the funding from the Countryside Commission for many of these schemes has been withdrawn. In a recent unpublished telephone survey of 678 farmers in Great

Table 9.3 Free conservation advice provision in the United Kingdom, 1993/94

	Total number of holdings	Total agricultural area hectares	ADAS Scottish Agricultural College (SAC) Department of Agriculture Northern Ireland (DANI)		FWAG		ADAS/SAC/DANI + FWAG				
			Adviser years	No. of holdings per adviser year	Adviser years	No. of holdings per adviser year	Total no. of holdings per adviser year	Total no. of hectares per adviser year	Adjusted total no. of adviser years	Adjusted total no. of holdings per adviser year	Adjusted total no. of hectares per adviser year
England	151,718	9,334,534	13	11,670	35	4,335	3,160	194,469	187	811	49,917
Northern Ireland[1]	28,775	1,006,632	5.5	5,232	0	0	5,232	183,024	9.5	3,028	105,961
Scotland	32,092	5,273,934	5	6,418	16	2,005	1,528	251,140	30.5	1,052	172,915
Wales[2]	29,903	1,490,322	6.2	4,823	0	0	4,823	240,375	49.2	608	30,291
UK	242,488	17,105,422	29.7	8,165	51	4,755	3,005	211,963	276.2	878	61,931

[1]Data refer to 1992/93
[2]FWAG has been treated as nil despite the fact that a small quantity of advice is provided by voluntary FWAG advisers
Source: Winter 1995: 89

Table 9.4 Number of trainees participating in selected ATB–Landbase agri-environmental courses, April 1993 to March 1994, Great Britain[1]

Course Title[2]	North England	Central England	South East England	South West England	Wales	Scotland
Conservation[3]	10	125	12	6	68	6
Conservation grants[4]	21				30	
Drystone walling	56			20	4	10
Fogging, misting and smokes			6			
Footpath conservation	8					
Hedge management[5]	140	44	43	105	74	
Ponds[6]		16	6	25		
Woodland management[7]	45	88	28	67	28	20
Soils[8]			20			
Region totals	280	273	115	223	204	36

[1]The definition of agri-environmental for training purposes is not easy and two important caveats need to be made about these statistics. First, we have no figures on pesticide or livestock pollution training, but these would probably be higher due to the issue of compliance with regulations. Second, there are other practical courses (e.g. on chainsaws) which have implications for conservation management but are not primarily conservation oriented. These have not been included in the analysis.

[2]Some courses on similar topics have been amalgamated to simplify the presentation. It should be noted though that if two courses occur within the same region, it is possible that the same farmer may have attended both. So there may be a small element of double counting in these figures.

[3]Includes the following courses: Conservation (North, Central, South East, South West, Wales, Scotland), Conservation Environmental Research by Institute of Hydrology (Wales), Looking at Environmental Conservation (Wales), Training Involved with Countryside (Wales).

[4]Includes the following courses: Conservation Grants (North), Conservation Grants for Farmers (North), Conservation Training Courses and Grants (Wales).

[5]Includes the following courses: Hedgelaying (6 courses – North), Hedgelaying (4 courses – Central), Hedge Management (South East), Hedgelaying (2 courses – South East), Hedgelaying (4 courses – South West), Hedgelaying (5 courses – Wales).

[6]Includes the following courses: Pond Creation and Management (Central), Pond Management (Central), Ponds (South East), Pond and Waterway Development and Management (South West), Pond Conservation (South West), Pond/Stream/Lake Conservation and Management (South West).

[7]Includes the following courses: After Care of Trees (Central), Beating up and Tree Identification (North), Coppice Management (South West), Hazel Hurdle Making (South East), Hedge and Tree Planting and Aftercare (Central), Planting, Looking at Derelict Hedgerow and Oak Woods (Wales), Practical Woodland Management (South East), Restoring Neglected Woodland (Central), The Woodland Scene (Central), Tree and Hedge Planting (Central, South West), Tree Identification (North), Tree Planting (South East, Wales, Scotland), Woodland (North, Central, South East), Woodland Management (North, Central, South West), Woodland Marketing (South West).

[8]Includes the following courses: Results of Eight Years' Experiments on Soil Structure (South East), The Quality of Soil and Future Cropping (South East).

Source: Winter 1995: data provided by ATB–Landbase

Table 9.5 Number of trainees participating in DANI agri-environmental courses in Northern Ireland

	1992/93		1993/94	
	Number of courses	*Number of trainees*	*Number of courses*	*Number of trainees*
Pesticides	136	969	80	615
Hedge/Tree Planting	14	206	9	120
Hedge Laying/Coppicing	−[1]	0	3	16
Farm Estate Maintenance[2]	12	117	12	134
Farm Conservation	3	34	−[3]	−
Organic Farming	14	120	−[4]	−
Vermin Control	1	30	13	215

[1]Eight courses were advertised but insufficient applicants.
[2]Includes pollution prevention.
[3]Several courses advertised but insufficient applicants.
[4]Several courses advertised but insufficient applicants.
Source: Winter 1995: data provided by Department of Agriculture Northern Ireland

Britain, undertaken for ATB–Landbase, it was discovered that 13 per cent had received an advice visit from FWAG and over 20 per cent from ADAS (including pollution advice).

An attempt to estimate the total provision available is given in Table 9.3 (for a full discussion of the data in this table see Winter 1995). The contrast between the four countries is striking, with Northern Ireland and Wales considerably less well covered than England and Scotland in terms of coverage of holdings, and Scotland and Wales the worse off in terms of area coverage.

If the advice situation is serious that of training is even worse. The ATB–Landbase survey revealed that only 3 per cent of farmers had undertaken any environmental training in the previous five years. Tables 9.4 and 9.5 provide evidence of the extraordinarily low levels of participation in environmental training in both Britain and Northern Ireland. Out of nearly 250,000 farmers in the UK just a few hundred seem to avail themselves of agri-environmental training provision each year, and yet as shown earlier in the chapter many millions of pounds are now being provided for farmers to manage their land in environmentally friendly ways.

The effectiveness of advice provision

It is important to consider the effectiveness of the advice offered to individual farmers in safeguarding and/or improving environmental quality on their farms. In a survey sponsored by the Countryside Commission in 1988 (Centre for Rural Studies 1990), combining an interview survey of farmers with field surveys to assess the effectiveness of advice implemented on the farms, the conclusions were generally favourable to the advisory agencies covered (4 county FWAGs, 1 national park and 1 county council). Advice was taken up to some extent by two thirds of the farmers interviewed and nearly one half had implemented all the advice received. The quality of the advice given was consistently high with a good level of detail. Field evidence showed that, on the whole, it was well implemented. However there were some problems, as some quotes from the summary of main findings reveal:

> Advisers were aware of the value of bringing existing conservation features into management, in relation to the creation of new ones.
> Their efforts in this direction were constrained both by the grant-aid available and by farmers' attitudes.
> The advisers' ability to adopt a whole-farm approach to conservation was hampered by the feature-orientated nature both of most grant-aid and of most farmers' thinking.
>
> (Centre for Rural Studies 1990: 79)

Following the CRS study, the DoE commissioned PIEDA to undertake a further study (PIEDA 1993). A telephone interview survey of 650 randomly selected farmers was undertaken. Of those in receipt of advice, 64 per cent had implemented it and 82 per cent had either implemented or intended to do so. Of those who had implemented advice, 51 per cent of farmers said they would not have done so or would have proceeded in more limited way without advice (so-called Quantitative Additionality).

The question of effectiveness, with regard to the balance between whole-farm advice and the creation or management of particular features, remains an important issue requiring further work. The most recent evidence available from FWAG (for the first three quarters of 1993) suggests that the whole-farm approach is adopted in the advice given in 62 per cent of new visits and 53 per cent of all contacts (see Table 9.6). Moreover, some long-standing clients of FWAG will have received whole-farm advice in the past and are now merely checking on particular aspects of implementation. This is encouraging, although farmers in receipt of such advice may or may not implement it in the manner intended. However, recent work by the author for MAFF suggests that FWAG operates with a very general view of whole-farm advice and that a continued reliance on feature specific advice is common, especially among some advisers.

Another important issue regarding effectiveness of advice is the type of farmers who receive and implement conservation advice. In some cases it has been suggested that those adopting conservation advice are already committed conservationists. Certainly this was suggested in the Countryside Commission's evaluation of the New Agricultural Landscapes project (1984), which found that advisers were largely preaching to the converted. A somewhat different argument has been utilised by Clive Potter who has claimed that many FWAG activists and farmers in receipt of conservation advice are older farmers who have already, in earlier years, done many of the things – such as hedgerow removal and the like – that FWAG attempts to remedy (Potter 1986). In either case there is concern over the fate of farms

Table 9.6 Summary of FWAG advice given in England, April 1992 to March 1993; and in Scotland, July 1993 to February 1994

ADVICE GIVEN ON:	ENGLAND No. of visits = 3,045		SCOTLAND No. of visits = 827	
	Frequency	% of cases	Frequency	% of cases
Landscape considerations	2,154	71	467	56
Woodland management	787	26	261	32
Woodland planting	764	25	281	34
Shelterbelt planting	338	11	111	13
Amenity planting	1,124	37	223	30
Hedgerow trees	1,201	39	142	17
Scrub management/planting	646	2	147	18
Pond management	770	25	92	11
Pond restoration	523	17	41	5
Pond creation	621	20	162	20
Ditch/watercourse management	765	25	213	26
Wetland management	594	20	281	34
Improved grassland management	555	18	80	10
Unimproved grassland management	1,092	36	239	29
Wildflower planting	438	14	58	7
Hedge management	1,488	49	129	16
Hedge planting	757	25	201	24
Field margin management	807	27	131	16
Pesticide management	657	22	105	13
Fertiliser management	870	29	143	17
Pollution control	281	9	112	14
Heather/moorland management	35	1	111	13
Shooting	207	7	45	5
Fishing	149	5	30	4
Species conservation	564	18	131	16
Drystone walls	143	5	147	18
Archaeological/historical	386	13	102	12
GRANTS ADVICE:				
Woodland Grant Scheme	827	27	253	31
Environmentally Sensitive Area	112	4	190	23
Countryside Stewardship	868	29	–	–
Farm and Conservation Grant Scheme	581	19	104	13
Hedgerow Incentive Scheme	439	14	–	–
Farm Woodland Premium Scheme	296	10	102	–
County Council Grant	1,236	41	102*	12
District Council Grant	194	6	–	–
Set-Aside	138	5	42	5
Other Grants	311	10	139	17

*Regional Councils in Scotland
Source: Winter 1995: 96

managed by those who do not fall into such categories. Similarly Eldon (1988) observed a noticeable correlation between economic status and awareness and use of conservation advice and grants. Farmers aware of and using advice/grant aid for conservation comprised 32 per cent of accumulator farms (those at the cutting edge of agricultural progress), 13 per cent of survivor farms and just 3 per cent of marginalised farms. Of farmers receiving advice solely from the private sector 49 per cent were totally unaware of conservation advice and grants and a mere 8.1 per cent had made use of them. In contrast 75 per cent of farmers receiving advice from the statutory network alone were familiar with advice and grants.

It is difficult to avoid the conclusion that in this aspect of policy greening, progress is slight, despite the claims made for both FWAG and ADAS. Resources for the advisory services are small and a relatively low proportion of farmers are touched by their efforts.

THE GREENING OF RESEARCH

Most of the issues considered so far in this chapter, as well as that of pollution discussed in chapter 10, show some evidence of a transformation in a policy sector, in terms of demonstrably greater influence being exerted by environmental interests. The world of agricultural research, by contrast, appears to have been remarkably insulated from direct influence by environmental interest groups. This is doubtless due, in part, to the Byzantine complexity of this policy sector. As in all mazes it is hard to find the centre, hard to identify policy goals and objectives. Yet, for all that, it is vitally important. Scientific research has been the engine of many of the changes affecting agriculture in recent decades. To turn the leviathan of agricultural research towards agri-environmental objectives would be to go a long way towards the redirection of agriculture that so many environmentalists seek.

In fact, the research world is a highly distinctive sub-sectoral policy community which, whilst linked to the agricultural policy community is, in most respects, entirely independent. Its independence derives from the fact that research priorities are determined by more than one government department with MAFF not necessarily the key player. Agricultural research was long oblivious to concerns sweeping other sectors of the agricultural and environmental worlds. Growing concern, in the 1960s and 1970s, at the environmental impacts of modern agriculture – from organochlorines to hedgerow removal – did little, if anything, to disturb or divert the underlying mission of the agriculture departments, the Agricultural Research Council (ARC) and their Research Institutes (RIs). 'Hard' production-orientated science remained their hallmark well into the 1980s. But during the early 1980s they came under increasing pressure, from parliamentary committees and in time from central government itself, to shift their priorities. This section focuses on agricultural research, seeking to examine how environmental issues have been treated within this sector. Far less attention has been focused on the work of non-agricultural biological science, in particular that associated with the Natural Environment Research Council (NERC), the Institute of Terrestrial Ecology (ITE), Scottish Natural Heritage (SNH), and English Nature. This is not an oversight, but merely a recognition that the agricultural research budget remains large and that agricultural research has a long-standing focus on practical applications within the farming industry.

This was the view taken in the 1980s by several parliamentary committees, resulting in the pressures which led the agriculture departments, the ARC and the RIs, to broaden significantly the scope of their research. One of the consequences of the countryside debate that emerged in the aftermath of the Wildlife and Countryside Act 1981, paralleled by the emerging scandal of CAP food surpluses, was an increased focus on the extent to which the agricultural research effort, much of it publicly funded, was reflecting the need for changed priorities. The matter was examined by the House of Commons Agriculture Committee in 1983, and again, with a more explicit focus on the environment, by the House of Lords Select Committee on Science and Technology in 1984. Consequently, MAFF was urged to increase its funding of strategic research and the Scottish RIs (SABRIs) and agricultural colleges were singled out for failing to respond to the need to spend a greater proportion of their research budget on agri-environmental research. A whole series of recommendations was made on how greater co-ordination between agricultural and environmental departments might be achieved.

Few would deny the huge importance of publicly funded research as a key element in the extraordinary advances in agricultural technology and output during the last fifty years. During that period the UK has been at the leading edge of agricultural science research. Agricultural research, as an identifiable activity, commenced in the middle of the last century with developments in the universities and colleges and the establishment of the first research institute at Rothamsted. The inter-war years saw the formation of most of the agricultural research institutes (RIs), mostly sponsored by the agriculture departments, which in the post-war period provided the launch pad for a rapid expansion of agricultural science research. In order to facilitate co-ordination of the RIs and to provide a focus for the combined efforts of the Development Commission and the agriculture departments, the ARC was formed in 1931. However, formal control of the RIs remained with the agricultural departments. Apart from a small budget of its own the Council's role was mainly advisory until, in 1956, it assumed financial control of the RIs in England and Wales. But the Department of Agriculture for Scotland remained the key funding authority north of the border, thus initiating a marked divergence between arrangements in the two parts of the union (the Northern Ireland model is close to the Scotland one).

Another key element in the research system was the network of experimental husbandry farms established to undertake work on the application of RI findings to practical farming. These came under the jurisdiction of NAAS, and provide the origins for the current research infrastructure of ADAS. Again arrangements were different north of the border with this aspect of R&D becoming an integral part of the unified 'Scottish system'.

For twenty years after the war the two existing research councils, the ARC and the Medical Research Council, were very much given their head, with little direct guidance from government. For the ARC this was a period for the expansion of the existing institutes and the establishment of more. By the early 1980s, the ARC was directly responsible for eight RIs and grant-aided a further fourteen in England and Wales and seven in Scotland. Most were tied very specifically to commodity sectors, such as the Grassland Research Institute, the National Institute for Research in Dairying and the Poultry Research Centre. In addition to the RIs, some twenty-seven smaller *units* were established up to 1980, many with fixed periods of life (Henderson 1981). These were mostly based at universities.

Although it would be foolish to suggest that topics such as plant physiology (a unit based at Imperial College from 1959 to 1971) or insect physiology (Cambridge University, 1944–1967) have no potential beneficial implications for the promotion of environmentally friendly agriculture, none of the units had a specific brief of this nature. For example no ARC unit was established to examine the effects of persistent organochlorines, apparent from the late 1950s, which was pioneered instead by Norman Moore and others at Monks Wood, an experimental station of the Nature Conservancy; and not without resistance from the dominant agricultural research fraternity:

> one of the main obstacles which confronted us at Monks Wood was the opposition of other laboratories and organisations. The members of at least one agricultural research laboratory felt that we were trespassing on their territory. We were studying the effects of agricultural technology on agricultural land. They, not we, should have been doing the work. In fact, they did some but not enough.
>
> (N.W. Moore 1987: 189)

Growing concern at the environmental impacts of modern agriculture – to organochlorines should be added, in the 1960s, many other concerns, including hedgerow removal – did little, if anything, to change the underlying mission of the ARC and its RIs until the 1980s or even 1990s.

In 1984 the Priorities Board for Research and Development in Agriculture and Food was established, issuing its first report in December 1985. Its remit was to advise the AFRC, the MAFF, the Scottish Office Agriculture and Fisheries Department, the Department of Agriculture for Northern Ireland and the Welsh Office Agriculture Department (see Figure 9.6). Its membership was small with an independent in the chair and independents in a majority. The message was clear – research had to be responsive to interests outside the research community itself. The Board as originally constituted had just eight members, comprising two farmers, the chairmen of Unilever and of Beecham Group, a professor of Biochemistry from Cambridge and one senior figure from each of three key agencies – AFRC, Department of Agriculture and Fisheries for Scotland, and MAFF. DANI was represented only by an assessor. The government expected that the Board's 'advice' would normally be acted upon.

Throughout the 1980s and early 1990s, the Priorities Board provided a

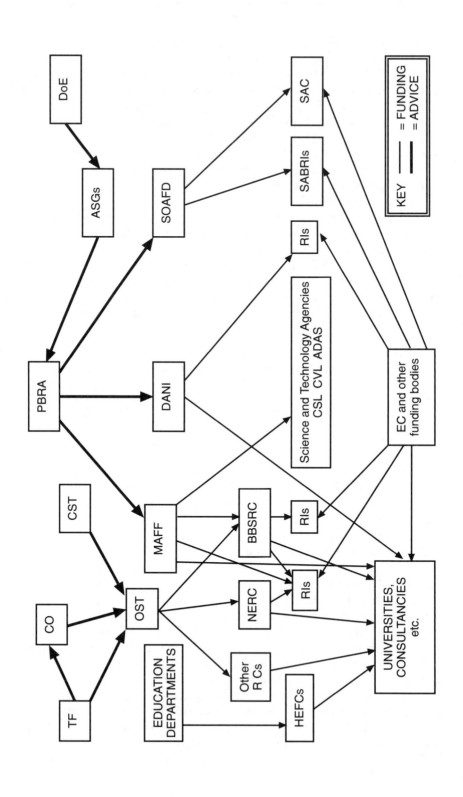

KEY
——— = FUNDING
—— = ADVICE

ASG   Advisory Sectoral Group
BBSRC   Biotechnology and Biological Sciences Research Council
CO   Cabinet Office
CSL   Central Science Laboratory
CST   Council for Science and Technology
CVL   Central Veterinary Laboratory
DANI   Department of Agriculture Northern Ireland
DoE   Department of the Environment
HEFC   Higher Education Funding Councils
MAFF   Ministry of Agriculture, Fisheries and Food
NERC   Natural Environment Research Council
OST   Office of Science and Technology
PBRA   Priorities Board for Research in Agriculture and Food
RC   Research Council
RI   Research Institute
SAC   Scottish Agricultural College
SABRI   Scottish Agricultural and Biological Research Institute
SOAFD   Scottish Office Agriculture and Fisheries Department
TF   Technology Foresight

Figure 9.6 The agri-environmental research network in the UK, spring 1995

steady stream of advice, much of it acted upon. It presided over a shift of R&D effort to food and the environment. But it is important to point out that there were severe limitations on the work of the Priorities Board. First, despite the government's strictures its role was purely advisory. This meant that both the sponsors and the customers have had some leeway in interpreting its advice. Much of the advice has consisted of establishing how the R&D budget might be allocated between broad budget headings. It is not beyond the wit of either research council or institutes to respond to suggestions of this nature, at least in part, by re-classifying on-going research programmes.

Moreover, opportunities for flexibility and manoeuvrability are enhanced by both the dual support system and the growth of private sector sponsorship of research. In 1983/84, 51 per cent of the AFRC budget was provided by MAFF and 46 per cent from the science budget (DES). By 1992/93 the corresponding figures were 20 per cent from MAFF and 62 per cent from the Office of Science and Technology (OST) science budget. The remainder of funding is from commercial sources or other government departments, a figure that rose from 4 per cent to 15 per cent during the same period.

The most recent completed exercise in establishing priorities was the Priorities Board's establishment of six advisory sectoral groups, one of which was the Advisory Group on the Environment Sector (AGES). After nearly three years' work, AGES submitted recommendations to the Priorities Board (AGES 1993b). It recommended a modest 5 per cent increase in the environment spend (following increases in the previous three years) but this was rejected by the Board which instead recommended increases in horticulture and food rather than the environment, although the environment sector was spared cuts. Of the ten members of the Priorities Board, not one represented any of the environmental NGOs, the Department of the Environment or its conservation agencies or NERC, although environmental interests were represented on AGES. This is despite a recommendation as long ago as 1984 by the House of Lords Select Committee on Science and Technology that *the Agriculture Departments' Priorities Board for agricultural research should have a member appointed to represent environmental interests.*

The work of AGES provides a comprehensive register of agri-environmental research currently taking place in the UK. The group estimated that the AFRC and Agriculture departments spent £44 million on environmental research in 1992/93, 17.5 per cent of their total research spend. This does not, of course, include additional large sums spent on relevant work by other agencies such as NERC, the NRA and the DoE. This non-sponsor work is estimated by AGES at an additional £23 million. These figures seem encouragingly high, but the definition of environmental research is broad (including, for example, some biotechnology research) and some projects are not centrally concerned with environmental issues (Winter 1995). Not surprisingly the research tends to build on the traditional 'hard science' bases of agricultural research. Twenty-three per cent of the AFRC and agriculture departments' environment spend (approximately £10 million) is on pests, diseases and weed control and 22 per cent on water pollution. This contrasts with just 8 per cent on wildlife conservation, which represents perhaps 1 per cent of the total agricultural research budget.

Clearly, the greening of agricultural research still has a long way to go. Environmental interests have had limited impact on the process of determining priorities and there are few examples of fundamental shifts in research priorities. On the contrary, progress has been slow and incremental, with innovation often limited by the highly bureaucratic and complex systems of policy formulation in the sector.

THE 1992 CAP REFORM

Hopes were high during the negotiation of the CAP reform that environmental management would, at last, become central to agricultural policy. Superficially, grounds for optimism remained even when the final package was agreed. The first five-year set-aside scheme caused considerable controversy within the environmental world. The CPRE has opposed it vigorously on the grounds of its detrimental impact on the landscape and the fact that it does nothing to discourage intensive farming on the rest of the holding. The RSPB, on the other hand, pointed to various advantages of set-aside as a habitat for bird populations. But the rules adopted for the 1992 'compulsory' set-aside scheme did little for the RSPB case. Rotational set-aside became the norm, preventing long-term diversion of land for conservation purposes. Moreover, some set-aside land can be used for non-agricultural, so-called 'industrial', crops. The environmental benefits of set-aside would appear to be strictly limited. Receipt of set-aside payments is conditional upon farmers not destroying landscape features and habitats, but this could be made far more comprehensive by including requirements for management of features (conditionality). Even more seriously, the arable area and set-aside payments are so generous that farmers with high arable acreage within an ESA are likely to be financially worse off entering an ESA scheme than if they remained outside the scheme. The same problem is likely to arise with the Habitat Scheme, which is seeking to encourage the long-term diversion of some five-year set-aside land for conservation purposes.

There are similar problems with the livestock sector, with limited environmental conditionality. The reduced stocking rates are set too high for any serious environmental benefits to accrue. Moreover the stocking rates are 'theoretical' rather than actual, relating to the number of animals claimed on farm, rather than the number actually there. Thus any inclined to herald the extensification premium as a potential environmental measure were rapidly disillusioned. The NFU was quick to pinpoint the weakness, although it was of course a benefit to their members, to whom the Union offered the following advice in an advisory leaflet: 'Having done your sums, if you discover that your CAP stocking rate is just over 1.4 GLU/ha, it may pay you not to claim on a few animals, in order to qualify for the bonus on all the rest.' Thus, environmental implications are not particularly encouraging, despite much of the early rhetoric. The bureaucracy to police the system revolves round claims for compensation rather than extensification *per se*. Moreover, the opportunities for adverse knock-on effects, agricultural commodity diversification, etc., may conspire to limit environmental benefits severely.

But what about the so-called Accompanying Measures? At the time of the launching of the reform package much publicity was given to accompanying measures, such as a programme of agri-environmental measures designed to ensure that agricultural production would become more compatible with the need to protect the environment and the countryside. Despite being heralded by some as a central element in the reform package, in practice the agri-environment programme has amounted to little more than a continuation and extension of existing programmes, such as ESAs and NSAs. In financial terms they take up a fraction of the total CAP budget. In 1993/94 total CAP funding in the UK amounted to £2,380 million, of which £840 million was devoted to arable area payments but just £43 million to the agri-environment package.

Moreover, there has been speculation that so generous are the direct payments under the commodity regime measures, that farmers will be disinclined to enter environmental schemes. For example, early signs are that subscription to the Moorland Scheme and the Organic Aid Scheme will be poor.

CONCLUSIONS

This chapter has shown how a number of 'green' policy initiatives have emerged within the agricultural policy community. However, far from providing a radical break with previous agricultural policy, as might appear to be the case at first sight, they have served to emphasise the durability of some of the underlying ideological features of the post-war agricultural policy community. Voluntarism in the way schemes are administered and the preservation of 'rights' of farmers to a degree of state support through the extension of their property rights to cover environmental goods are part of this story. So too is the survival of the Ministry of Agriculture and the close relationship between MAFF and its client groups, particularly the NFU.

Areas such as conservation advice and agri-environmental science remain poorly resourced alongside the mainstream agricultural work on advice and research. And the environmental land-management schemes receive finance that pales into insignificance alongside mainstream agricultural support. Progress towards cross-compliance in agriculture, whereby farmers would only receive payments if they were to comply with environmental conditions, still seems a long way off.

Consequently those who look for evidence of the greening of agricultural policy will find much at the level of rhetoric and of policy initiative. The environmental groups are now fully involved in the agricultural policy debate, but much of their effort is marginal to the core policy process. In some ways, the extraordinary complexity of the reformed CAP makes this almost inevitable. A small number of environmental groups, or individuals, have acquired enough knowledge to inhabit this world with ease, but for many, especially when they are concerned with a wider set of environmental issues, the task is beyond them. Much in the same way that the agricultural policy community used to hide behind the mathematical complexities of the annual review, so in the mid-1990s it is consumed by the mind-numbing details of the regulations that form the basis of the Integrated Administration and Control System (IACS) of the reformed CAP. Green issues are easily squeezed out or lost sight of in such a context.

10

AGRICULTURAL POLLUTION

INTRODUCTION

Pollution of the environment is defined in the Environmental Protection Act 1991 as 'the release (into any environmental medium) from any process of substances which are capable of causing harm to man or any other living organisms supported by the environment'. Essentially there are four types of agricultural pollution:

- pollution of the land or soil
- air pollution
- water pollution
- pollution of food, that is, of farm crops and animals (for example through pesticide residues).

This chapter focuses in particular on the issue of water pollution as that is the one that offers the most immediate and obvious example of pollution with damaging effects on a key element of the countryside – aquatic environments.

But it should not be forgotten that, in addition, soil pollution, alongside the important issue of soil erosion, is an emerging concern and one that has been highlighted in the Fifth Environmental Action Programme of the EU, in a draft EU directive, and at the time of writing is the subject of an investigation by the Royal Commission on Environmental Pollution. The UK government has responded with a code of good practice for soil management (MAFF 1993b). Similarly air pollution has been the cause of some concern over the years – in terms of airborne spray drift, smells and the smoke from burning crop residues. It is estimated that methane from farm animal waste accounts for about 30 per cent of the UK's total emissions of this greenhouse gas, and about 50–60 per cent of ammonia (MAFF 1994). A code of good agricultural practice covering air has also been issued (MAFF 1992) and a ban on straw burning was imposed in 1993.

Space prevents us from dealing with all these issues, so the chapter concentrates on two main pollution issues: pollution from nitrogenous fertilisers, and slurry and silage effluent (for an excellent account of the pesticide issue see Ward *et al.* 1993). First, though, it is important to establish the political and institutional parameters of the pollution debate as it has developed in the UK.

THE POLITICS OF POLLUTION

Probably the most important fact to grasp with regard to the politics of agricultural pollution is that industrial pollution became a policy issue long before it was recognised that agricultural pollution might also be a serious problem. In many ways pollution became *the* environmental problem even during the last century and certainly in the 1960s and 1970s. In Britain public concern was particularly promoted by the Torrey Canyon disaster off the coast of Cornwall in 1967 when 120,000 tonnes of crude oil were spilled. Domestic agriculture seemed so far removed from the issue that it was scarcely even considered, although this is something of a paradox arising, perhaps, from the British preoccupation with the appearance of the country-side, its landscape. On a global scale, pesticide pollution was very much to the fore in the emergence of the new environmental movement in the 1960s, particularly following the publication of Rachel Carson's *Silent Spring* in 1963. And in Britain pesticides provided the focus for a widely read book by Kenneth Mellanby (1967; for discussion of the origins of pesticide concern see Sheail 1975). Perhaps as a consequence of the action taken to curb the use of the most dangerous pesticides, for example the banning of most uses of persistent organochlorine pesticides between 1969 and 1973 (N.W. Moore 1987), the focus on agriculture waned – but not the focus on pollution generally which gathered pace in the 1970s. In the wake of the Torrey Canyon disaster, the standing Royal Commission on Environmental Pollution was established in 1970. With a dedicated permanent secretariat the Commission published a spate of authoritative reports putting pressure on government to improve its integration of policy.

Critically as far as political science is concerned, MAFF, with few exceptions, was not a central agency in the development of pollution policies, primarily because it did its best to prevent the application of regulations to agriculture. A fierce battle was fought by agricultural interests to protect farmers from regulation once pollution policies began to impinge on the farming community. However, their inability to contain the issue within the tight confines of the agricultural policy community meant that their efforts were inevitably only partly successful. This is not to imply that pollution regulation in the UK has been particularly tough, but it has become tougher in recent years and the agricultural industry faces a degree of regulation that many farmers would have preferred to avoid. The origins of pollution policy date back to the last century, when efforts were made to curb some of the worst excesses of pollution on public health grounds (Public Health Act 1875). Consequently,

> Britain can boast what is normally considered the world's first national public pollution control agency, the Alkali Inspectorate, which was established by the Alkali Act 1863 to control atmospheric emissions primarily from the caustic soda industry. Water pollution controls followed in the Rivers Pollution Prevention Act 1876, although these proved to be virtually unenforceable in practice
>
> (Ball and Bell 1991: 7)

However, the Control of Pollution Act 1974 provides the first example of modern comprehensive pollution legislation. Notwithstanding this, many of its provisions were not implemented for many years (for example most of Part 2 was held in abeyance until 1986: W. Howarth 1989) and enforcement difficulties remained a significant factor, the problem of implementation deficit referred to in chapter 3 (Weale 1992: 17). A major problem, arising from the incremental style with which environmental issues had been addressed in Britain, lay in the piecemeal and fragmented nature of pollution regulation. As late as the mid-1980s the Department of the Environment, the Health and Safety Executive, environmental health departments of district councils, county councils, local planning authorities, regional water author-ities (RWAs) and MAFF all held separate duties for elements of pollution control. The situation was complex; for example, whilst RWAs were responsible for emissions to water, responsibility for emissions to sea water outside a three-mile zone rested with MAFF. As O'Riordan and Weale (1989) have pointed out, such disjointed developments had two main consequences:

> Firstly, they led to a situation in which there was no statutory or procedural basis for examining the effects of emissions displaced into one medium from another. The piecemeal development of pollution control had created a situation in which no one authority had responsibility for looking at pollution in the round. The second feature was the variability of decision rules through which pollution abatement took place within and between the plethora of regulatory agencies.
>   (O'Riordan and Weale 1989: 283)

These shortcomings were brought to the government's attention in 1976 when the Royal Commission on Environmental Pollution (Cmnd 6371) urged the need for a unified inspectorate to provide a comprehensive approach to pollution control, an inspectorate firmly located within the DoE. Over six years later, the government rejected this relatively modest proposal, despite acknowledging the force of the Commission's logic (Weale 1992). One of the difficulties faced by those anxious to implement the Commission's recommendations was the extent to which such a new agency would cut across existing departmental responsibilities, particularly, for Labour, cherished tripartite bodies such as the Health and Safety Executive (O'Riordan and Weale 1989).

The advent of a Conservative government in 1979, committed to a close examination of regulatory arrangements in all sectors, somewhat para-doxically set in motion the changes that eventually brought about a strengthening of regulation and its co-ordination. A key figure was William Waldegrave, whose spell in the Department of the Environment was marked by a championing of the case for a new agency (O'Riordan and Weale 1989). However, Waldegrave and others resisted any suggestion that a new environmental protection ministry might be the answer. Another report by the Royal Commission on Environmental Pollution (Cmnd 9149, 1984) provided more pressure on government as did the public disquiet over the

Sellafield nuclear reprocessing plant (Weale 1992). Eventually arguments for improved government efficiency provided a means of breaking the political log jam and Her Majesty's Inspectorate of Pollution (HMIP) was formed in 1987 (Weale, O'Riordan and Kramme 1991). The growing influence of EC environmental directives should not be discounted either, in particular their emphasis upon an integrated approach with specific quality objectives (Ball and Bell 1991). Key directives include the Directive Relating to the Quality of Water Intended for Human Consumption (81/971/EEC), the Dangerous Substances Directive (76/464/EEC), and the Bathing Waters Directive (76/160/EEC).

At the outset, HMIP was relatively weak, indeed little more than an amalgam of existing inspectorates (Owens 1990), and lacked the necessary legislative basis to develop the concept and practice of Integrated Pollution Control (IPC). However this was remedied by the Environmental Protection Act 1990 which referred to HMIP as 'the enforcing authority' for IPC. The main objectives of IPC, as set out by the DoE, are:

- to prevent or minimise the release of prescribed substances and to render harmless any such substances which are released;
- to develop an approach to pollution control that considers discharges from industrial processes to all media in the context of the effect on the environment as a whole.

(Ball and Bell 1991: 218)

It would be a mistake, however, to see the birth of HMIP or the enshrinement of IPC in legislation as the abandonment of all aspects of the UK's regulation style; rather a hybrid is emerging combining the thinking and practice of the UK and of continental Europe (A. Jordan 1993).

For example, HMIP is not the lead agency for all aspects of pollution, as might have been expected from a rigorous implementation of the IPC approach. For example, some aspects of air pollution are regulated by local authorities. While the Environmental Protection Act 1990 covers air pollution, waste disposal, litter and noise, water is covered by separate legislation. The Water Act 1989 and the Water Resources Act 1991, which superseded the Control of Pollution Act 1974, established another new agency: the National Rivers Authority (NRA) (G. Hill 1993).

THE POLITICS OF WATER

Since the 1970s, according to one recent commentary, water policy *has been transformed from an area of extremely low to extremely high political salience* (Maloney and Richardson 1994: 111). Moreover, it has moved from being a relatively closed policy-making community, primarily comprising technocratic professionals, to being an open and diffuse issue network with hundreds of participants (Maloney and Richardson 1994).

Much of this change can be linked to legislation. The Water Act 1973, largely motivated by internal pressures and internal agreement by the water professionals on the need for change (Richardson and Jordan 1979), established ten unelected Regional Water Authorities in England and Wales, which took over what had been local government functions. The main losers were the local authorities, preoccupied with local government reform:

> In practice, the 1973 Act created a RWA membership dichotomised along two lines of legitimation, each rooted in competing ideologies – technical and electoral. Within this arrangement the domination of professionals, whose legitimacy derived from their technical competencies, sat some what incongruously with local authority nominees whose legitimacy derived from popular election.... in reality local authority members were swamped with the technical details during RWA meetings.... In terms of network analysis, a member of the core policy community had been pushed to the periphery.
>
> (Maloney and Richardson 1994: 114–115)

The Water Act 1983 abolished the role of the local authorities altogether, signalling not only the Conservatives' disenchantment with local authority involvement in the policy process, but also the view that water should be treated as a commodity rather than a service. It also made the RWAs directly responsible to the DoE rather than indirectly, through a National Water Council. This was a reorganisation which reflected the dominance of engineers in decision making and a consequent emphasis upon technical rather than political factors. Linked to the technical approach was a tendency to see water as an economic good rather than a service (Sewell and Barr 1978). The NFU, representing a major water customer and with an empathy for the technocratic approach, was well able to cope with these changes and retained an important stake in the water policy community throughout the 1970s and 1980s (Maloney and Richardson 1994; Maloney *et al.* 1994; Saunders 1984; Parker and Penning-Rowsell 1980).

Further policy change occurred in the mid- and later 1980s, which resulted in the formation of the NRA. The genesis of the NRA, like that of HMIP, involved a combination of policy imperative and political expediency. In the late 1980s, a number of developments combined to provide a fresh impetus for tighter controls of water pollution. Of great domestic political significance were the politics of water privatisation. The measure was heralded in 1986 during Mrs Thatcher's second administration, and became a key issue for debate during the 1987 general election campaign. In that campaign and during the passage of the Water Bill itself, the issue of environmental protection loomed large. Indeed environmental pressure groups secured some notable advances in environmental protection as the government sought to distance itself from any suggestion that privatisation would lead to a dereliction of environmental responsibility by the erstwhile water authorities faced with new market opportunities.

It is fair to say that, while water privatisation might at first sight appear as a classic example of New Right policies, its genesis in fact lies in the earlier changes already discussed. Indeed a key concept, 'integrated river basin management', lay in the control given to RWAs over every aspect of water management in their region (Richardson *et al.* 1992).

The 1986 proposals for water privatisation caused a furore, primarily because of the proposal that the water companies would be regulated not by a government department or agency, but by themselves. Pressure groups, such as the CPRE, objected strenuously that this was no way to promote effective regulation in the public interest. Even business organisations such as the CLA and the CBI objected. Much of the argument concerned a legal issue as to whether private companies could be construed as 'competent authorities' for implementing EC directives, in terms of European law. The government appeared to be isolated in the face of the opinions of legal experts, a wide range of pressure groups and eventually the European Commission itself. After the 1987 election the new minister responsible (Nicholas Ridley had replaced Kenneth Baker as Secretary of State for the Environment) moved quickly to act upon the clause he had managed to insert into the Conservative election manifesto, that a regulatory authority would be established:

> The government's shift was, however, not simply due to the consider-able external pressures being exerted upon it. The change reflected much more closely the new minister's views.... Mr Ridley was concerned by the notion of one private company having the power to prosecute another. The decision to slow down the policy process and eventually to establish the National Rivers Authority was his alone.
>
> (Richardson *et al.* 1992: 167)

Consequently, the establishment of the National Rivers Authority (NRA) became a central plank of the Water Act 1989. Other features or outcomes of the Act included a tougher prosecution policy and the removal of the 'good agricultural practice' defence against prosecution (although the NRA was to take into account contravention in determining how to exercise its powers). But in other ways the Act was not particularly tough and the environmental groups, in particular, voiced concern at the lack of clarity with regard to the relationships between the many organisations to be involved in the new system (Bowers *et al.* 1988).

In particular, while environmental regulation required the establishment of the NRA, consumer protection (in terms of pricing) required another new regulatory body, the Office of Water Services (OFWAT). Increasingly the NRA's imperative to improve water quality (which costs money) was set on a collision course with OFWAT's duty to regulate the prices charged by the monopolistic privatised water companies. At the same time, despite having some responsibility for making appointments to the NRA (with the DoE and Welsh Office), MAFF faced increasing difficulties in maintaining a powerful voice within the policy sector (although the expansion of the Nitrate Sensitive Area policy gives it a continuing policy stake).

One of the key outcomes of the formation of the NRA was to open considerably the policy community to the influence of environmental pressure groups. The NRA identifies environmental groups as part of its policy community in a manner that had not been the case under the former regional water authorities (Lowe and Ward 1993). The NRA assumed responsibility for water management and some water pollution (including agricultural pollution), although HMIP retained some powers over industrial pollution (Bowman 1992).

> It was clear that given this remit, there was an inherent ambiguity in the relationship between HMIP and the NRA, an ambiguity not made any easier by the NRA being some ten times larger in terms of staffing than HMIP.
>
> (Weale 1992: 105–106)

Moreover, despite the new powers of the NRA, a separate Drinking Water Inspectorate exists within the DoE. In Scotland, it should be noted, the functions of the NRA are largely discharged by seven river purification boards which remain closely linked to local authorities (Ball and Bell 1991). In Northern Ireland the Department of the Environment's Environment Service discharges the water pollution regulation functions. In the light of the continuing administrative complexity and uncertainty, Prime Minister John Major announced in 1991 that yet a further round of reorganisation, involving both the NRA and HMIP, was contemplated for England and Wales, and that a new agency would be created for Scotland; this, despite having rejected such proposals just ten months earlier. At the time of writing, legislation to usher in a new environmental protection agency is before Parliament.

The advent of the NRA heralded a new era for the agricultural policy community insomuch as a new agency was established with powers which, potentially, had implications for farms across the whole of the agricultural estate, and not just in specially designated areas, such as SSSIs. Under Section 16 of the Water Resources Act 1991 a duty is placed upon the NRA 'to conserve and enhance the natural beauty and amenity of inland and coastal waters, and of land associated with them'. Under Section 84 it has responsibility for water quality covering all groundwaters and fresh waters. It has tough prosecution powers and 'may, if it considers it necessary, take direct action to prevent and remedy pollution and recover its costs where it can identify those responsible for the pollution' (Bowman 1992: 567). The NRA is required to advise the Secretary of State for the Environment on Water Quality Objectives (WQOs) which are, if approved, set as statutory objectives. WQOs are to be related to Water Quality Standards (WQSs) also to be developed by the NRA.

In the light of the implications for agriculture of the NRA, it is scarcely surprising that stories have emerged that MAFF, in the run up to the decision to establish a new environmental agency, attempted to assume some of the NRA's responsibilities, an indication that IPC is, as yet, but shallowly rooted in the UK system:

The decision [to form a new agency] followed an intense Whitehall battle over the size and structure of the NRA, with certain constituencies arguing very forcefully for the Authority to be carved up, the operational functions being transferred to the Ministry of Agriculture, Fisheries and Food thereby leaving the proposed agency with only a narrow regulatory ambit.

(A. Jordan 1993: 422)

The environmental lobby fought hard to protect the NRA from such an erosion of its powers (Carter and Lowe 1994). Carter and Lowe identify five main reasons, in terms of the historic nature of environmental politics in the UK, as to why the further rationalisation of pollution might have failed during the intensive lobbying of 1991–1993, reasons briefly summarised as follows:

- the fragmentation of policy arrangements and legislation;
- the decentralisation of regulation;
- the informal, accommodative and technocratic approach to environmental administration;
- the voluntarist approach;
- the domination of producer interests within a closed policy community.

It should be apparent already that many of these features were under pressure by the early 1990s. The MAFF attempt to wrest control of some functions back from the NRA before a new agency was formed represented an attempt to reassert many of these principles, safeguarding the cherished freedoms of the farming community. However, Carter and Lowe (1994) explain how the intellectual case for greater integration and tighter regulation had become almost unassailable, and was widely accepted in the DoE. Moreover, the rejection of the MAFF bid was hardly surprising in the light of a new proposed EC directive on integrated pollution prevention and control. Increasingly the politics of water reflect the Europeanisation of environmental politics.

THE NITRATE ISSUE

It was another EC directive that first, albeit somewhat belatedly, focused MAFF's attention on the question of agricultural pollution. The EC Directive Relating to the Quality of Water Intended for Human Consumption was first proposed in 1975, notified in 1980, and brought into force in 1985. The directive defined nitrate pollution of drinking water at 50 mg per litre. Prior to that the UK had worked to the World Health Organisation's standard set in 1970 of 100 mg per litre. The reduction of the permissible levels so dramatically placed the UK government in a particularly difficult situation, for it was clear from the mid-1970s that increasing fertiliser applications in arable agriculture over the previous forty years (see Figure 10.1) had resulted in a significant increase in nitrate levels in groundwater

(Department of the Environment 1974; Cmnd 7644, 1979; Wilkinson 1976). The increase had been dramatic – estimated at 650 per cent in the case of nitrogenous fertiliser between 1955 and 1982 (Jewell 1991; Department of the Environment 1988). Consequently the maximum admissible concentrations were being exceeded in some parts of Britain, primarily areas of intensive arable agriculture where use of fertilisers was dramatically higher than in other agricultural regions (Figure 10.2). In 1989, 154 UK sources exceeded the limit and by 1990 the number had increased to 192 (MAFF 1993c). In order to comply with the directive the water companies were obliged to introduce expensive blending or treatment programmes.

Chalk aquifers are particularly at risk and because nitrate moves through the chalk so slowly (estimated at 0.5 metres per year in the Isle of Thanet for example), then the results of fertiliser applications and ploughing many decades earlier are only now being felt. The study in the Isle of Thanet discovered concentrations of 30 mg/l under fertilised arable land compared to 10 mg/l for fertilised permanent grassland and only 3 mg/l for unfertilised pasture (NRA 1992). In some instances remedial action lies in the gift of agencies owning and leasing water catchment land in the vicinity of reservoirs. Thus in the mid-1980s evidence of increasing nitrate concentrations at the Batheaston and Monkswood reservoirs led to restrictions on arable agriculture on land owned by the water authority, Wessex Water. The subsequent increase in permanent grassland has led to a reduction in nitrate levels (NRA 1992; Tuckwell and Knight 1988).

However, in most instances such measures are not so easy, either because the physical features of the aquifer prevent such an immediate response to changed farming practices or because controls over farmers' actions are not

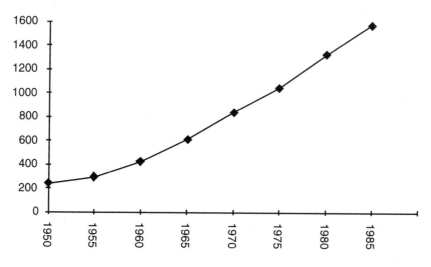

Figure 10.1 Annual use of nitrogen in fertilisers in the UK (thousand tonnes)

Source: Fertiliser Manufacturers' Association annual reviews

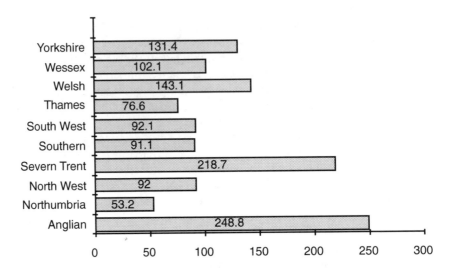

Figure 10.2 Estimated tonnage of inorganic nitrogen applied (thousand tonnes per year) in the late 1980s in NRA regions

Source: NRA 1992: 137

so easy to implement. Instead government action is necessary; yet, for a variety of reasons, the UK was so slow to act that the European Commission eventually took successful court action against the UK government for its failure to comply with the drinking water directive. Thus one of the messages of this case-study is that the translation of EC environmental policies into national policies may be fraught with difficulty. Writing in the late 1980s, one group of commentators was moved to cite the UK's apparent intransigence as an instance of 'non-decision making' (Hill *et al.* 1989).

Why this was so and the extent to which subsequent developments mark any change in the UK's approach is the main concern of this part of the chapter. In the early 1980s, the UK government defended its position by emphasising the efficacy of five main methods available to the RWAs for dealing with the problem without the requirement to reduce pollution at source:

- *Blending and source replacement.* Usually this is the cheapest way of dealing with the problem, and is already widely practised. As nitrate levels rise this becomes increasingly less useful as a technique. It becomes more difficult and more costly the further away the cleaner sources are, and may involve deterioration in respect of other water quality parameters.
- *Storage.* This allows microbiological denitrification to occur, but other treatment problems arise as a result of storage. Large costly reservoir schemes would be necessary in areas where currently the use of ground water obviates the need for them.

- *Denitrification.* In the UK two denitrification processes are favoured: ion exchange and microbiological denitrification. Denitrification is an expensive way of providing low-nitrate water supplies and can lead to a reduction in water quality in other respects.
- *Management of boreholes.* The re-siting and deepening of boreholes and changing of purifying procedures can help to reduce nitrate levels but this is an expedient of largely interim limited value.
- *Provision of bottled and tankered water.* This has been adopted as an emergency measure in some areas in the past, to provide lower nitrate level water temporarily for babies and pregnant women. As a long term expedient for a large number of people it is far and away the most costly alternative option.

(Hill *et al*. 1989: 229)

In the most seriously affected region, East Anglia, the strength of the farming interest within the policy community meant that as late as the early 1980s the Anglian Water Authority was refusing even to consider the possibility of tackling the problem by limiting applications of fertiliser. In 1985 the consultancy group, Lawrence Gould, was contracted by the DoE to report on the issue and concluded that the cheaper of the remedial methods represented a more cost-effective approach to the problem than cutting applications of fertiliser, a view consistent with that of the Royal Commission on Environmental Pollution (1979).

In later work, however, the consultants suggested that in some zones fertiliser control might be a cheaper option (Hill *et al*. 1989). In such deliberations neither the consultants nor the DoE appeared to place a great deal of emphasis upon the significance of those two underlying principles of EC policy – the desirability of preventing pollution at source and the polluter pays principle.

In addition to the apparent reluctance in the mid-1980s to tackle the nitrate pollution problem at source, there was also considerable reluctance to accept the standards established by the EC, which were seen as unacceptably stringent. This was certainly the position, as might have been expected, of key interest groups such as the NFU and the Fertiliser Manufacturers Association, and of individual fertiliser companies such as ICI which complained that the Drinking Water Directive contained no medical evidence in support of the level chosen (ICI 1986). In 1987 the House of Commons Select Committee on the Environment, not renowned for back-pedalling on environmental issues, urged the DoE to seek a re-examination of the 50-mg level. The DoE's own Nitrate Coordination Group discussed alternative standards when considering the issue (Department of the Environment 1986) and the government's own advice and interpretation of the directive indicated *at best ambivalence about the importance of the limit* (Hill *et al*. 1989: 230). For example, the government's Chief Medical Officer regarded the limit as arbitrary (Seymour *et al*. 1992), as indeed did the public sector Water Authorities (Baldock 1988).

Indeed, while there was some medical suggestion that infant methaemo-globinaemia (blue baby syndrome) might be a risk at levels above 50 mg, the evidence regarding stomach cancer remained lacking. It was left to Friends of the Earth (FoE) to take the lead on urging caution, not merely because of inconclusive evidence but because of lack of evidence. In urging caution, and indeed precaution, FoE was allying itself closely with EC legal opinion and, in particular, the principle of *Vorsorgeprinzip* or precaution, developed in Germany in the 1980s, which suggests that policy makers are compelled 'to go "beyond science" in the sense of being required to make decisions where the consequences of alternative policy options are not determinable within a reasonable margin of error and where potentially high costs are involved in taking action' (Weale 1992: 80). If this principle was taken up with enthusiasm by campaigners and, in the Fourth Environmental Action Programme, by the European Commission itself, it has certainly never been accepted widely by the agricultural policy community or within UK government. Both continue to argue that scientific evidence is required prior to action (see, for example, the 1990 White Paper Cmnd 1200). The requirement for scientific evidence, whilst used to justify its failure to embrace wholeheartedly certain restrictive EC policies which might limit economic activities, has not been pursued with the same vigour by the UK government when considering business policy decisions, which may be based on equally slight scientific evidence. As late as June 1992, no less a body than the Royal Commission on Environmental Pollution continued the challenge to the EC limit:

> With regard to human health, a substantial intake of nitrate can contribute to the development of methaemoglobinaemia in infants under six months but other factors are important and the disease is extremely rare in the UK. There appears to be no epidemiological evidence that the concentrations of nitrate in water in the UK are associated with an increased risk of cancer. We have not been convinced that the strict limit in the EC drinking water directive is needed to safeguard health in the UK.... We regret that the EC limit has created anxiety in the minds of consumers whose supplies are known to be marginally in breach of the limit, has removed from use several water sources which were regarded as secure, possibly diverted resources and attention from more deserving objectives, and created some environ-mental problems in disposing of nitrate removed from supply.
>
> (Cmnd 1966, 1992: 185–186)

Whilst the agricultural policy community fulminated over the lack of scientific evidence of human health problems arising from the application of nitrogenous fertiliser, they were able to ignore completely ecological problems associated with nitrate pollution, about which there was plenty of scientific evidence. On the issue of eutrophication (nutrient enrichment) the evidence was clear, and even the new tighter EC standard was set far too high for wildlife according to one NCC official (Association of Agriculture 1988;

Seymour *et al.* 1992). A Royal Commission report in 1992 emphasised the vital importance of reducing the nutrient enrichment of freshwater and the sea, an enrichment which was accelerating as a result of the continued application of nitrates.

In the light of this aspect of the debate, and in the context of fears that inaction might lead to a case being brought against the UK in the European Court, some steps to influence agricultural practices became inevitable. Thus it was that in 1985, the date set by the 1980 Directive for countries to comply with the standards, MAFF launched its *Code of Good Agricultural Practice.* The Code advised farmers on appropriate practices associated with organic and inorganic fertilisers but suggested little in the way of serious restrictions on their application:

- Care should be exercised in the handling and application of solid and liquid fertilisers, particularly to avoid polluting relevant waters either directly or indirectly.
- Application rates of fertilisers should take account of crop requirements and the nitrates provided by any organic manure and the soil. To reduce the danger of nutrients being leached out and polluting relevant waters, fertilisers (particularly nitrogenous fertilisers) should not exceed maximum recommended rates.
- Nitrogenous fertilisers should only be applied at times when the crop can utilise the nitrogen. In autumn and winter application should be avoided except when there is a specific crop requirement.

(MAFF 1985: paras 1.5–1.7)

The Code was long awaited, for it had been heralded in the Control of Pollution Act 1974. But as a step along the path towards stricter controls over agricultural pollution, the Code was, at best, a mixed blessing. Under the terms of the 1974 Act, farmers, along with all citizens, were liable to prosecution for water pollution unless the entry to the water system was authorised or was *attributable to an act or omission which is in accordance with good agricultural practice.* Thus, in effect, the Code actually offered *legal protection* to farmers, if they could demonstrate that pollution had occurred even when they had adhered to the principles of *good practice* as defined not by the DoE (the Department responsible for enforcement of the Act) but by the Ministry of Agriculture and other UK agriculture departments. In the case of nitrate pollution, attributing blame to any specific farmer was difficult enough in any case without also having to demonstrate that actions had been taken which were inconsistent with the Code.

Thus, the Code had as much to do with protecting farmers as with promoting good practice. It also demonstrated yet again the ability of the agricultural policy community to respond to a new policy issue in a manner designed to preserve the freedoms of farmers. However, in the case of pollution the policy community faced considerable difficulties in *domesticating* the issue as they had done with such conspicuous success with regard to habitat and landscape protection. In the case of non-agricultural threats to habitats/landscape, then, the planning system was in place; elsewhere the agricultural policy community had a firm hold on the agenda in its promotion of the voluntary principle over the wider countryside and the commitment it had secured to generous compensation payments in operation elsewhere. In ESAs environmental policy was implemented by MAFF itself, and other policies were implemented by the NCC which was not a central government department and had shown itself reluctant to take an aggressive stance on policy issues which affected the agricultural industry.

But pollution, potentially, was different. Here a body of law, and agencies, had grown up to tackle pollution in general as we saw in the first section of this chapter. Thus, whilst it might be possible for MAFF to find ways of limiting the impact of pollution policy developments upon the agricultural community, it could not easily hive off the issue and bring it under its own jurisdiction, or certainly not without making substantial commitments to a high degree of regulation and control. To do so would require amendments to much extant legislation. Moreover, by the mid-1980s, agriculture itself was under increasing scrutiny for over-production. It was hardly an opportune moment to seek an extension of agricultural 'exceptionalism'. The policy agenda had shifted too far. Indeed in some quarters it was even suggested that nitrogen limitation might be a possible means of achieving required cuts in surplus production. None the less, the administration of any controls over agriculture clearly required some involvement of MAFF. Thus when the Water Act 1989 established Nitrate Sensitive Areas, MAFF was given responsibility for their administration, but the NRA remained very much a power behind the scenes both in terms of a continuing critical scrutiny of how the scheme might work and in its powers to take further action if that proved necessary.

The 1989 Act removed good agricultural practice as a legal defence against a water pollution charge, but retained the Code of Good Agricultural Practice requiring the NRA to take into account contraventions of the Code in the exercise of its powers (W. Howarth 1992).

Nitrate Sensitive Areas (NSAs)

Early in 1989 a new draft EC directive, the Nitrate Directive, was issued which proposed that states should designate all areas where waters might be polluted by nitrates and to impose therein very strict limits on fertiliser use and stocking densities. This resulted in an amendment to the Water Bill, incorporated as Section 112 of the Water Act 1989, which established a framework for Nitrate Sensitive Areas. *Mandatory* restrictions on the use of nitrates in NSAs were provided for but such powers were to be held in reserve. NSAs, ten of which were established in 1990, would, in the first instance at least, be based upon voluntary compliance backed up by management agreement payments on an annual basis.

The ten experimental NSAs covered just 15,000 hectares and some well-known high nitrate aquifers were omitted. In addition to the NSAs, a further 23,000 hectares were designated as Nitrate Advisory Areas, where an intensive advisory campaign would be run to encourage farmers to adopt certain management techniques so as to minimise nitrate pollution.

There are two types of NSA agreement: basic scheme and premium scheme. Under a basic scheme agreement the farmer is restricted in the use of chemical fertilisers and natural fertilisers such as farmyard manure, but the scheme is designed to allow a continuation of commercial farming operations, without dramatic alteration to the existing crop rotations. The main measures are as follows:

- Adherence to economic fertiliser recommendations (including full allowance for nitrogen from manures and previous crops).
- Reduction below the economic recommendation of 25 kg/ha N for winter cereals, 50 kg/ha for winter oilseed rape.
- Requirement for cover crops on land which would otherwise be bare over winter.
- No more than 175 kg/ha N as organic manure to be applied per year.
- Poultry manure and slurries not to be applied during July to October inclusive (September to October for grassland).

(MAFF 1993d)

The premium scheme offered higher payments to farmers prepared to convert some of their arable land to low-input grassland. Participation in the scheme was voluntary, but once committed to the scheme all the land farmed by a participant that lay within the NSA had to be entered either at basic or premium rates. Initially the NSA scheme was established for an experimental period of five years. Participation was high with 87 per cent of land entered into the basic scheme, ranging from a low of 52 per cent in the Boughton NSA in Nottinghamshire to 98 per cent of the Kilham NSA in the Yorkshire Wolds, although uptake of the premium scheme was limited outside one of the NSAs, Sleaford (MAFF 1994). The evidence so far available suggests that the pilot NSAs have been successful in reducing both the amount of fertiliser applied and the concentration of nitrates leaving the soil as shown in Figure 10.3.

One of the problems with the NSA approach, in common with so many site-specific schemes in the UK, is the danger of displacement of problems to land lying outside the NSAs, either through an intensification of production by a farmer with land in and land without an NSA or a generalised intensification by other farmers able to exploit market opportunities presented by lack of restrictions. This was tackled to a considerable extent by the water catchment area approach, but in response to continuing fears the government has continued to stress the importance of the voluntary approach within the wider farmed countryside. In July 1991, for instance, its revised *Code of Good Agricultural Practice for the Protection of Water* was issued.

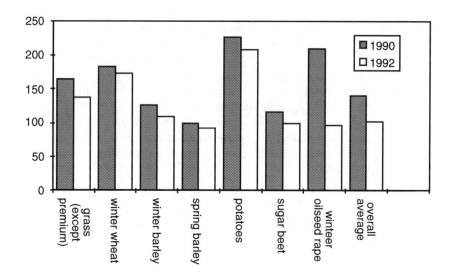

Figure 10.3 The average amount of nitrogen fertiliser applied to crops within the NSAs before and after NSA agreements (kgN/ha/year)

Source: MAFF 1993d

The 'Directive concerning the protection of waters against pollution caused by nitrates from agricultural sources' (the Nitrate Directive) was approved by the Environment Council of the EC in 1991. The directive requires member states to:

• designate as Nitrate Vulnerable Zones all known areas of land that drain into waters where the nitrate concentrations exceed or are expected to exceed 50 mg/litre or where there is evidence of nitrate limited eutrophication;
• establish action programmes which will become compulsory in these zones at a date to be agreed between 1995 and 1999;
• review the designation of NVZs at least every four years.

Full compliance by all farmers is expected by 1999. Farmers in NVZs will be required to maintain records of all fertilisers and manures used on the farm on a field-by-field basis. The regulations covering enforcement procedures will be available in 1995. A crucial aspect of the NVZs is that compensatory payments are not to be paid, as the agriculture departments have argued that compliance merely requires the farmer to conform to the codes of good agricultural practice. This would appear to represent a victory for those advocating adherence to the polluter pays principle and a defeat for the NFU's assertion of rights to payment for the production of environmental goods. However NSA payments, all of which are located in the new NVZs, will also continue. As the NSA basic scheme in most cases differs little from the Code of Practice requirement there would appear to be an anomaly here with some farmers eligible for payments and others not. The NFU has been vociferous in its complaint that this is contrary to principles of natural justice.

Whereas nitrate pollution is, on the whole, a slow process due to gradual leaching, the direct pollution of watercourses by animal slurry or silage effluent can be much more immediate and direct, although, of course, there may also be relatively slow and continuous pollution, contributing either to eutrophication or high concentrations of nitrates in aquifers. But this section concentrates on *pollution incidents* in much the same way as with industrial pollution accidents. Inadequate storage leading to a spillage or overflow, or spreading of manure in inappropriate conditions (for example when the ground is waterlogged), can lead to widespread loss of aquatic life and long-term environmental degradation. On occasions it can also lead to a considerable clean-up cost as polluted water is pumped from rivers and/or hydrogen peroxide is added to the water in an effort to maintain oxygen levels. Fish may even be stunned and removed to safer waters. The NRA may seek to recover the costs of such actions from those guilty of the pollution offence. As with the nitrates issue, the problem has been caused by the rapid intensification of agriculture in the post-war years, and the rising tide of incidents in the 1980s is illustrated in Figure 10.4.

Not only have stock numbers increased but so has the risk of slurry or silage pollution as a result of the concentration of livestock production in certain regions of the country and, within those regions, on particular farms. Consequently the size of livestock units has increased dramatically. Eighty

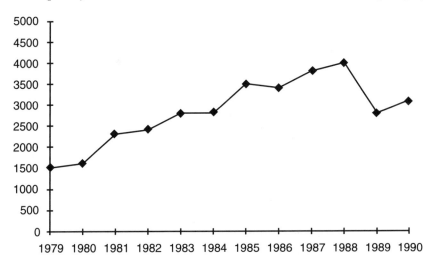

Figure 10.4 Number of farm pollution incidents in England and Wales

Source: NRA 1992: 22

per cent of farm pollution incidents in the 1980s were from dairy or beef units (Ward *et al.* 1993) and in 1992 the NRA estimated that in many catchments 40 per cent of farms were polluting or in danger of doing so (NRA 1992). Not surprisingly the number of reported farm pollution incidents has also increased, although much of the increase is put down to a higher incidence of reporting (data was not even published on an annual basis until 1985).

Increases in stocking rates due to advances in grassland productivity have meant that, in the case of dairying, there is less land on which to spread the slurry unless arrangements are made with other farms. The situation is compounded by other developments such as the increasing proportion of slurry-based rather than straw-based livestock housing systems and the switch from hay to silage as the normal winter feed for cattle (Lowe *et al.* 1992), as illustrated in Figure 10.5 – a process which accelerated after the imposition of milk quotas in 1984 as farmers strove to hold down the costs of bought-in food by improving grassland production. A dairy farm in Devon of 53 cows (average for Devon but smaller than the national average) has

> a pollution potential load equivalent to that of a community of 465 inhabitants. If silage is made for winter storage, which is likely, then an average crop for a herd of this size would be 650 tonnes. If this crop had been wilted there would be 145,000 litres of silage effluent to be disposed of at the rate of 19,000 litres per day. The potential pollution load of this effluent is equivalent to that of a community of 10,800 inhabitants.
>
> (South West Water Authority 1986; quoted in Lowe *et al.* 1992: 1)

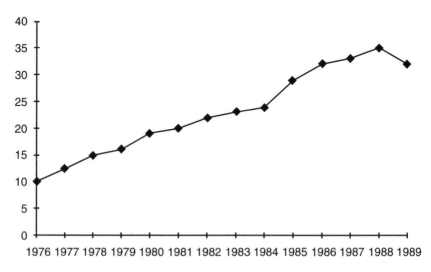

Figure 10.5 Silage production in England and Wales (million tonnes per year)

Source: NRA 1992: 119

Despite the concentration of production and the capital-intensive nature of modern dairying, many dairy farms are relatively small businesses which have been slow to make the necessary investments in adequate storage and handling facilities. In 1989 a third of reported pollution incidents involving cattle wastes were a consequence of inadequate facilities (NRA/MAFF 1990). The policy focus has therefore been towards improving design, construction and siting of storage facilities. But, as Lowe *et al.* point out, 'the disposal problem, which is fundamentally constrained by the availability of land, is not thereby diminished' (1992: 2).

Although the issue did not come into focus as a major policy concern until the late 1980s, the rudiments of the problem and the need for policy initiatives were understood some time earlier. As with nitrates, the issue can be characterised as an example of non-decision making for a considerable period of time. MAFF itself recognised that a problem was emerging when it encouraged ADAS to set up a Farm Waste Unit in the 1970s, but this was primarily in response to concerns about smell rather than water pollution. A key development in the emergence of the issue on to the political stage was an investigation by the House of Commons Environment Committee (1987) on the pollution of rivers and estuaries. The Committee spoke of *a growing tide of farm pollution incidents* and urged MAFF and DoE to take *a far more interventionist and regulatory approach to farm pollution*. The Committee was clearly pushing at an open door, as the government's own response to the Committee's report showed (House of Commons 1988).

Government policy to tackle livestock pollution incidents has had four main elements:

- a tougher enforcement policy
- a major advisory campaign
- capital grants to encourage the improvement of facilities
- expenditure on research associated with the problem.

The tougher enforcement policy fell to the NRA. The number of prosecutions and the size of fines increased with the annual number of convictions for farm pollution offences rising from below 100 between 1973 and 1983, to 225 in 1987 and between 150 and over 200 since then (Ward *et al.* 1994).

The last chapter discussed the importance attached to giving advice to farmers in the greening of agricultural policy and, indeed, some of the shortcomings in the resources devoted to this aspect of agricultural policy. Advice on the control of pollution has certainly been seen as one of the main ways in which ADAS can meet its obligation to provide public-good advice, with some 56 staff years devoted to pollution advice in 1992/93, representing the major proportion of its public-good work:

MAFF's intention is that a free visit should supply general advice geared to the farmer's immediate waste control problems. Visits themselves tend to involve discussion of quite complex technical matters, and the advice frequently indicates what options might be appropriate. It may even include recommending a particular type of system. But although individual advisors can exercise considerable discretion, free advice always stops well short of providing an actual design or provision of a farm waste management plan, both of which are available under ADAS's commercial services.

(Lowe *et al.* 1992: 11)

Such was the position in 1992, but in 1993 the free advisory system was adapted so as to offer farmers a choice between a general pollution visit and assistance with the preparation of a farm waste management plan. The NRA does not seek to offer technical advice on waste management although it will advise on whether a danger of pollution exists. Despite the efforts of ADAS there are some serious shortcomings in the coverage. In 1991/92 MAFF and the Welsh Office set an objective of 5,000 free pollution advice visits in England and Wales, allocated between regions. Lowe *et al.* point out that the 640 allocated to Devon and Cornwall represent just 5 per cent of the total number of livestock holdings in the two counties, and that the number is 'determined in the light of previous demand and staff resources, not by reference to the scale of the problem' (Lowe *et al.* 1992: 12). Not surprisingly, in the light of the tougher enforcement policy adopted by the NRA, the demand for ADAS free advice has been high, with reports in some regions of long waiting lists. Linked to the advisory campaign are capital grants to encourage the improvement of facilities. These were set, under the terms of the Farm and Conservation Grant Scheme introduced in 1989, at a rate of 50 per cent (an increase from the existing grant of 30 per cent) for the provision, replacement or upgrading of storage, treatment and disposal systems for agricultural wastes. In addition regulations have been adopted setting minimum construction standards for new or substantially modified storage facilities. The take-up of grants has been high.

Finally, research effort has been directed primarily towards effective spreading methods and storage, rather than towards more fundamental research on the capacity of soils to cope with waste, still less with socio-economic research on how to tackle the problem at source (Lowe *et al.* 1992). There is, however, an increasing and welcome emphasis upon the implications for air pollution of some of the measures that have been taken to prevent water pollution. MAFF's 1994–1996 Environmental Protection Research Strategy identifies seven main objectives for a proposed research programme on farm waste pollution at an annual cost of nearly £2¼ million:

- to develop practical and economic methods to reduce silage effluent and slurry production
- to develop safer methods to contain wastes and prevent accidental release at low cost
- to develop disposal outlets that are environmentally sound, result in beneficial recycling and do not impose excessive costs
- to provide improved information on the extent and fate of gaseous

ammonia emissions associated with farm wastes and develop means of control which do not impose excessive burdens on the industry

- to control odours associated with farm wastes
- to provide reliable estimates of greenhouse gas emissions from farm wastes under different conditions and evaluate ways of reducing them
- to provide a sound scientific basis for the Codes of Good Agricultural Practice to Protect Water and Air.

(MAFF 1994: 27)

The continued emphasis upon control and upon measures that do not adversely affect the economy of the industry is apparent from this list. The research philosophy is very much that of trying to discover a series of 'technical fixes' to solve particular problems rather than grappling with some of the underlying socio-economic problems of production concentration. Research covering both the socio-economic structure of agriculture and the linkages between environmental and socio-economic sustainability appears to be lacking. This is an almost inevitable consequence of a research strategy led by the requirements of different policy sectors within MAFF (Winter 1995).

CONCLUSIONS

Ward *et al.* (1994) have argued powerfully that pollution represents an example of how the restructuring of the countryside, consequent upon population shifts, is now leading to major revisions to the former political order in the agricultural sector. Agricultural pollution had been so neglected during the 1970s and early 1980s that MAFF had failed to develop the proactive lead in this area that might have preserved the integrity of the policy community. Instead its purely defensive posture was quite inadequate to cope with the changes wrought in the late 1980s and 1990s when the issue came to prominence. As regards the farming lobby, having failed to engage with the issue in the 1970s and 1980s when it might have taken the kinds of action which would have secured it a central place in current policy deliberations, it too has been out-manoeuvred by an increasingly sophisti-cated environmental movement. But as Maloney and Richardson (1994) have pointed out the community has been fundamentally destabilised, with conflict rather than consensus becoming the chief characteristic of the policy sector.

11

FARMING'S RICH RELATION: FORESTRY POLICY AND POLITICS

INTRODUCTION

On the grounds that forestry occupies such a small proportion of rural land in Britain, it was the original intention to entitle this chapter 'Farming's poor relation'. To have done so would have been singularly inappropriate, for private forestry has been, for most of the period under review, the preserve of rich – and often extremely rich – men and women. Forestry is a minority interest but, for some of those involved, a consuming one. In marked contrast to many other European countries, UK forestry occupies relatively small areas of land and is a minor player in the rural economy. Forested land accounts for just 10 per cent of the land surface of Great Britain, supplying 14.3 per cent of timber consumption needs in 1991 (Grayson 1993). Earlier in the century the coverage had been even smaller. The ravages of early industrialism, when timber was extensively used in mining, construction and as a fuel, combined with the periodic demands of war, served to reduce forestry to such a point that woodland covered just 5.6 per cent of the land area in the immediate aftermath of the 1914–1918 war (Grayson 1993).

In the light of this history, it is hardly surprising that a campaign to afforest significant tracts of (mainly upland) Britain with conifers was launched in the early decades of the century. With some marked changes of emphasis this campaign has been sustained ever since. This chapter examines the genesis of forestry policies and their progress.

THE DEVELOPMENT OF FORESTRY POLICY TO 1919

By the nineteenth century, private forestry in Britain had become connected primarily with sporting pursuits rather than with timber production. Coverts for shooting and fox hunting abounded in lowland England, but these were generally small and, in terms of timber production, insignificant, although of considerable importance to the landscape. The Crown forests, however, were managed for commercial timber production. Based largely on ancient royal hunting forests, such as the Forest of Dean, the New Forest and Sherwood Forest, the management of the Crown forests was transferred to the Commissioners of Woods and Forests in 1810 (N.D.G. James 1981). During the nineteenth century, especially the early decades, the Commissioners made strenuous efforts to rejuvenate existing woodland areas within the

Crown forests, to plant open areas and to acquire new land for planting. Their aim was to ensure the availability of 100,000 acres of productive oak woodland which had been calculated as the requirement for the provision of a regular supply of timber for the Royal Navy. Consequently almost all suitable Crown land had been planted by 1850 but the target of 100,000 acres remained elusive (N.D.G. James 1981). The Commissioners were beset by the problems of unenclosed land within the hunting forests where the customary rights of forest residents to grazing, estover (firewood) and turbary (peat for fuel) were exercised. The erosion of these rights by the Crown, intent on increasing its commercial timber production, led to bitter disputes in the middle years of the century. A government inspector sent to examine the New Forest in 1848 was burnt in effigy (E.P. Thompson 1993) and the Deer Removal Act 1849, which empowered the Crown to enclose a further 10,000 acres in the New Forest for planting, caused such outcry that it had, eventually, to be amended (Pasmore 1976).

During the 1860s the introduction of iron-clad ships led to a dramatic about-turn in the priority, established over several centuries, of safeguarding and replenishing the oak woodlands of Britain. With the demise of demand from the shipyards, British forestry was thrown into disarray. In particular the hallowed place of the English oak was seriously undermined. The special requirement of wooden shipping was for hardwood, and oak was the most suitable of all the hardwoods. Many other requirements for timber were less specific and the future now lay with the faster growing softwoods. A tradition of planting Scots Pine (as a nurse crop) with oaks, heralded the era of coniferous planting, which gathered pace with the introduction from North America in the mid-nineteenth century of fast-growing exotic species such as the Sitka Spruce, Douglas Fir and Lodgepole Pine. By the close of the century most new planting was of conifers, but the rate of planting remained modest. In 1902, a Departmental Committee was appointed by the Board of Agriculture with the following terms of reference:

> to enquire and report as to the present position and future prospects for forestry, and the planting and management of woodlands in Great Britain, and to consider whether any measures might with advantage be taken, either by the provision of further education facilities, or otherwise, for their promotion and encouragement.
>
> (Cmnd 1319; quoted in N.D.G. James 1981: 197)

The report castigated the neglected state of forestry in Britain and urged a programme of education and practical demonstrations to encourage better management of the nation's woods and fresh planting. This was essentially an exercise in piety, with forestry seen as a good thing but with little attempt to justify such a position or suggest how the objectives might be achieved. With the naval argument obsolete, forestry seemed destined to remain the minority interest of a few large landowners and retired officials from the Indian Forestry Service. But this was to change in the first decade and a half of the twentieth century for two reasons: welfare and warfare.

Welfare

As indicated in chapter 4, the early years of the twentieth century were a period of dramatic political development as the two major parties strove to accommodate themselves to the challenges posed by the expansion of democracy and the emergence of the Labour Party. No longer could major social problems such as unemployment be ignored or marginalised within the political process. The welfare state may not have been born until 1946, but welfare politics pre-dated this by some fifty years. The switch of many rural seats from Tory to Liberal in 1906 and the emergence of Lloyd-George, from rural Wales, as a key Liberal and radical politician helped to ensure that the new welfare politics had a rural dimension.

Between 1881 and 1911 the number of agricultural workers in Britain declined by 150,000. In 1907 an Afforestation Conference, presided over by the President of the Board of Agriculture, Lord Carrington, was held to consider the potential role of forestry in reducing the growing level of urban and rural unemployment, a theme first promoted by the Association of Municipal Councils (N.D.G. James 1981). Two years later, the Royal Commission on Afforestation (Cmnd 4460) reported that 'a national scheme of afforestation would contribute to the solution of the unemployed problem' (quoted in N.D.G. James 1981: 201). Its main conclusions were that:

- Nine million acres of afforestation was desirable in UK without substantially encroaching on agricultural land. Assuming a sixty-year rotation this would mean planting at a rate of 150,000 acres each year.
- Permanent employment would be provided at one man per 100 acres.
- The net revenue after eighty years should show a return of 3.75 per cent on the net cost.
- A special Board of Commissioners, having powers of compulsory purchase, should be set up to deal with the work of afforestation.

In response, the Chancellor of the Exchequer, Lloyd-George, introduced the Development and Road Improvement Fund Act 1909 which established the Development Commission. In his 1909 budget speech, Lloyd-George highlighted the importance he attached to forestry, hoping that grants would be made available for 'schools of forestry', the acquisition of land for planting, and the establishment of experimental forests (N.D.G. James 1981). In order to take this work forward, the Development Commission set up a Forestry Committee, comprising four commissioners, to deal with this side of its work. In practice the Commission was limited in what it could do as it was not permitted to help profit-making organisations, nor could it initiate action itself, but only offer advice and aid (Minay 1990). Its first report on forestry in 1911 urged a programme of education and research, including an examination of what would be the most suitable sites for afforestation, prior to any action.

Responding to the commissioners' report in 1912, the President of the

Board of Agriculture appointed an Advisory Committee on Forestry to examine the main issues raised in the report, in particular:

- to consider and advise upon proposals for a forestry survey;
- to draw up plans for experiments in silviculture and to report upon questions relating to the selection and laying out of forest demonstration areas;
- to advise as to the provision required for the instruction of woodmen.

In its report (Cmnd 6713), the Committee proposed modest means of proceeding on each of the above, including firm recommendations for a survey covering the location of land suitable for afforestation (N.D.G. James 1981).

Warfare

The second, and most important, of the factors which prompted the re-emergence of forestry as a significant concern for policy makers was the 1914–1918 war. Paradoxically, revolution in naval warfare, which had prompted the neglect of forestry fifty years earlier, was now the source of escalating concern over the state of the nation's woods. The deadly effectiveness of attacks by German submarines on the British merchant fleet meant a severe disruption to timber imports. At the outset of the war, the UK was importing the bulk of its timber requirements, amounting to approximately 400 million cubic feet of timber per annum, mostly from northern Europe and Russia (N.D.G. James 1981). Pitprops were among the most important uses for timber, with coal remaining the key source of power for nearly all sectors of the economy including, of course, the Royal Navy.

In order to increase home supplies, a Home-Grown Timber Committee, appointed by the President of the Board of Agriculture in 1915, was empowered to organise timber supplies and establish and staff its own sawmills. In 1917 the Committee was replaced by a Directorate of Timber Supplies, which was recast a few weeks later as the Timber Supply Department of the Board of Trade. The Department introduced fixed prices for timber and, equally importantly, felling licences. In future landowners would have to apply for a licence to fell timber, thus allowing the Department to determine the appropriate rate of exploitation of the country's timber resources as a whole. Additional powers of compulsory purchase were rarely used.

Thus, as with agriculture, wartime provided a context in which reliance on the operation of private sector markets was set aside and replaced by state direction and controls, in which private economic concerns became secondary to the national interest. But also, as with agriculture, these developments presented a unique opportunity for the private sector and its representatives to demonstrate their preparedness to work in tandem with the state and to make the necessary investments – both political and economic – for the future.

The initial impact, however, was dramatic and there was little opportunity for assessing the requirements of the future. No less than 450,000 acres of woodland were felled in Great Britain during the 1914–1918 war. Much of this was oak woodland planted during the eighteenth and early nineteenth centuries; and some of it was ancient broadleaved woodland. Watkins (1990) characterises ancient woodland as a mixture of primary woodland (i.e. woodland areas that have never been cleared for agriculture or other uses) and secondary woodland of medieval or earlier origin. Many of these sites were subsequently replanted with conifers. The war, whilst it rejuvenated forestry, did so at a tremendous ecological cost with secondary coniferous plantations replacing ancient semi-natural deciduous woodland.

Warfare, as well as transforming current production levels and techniques, also tends to inspire visions of a happier future. A forestry sub-committee of the Selborne Committee (see chapter 4), chaired by the Rt. Hon. F.D. Acland (Parliamentary Secretary to the Board of Agriculture), reported in 1917 (Cmnd 8881). Its report was a virtual blueprint for the new forestry policy that was to emerge after the war and that was to dominate British forestry for the following seventy-five years. Aware of the need to reduce reliance on timber imports, but aware also that similar arguments were being applied to domestic agriculture, the Committee effectively abandoned the lowlands, focusing all its attention on the potential for coniferous afforestation of the uplands. They estimated that 2 million acres of rough grazing could be made available for growing timber without reducing the home production of meat by more than 0.7 per cent, and at the same time giving employment to ten times more people than if the land remained in agricultural use. The sub-committee calculated that 1.7 million acres were needed to ensure independence of imported timber for up to three years given an emergency. To achieve this a planting programme over eighty years should be instigated, with 250,000 acres planted in the first decade, 150,000 acres by the state and 100,000 by the private sector (Pringle 1995). So was born the idea of a public body with twin aims – on the one hand to regulate the private sector and provide advice and grant aid; on the other, to act as a commercial enterprise in its own right, in effect a nationalised industry.

A key issue to be addressed was to whom the new body would be responsible. The links between forestry and agriculture, as the two major rural land users, were clear, and suggested the need for a direct link with the agriculture departments. However, there were three independent agriculture departments in the UK – the Board of Agriculture and Fisheries for England and Wales, the Board of Agriculture for Scotland, and the Department of Agriculture and Technical Instruction for Ireland. The sub-committee pressed the case for a national body covering the entire British Isles, and not tied to any one of the existing agricultural departments or, indeed, jointly to all three (Pringle 1995). The sub-committee's findings were generally accepted, and in 1918 an Interim Forest Authority was established. This was a precursor to the Forestry Commission itself, which was formed in September 1919, following the successful passage through Parliament of the first of many Forestry Acts.

THE FORESTRY COMMISSION TO 1951

The Commission was chaired by Lord Lovat and comprised eight commissioners who would operate at arm's length from government, and 'not less than two of whom were to have special knowledge and experience of forestry in Scotland; at least one was to have scientific attainments and a technical knowledge of forestry and one was to be a member of the House of Commons' (N.D.G. James 1981: 215). The forestry powers of the existing agriculture departments were transferred to the Commission. In 1923, control of the Crown forests was also transferred to the Commission, thus completing the establishment of a major national body and landowner.

In terms of its future development, two key features of its constitution need to be stressed. First, the emphasis on technical forestry expertise, whilst a natural and innocent-sounding requirement, meant in practice that most commissioners were not only enthusiasts for forestry but in many cases major landowners too. In the case of agriculture, despite being so heavily influenced by the farming lobby, policy has always remained formally in the hands of government ministers and civil servants. In forestry, however, the establishment of the Forestry Commission meant that the implementation, and to a considerable degree the formulation, of forestry policy were placed in the hands of the foresters themselves.

Following the first meeting of the Commission, Lord Lovat and another commissioner, Lord Clinton, rushed home, both anxious for the privilege of being the first to plant Forestry Commission trees. Devon, being closer to London than Scotland, Lord Clinton won. The story is told as an amusing incident in the standard accounts of forestry history by N.D.G. James (1981) and Pringle (1995). What their accounts fail to highlight is the significance of their lordships rushing off to plant on their *own* estates, on land to be sold or leased to the newly formed Commission. As large landowners, several of the commissioners stood to benefit from sales or leases of land or from private planting grants. Nor, in a market where imports predominated, did they fear competition from the productive potential of a state industry. On the contrary, they stood to gain from infrastructure developments, particularly the increased saw-milling capacity, that would be the inevitable corollary to an expansion in home production. This is not to suggest that such commissioners were motivated merely by narrow self-interest – they were convinced that afforestation was in the national interest – but in the years to come they could hardly be expected to adopt a particularly detached or critical approach to forestry policy. There was even less opportunity in forestry than in agriculture for the emergence of radical or heretical voices within such a closed policy community.

A second key feature of the newly formed Forestry Commission was the extent to which it formalised a separation from agriculture. Administrative arrangements cannot be blamed for everything, but in this case the arrangements did little to bring together two rural land uses in potential conflict, and merely aggravated an existing dichotomy peculiar to Britain.

Whereas forestry and farming went hand in hand in many countries of Europe, in Britain the landlord–tenant system was erected on a strict division of responsibility, with the landowners controlling estate woodlands and tenants farming the land. Tenants had little or no responsibility even for small areas of wood within the boundaries of their farms, where the sporting rights, as well as the timber, were reserved for the landlord. In many cases the reserved rights even extended to large hedgerow trees. The way in which the Forestry Commission was constituted held out little promise for any fundamental examination of the relationship between forestry and agriculture and may be partly responsible for the lack of interest in woodland matters still displayed by many farmers (Lloyd *et al.* 1995; Watkins *et al.* in press; Williams *et al.* 1994). All that the Forestry Act 1919 did to alleviate this difficulty was to establish Consultative Committees for England, Wales and Scotland (replaced by National Committees in 1945) which, in addition to facilitating representation of a range of forestry interests, stipulated a representative from the agriculture department.

The Commission faced its first serious threat in 1922. The financial crisis which saw the repeal of the Agriculture Act 1920, prompted the Geddes Committee (Cmnd 1582) to call in 1922 for the Forestry Commission to be wound up and for state-sponsored afforestation to cease. There are probably two reasons why the Commission was able successfully to head off the challenge, where agriculture had failed. First, in contrast to farming (see chapter 4), forestry presented a united front. Lord Lovat, in effect, led a small and tightly knit elite, able to marshal a powerful set of arguments, based mainly on the issue of national security, to support the continuation of state afforestation. In contrast to NFU leaders at the time, Lovat was well connected and he did not have a highly disparate constituency to deal with. Second, as far as the government was concerned, to withdraw support for forestry would involve not only reneging on its forestry policy, but also mean a volte-face on commitments made to seek alleviation of unemployment through forestry. Few claims could be made for agriculture as a means of soaking up surplus labour. Late in 1921 the Commission had been instructed to acquire additional land for planting in high unemployment areas (Ryle 1969). It seemed scarcely credible for the government to alter track so rapidly. None the less the Commission faced some economies, including a reduction in its planting rate, the redundancies of ten forest officers and a reduction in training places in the new forestry schools.

Having survived such an early threat, the Commission resumed its key task of planting. By 1929, it had planted 138,279 acres and, with the Crown forests as well, was managing an estate of 600,000 acres. In addition, it had grant-aided the afforestation of 76,736 acres by private landowners and local authorities. But there was more to a national forestry programme than merely growing trees. In time, the timber would have to be harvested, processed and sold. In an attempt to remedy what was widely perceived to be a chaotic system of timber marketing, a number of marketing initiatives were launched in the 1930s, as set out in Table 11.1. The Home-Grown

Timber Advisory Committee, formed in 1939, became a permanent feature of the forestry policy community. It consisted of representatives of the timber trade, woodland owners, the Board of Trade and the Commission itself. The marketing thrust continued in the immediate post-war period, with the Commission expressing, in 1948, its willingness to assist co-operative marketing schemes, and in 1949 establishing the Advisory Committee on Utilisation of Home-Grown Timber, with representation from the Commission itself, the Board of Trade, the Forest Products Research Laboratory, the Rural Industries Bureau, the United Kingdom Forestry Committee, the Timber Development Association, the Home Timber Merchants Association of England and Wales, and the Home Timber Merchants Association of Scotland.

The most important characteristic of these initiatives was the closeness of the relationship that was emerging between the private sector, state forestry and the forest authority. There have been many instances of individuals sitting astride all three functions. To take just one example from the 1930s, Lord Clinton served on the National Home-Grown Timber Council, not in his capacity as a forestry commissioner (and past chairman) nor yet as chairman of the board of governors for the Imperial Forestry Institute at Oxford, but as a representative of the Home-Grown Timber Marketing Association representing timber growers, of which he was president.

With only a maximum of twenty years' growth, the Forestry Commission plantations played only a limited role in timber provision when western Europe was once again plunged into war in 1939. During the 1939–1945 war, controls were exercised over prices, sales and felling by the Timber Control Department of the Ministry of Supply, which heavily circumscribed the work of the Forestry Commission. As in the 1914–1918 war, extensive tree felling took place, amounting to a total of over 500,000 acres, the vast majority of which was in private hands. Broadleaved woodland planted in the mid-nineteenth century, which had survived the First World War as not yet mature, now came under the axe. Unfortunately, by no means all of it was replanted in the post-war period, at a time when agricultural demands for land seemed inexhaustible.

During the war two reports were published on post-war forest policy (Cmnds 6447 and 6500). Endorsing the Acland Report's target of 5 million acres, based on new planting of 3 million acres and the better management of 2 million, the reports came up with firm recommendations on how this might be achieved. Acquisition of bare land by the state remained a key policy plank but, in addition, attention was now focused firmly on the private sector. The encouragement of private planting and management by the Commission had been less successful than its own planting programme during the inter-war years. In much the same way as recalcitrant farmers faced the threat of supervision orders or eviction in and immediately after the war, so it was proposed that woodland owners should be required to dedicate formally their woods to forestry or face compulsory acquisition by the Commission. The first report proposed that to enter a Dedication Scheme,

Table 11.1 Timber marketing initiatives in the 1930s and 1940s

Initiative	Comments	Date
FC Inter-Departmental Home-Grown Timber Committee	Resulted in the issuing of FC booklets specifying nature of demands for timber in mining, shipbuilding, packing-cases and woodturning.	December 1931 to September 1933
Home-Grown Timber Conferences	Sponsored by the CLA, the second conference led to agreement on the need for a marketing association.	December 1931 and November 1932
Formation of Home-Grown Timber Marketing Association	Produced bulletins on timber prices, etc. Acted as a representative body for timber producers on marketing matters.	January 1934
National deputation to House of Commons	Urged increased state support for home-timber production.	February 1934
FC sub-committee	Formed to examine and report on marketing matters; recommended a timber council.	May 1934 to February 1935
Formation of Home-Grown Timber Council	Grant-aided by FC and formed as a limited company but not to be involved in commercial operations. Functions: collection and publication of statistics, trade information, publicity for marketing, and forest products research.	March 1935 to 1939
Formation of Home-Grown Timber Advisory Committee	Formed initially as a sub-committee of the FC, its initial concern was to draw up maximum prices for timber for use if war was declared. Later it became the principal conduit for advice for the FC from the timber trade.	January 1939 to present day
Formation of Advisory Committee on Utilisation of Home-Grown Timber	To advise the commissioners on measures designed to promote the utilisation and sale of produce from British woodlands.	December 1949

Source: after N.D.G. James 1981

for which financial assistance would be available, the owner would have to undertake to:

- use the land in such a way that timber production is the main object;
- work to a plan, to be approved by the Forestry Authority, which would lay down the main operations to be undertaken;
- employ skilled supervision;
- keep adequate accounts.

The supplementary report was produced following consultations with private sector interests and was chiefly concerned with establishing appropriate levels of financial incentive (N.D.G. James 1981). The financial details, based on loans and maintenance grants, were complex but they certainly acted as effective inducement to many landowners once the Dedication Scheme was enshrined in statute under the terms of the Forestry Act 1947.

In the immediate aftermath of the war, the Forestry Act 1945 marginally reduced the independence of the Forestry Commission, by placing forestry under direct ministerial responsibility, either through the Minister of Agriculture or the Secretary of State for Scotland (in 1964 responsibility was transferred to the newly formed Ministry of Land and Natural Resources but reverted to MAFF in 1967; and later the Scottish Office assumed lead responsibility). Henceforth, a minister would be answerable to Parliament on forestry matters and land would be acquired in the name of the minister who would place it at the disposal of the commissioners. But once land was under the commissioners' control, the government's power was strictly limited. A minister could only manage or sell Commission land which 'in their opinion, was not needed, or not to be used, for the purpose of afforestation or any other forestry purpose' (Pringle 1994: 32). In addition, broad policy direction would be provided by the minister.

The thinking behind these changes had less to do with any desire for political accountability to ensure that forestry might be seen to serve a wide range of publicly acceptable purposes, than with a need to balance the land-hunger of forestry and agriculture at a time when the nation had great need of both timber and food. Crucially, the Commission was not directly answerable to the Ministry. The changes meant that a potentially sympathetic minister in Cabinet would be empowered to argue the forestry case, rather than the Commission having to deal directly with the Treasury. But the Commission's independence of operation would be largely preserved. Parallels might be drawn with the freedoms of operation enjoyed by Her Majesty's Stationery Office (HMSO), the National Savings Bank and the Crown Estate Commission, the only difference being that these three organisations might reasonably be expected to be financially self-supporting while the Commission remained heavily dependent on state funding. In one respect, the potential influence of government was reduced by the 1945 Act, which removed the right of members of the House of Commons to serve as commissioners. However, members of the Upper House suffered no such diminution of their potential influence on forestry policy.

Another important aspect of the 1945 Act was the establishment of National Committees for England, Scotland and Wales, to which the Commission could delegate any of their functions. In part, this was purely an administrative reform in recognition of the increasing scale of the Forestry Commission's activities. A degree of decentralisation was required to ensure efficient management. But, in addition, the national committees represented the first in a series of moves which, in effect, deepened the Commission's links with the world of private forestry. Hitherto, a relatively small national elite of large landowners, often commissioners themselves, had sufficed to deliver appropriate policies.

Now that the Commission had ceded, however marginally, some of its power to MAF, it needed to be buttressed from below. In 1946, Regional Advisory Committees (RACs) were established to provide channels of communication between the Conservator, in charge of each Forestry Commission Conservancy (i.e. region) and those within the Conservancy interested and involved in forestry. As if these changes were not enough to fully incorporate private interests, two further committees specifically to represent private forestry interests in England and Wales and in Scotland were formed in 1946 and known as Private Forestry Committees. The justification for the new bodies was innocent enough – the need to consult with the private sector to ensure the success of a new policy to boost private sector forestry:

> In putting forward their proposed scheme for the Dedication of Woodlands, the Commissioners had expressed the view that it 'should consult freely with the Central Landowners' Association, the Scottish Land and Property Federation and the Royal Scottish and Royal English Forestry Societies'.
>
> (N.D.G. James 1981: 238)

In 1948, the Private Forestry Committees were merged to form the UK Forestry Committee which, together with the RACs, became a formidable champion of forestry interests.

Other changes in the immediate post-war period strengthened the regulatory structures surrounding forestry and trees. The Town and Country Planning Act 1947 empowered local authorities to make Tree Preservation Orders (Watkins 1981, 1983). These became an important instrument for the protection of individual trees on amenity grounds and, less frequently, for the protection of small woods. TPOs could not apply where dedication covenants with the Forestry Commission had been entered into by land-owners (or where financial grants had been made to landowners by the Commission prior to the 1947 Act). This was an early sign, however implicit, of the expectation that the Commission would take amenity factors into consideration with the overriding concern for timber production. It also had the unintended consequence of encouraging landowners to think rather more positively of the opportunity of entering into a dedication scheme.

A further-reaching regulatory innovation was the retention of war-time felling licences. Initially, these were operated by the Board of Trade, but in 1950 they passed to MAF to be operated by the Forestry Commission. The Commission also assumed responsibility for regulating felling proposals at a national level, a task delegated to the Home-Grown Timber Advisory Committee (HGTAC). The Forestry Act 1951, which received Royal Assent in October, just weeks before Labour lost the 1951 election, established and clarified the framework for felling controls. In addition, it gave the Forestry Commission the duty of 'promoting the establishment and maintenance in

Great Britain of adequate reserves of growing trees', firmly clarifying the Commission's central role in the post-war rural land-use settlement not only as a forest enterprise but, additionally, as forest authority. The same legislation saw the reconstitution of the HGTAC and the RACs as statutory bodies.

THE POST-WAR SETTLEMENT IN PRACTICE: FORESTRY AFTER 1951

The immediate post-war period had been a momentous time for British forestry, with the establishment of a system of governance to cover all aspects of forestry from planting to felling in both the public and private sector. On the surface strongly state-led, in practice the private sector was deeply entrenched in a tightly knit and closed policy community that continued to be dominated by an elite comprising large landowners and professional foresters. The stage seemed set for a peaceful expansion of forestry in a quiet, undisturbed and uncontested backwater of government. In the event the political peace was to last for little more than decade but, at first, the signs were indeed propitious. The newly elected Conservative government's desire to shed unnecessary state controls in agriculture, and elsewhere in the economy, did not extend to forestry. With firm state backing, the 1950s are now remembered as 'boom years for forestry' (Pringle 1994: 42). By the end of the decade, the Commission had planted over 1 million acres and 583,000 acres in the UK had been entered under the dedication scheme.

Despite the plethora of committees, private sector associations and other initiatives on timber marketing, there was still widespread concern that, while trees were steadily being planted, the timber industry as a whole was ill-organised. In 1953, the UK Forestry Committee voiced its concerns to the Secretary of State for Scotland. As a consequence another committee was set up. The Committee on Marketing of Woodland Produce, under the chairmanship of H. Watson, deliberated between 1954 and 1956. Its terms of reference were 'to consider what measures might be taken within the home timber industry to improve the arrangements for marketing produce from privately owned woodlands'. Its report recommended a central independent consultative body to replace the HGTAC and the formation of a woodland owners' association:

- to represent owners
- to publicise financial assistance and facilities available on silviculture and marketing
- to collect information on prices, markets and supplies.

(Watson Committee 1956)

Marketing associations existed for timber; a number of statutory or semi-official committees provided opportunities for woodland owners to air their views, and, of course, some major woodland owners were themselves commissioners. However there was no single organisation which any woodland owner might join, in order to participate in the policy community. A need for this was perceived, not least because of the Conservative government's desire to ensure successful private forestry alongside the state sector. The dedication scheme was still encountering resistance from some owners. The Watson Committee believed that successful policy implementation required a mechanism for conveying the practical views of owners on how grant schemes should be constituted.

The drive to create an effective woodland owners' organisation was taken up by government. In July 1958, the Minister of Agriculture made it clear that new and improved grants for woodland owners would be dependent upon 'the formation of an effective woodland owners association' (N.D.G. James 1981: 252). Here, then, is an example of government promoting and sponsoring the formation of a pressure group. It was a move which led to a highly effective lobbying organisation at the very heart of the forestry community. In December 1958, the Timber Growers' Organisation Ltd was formed, for its first year under the auspices of the CLA. The following year saw the formation of the Scottish Woodland Owners' Association, which later became Timber Growers Scotland (the two merged to form Timber Growers UK in 1983). Co-ordination was effected through the Forestry Committee of Great Britain, which replaced the UK Forestry Committee.

The first hint that all might not go well for the forestry lobby in future forestry policy deliberations, came with an independent report on forestry in the uplands, commissioned by government and published in the summer of 1957 (Zuckerman 1957). The report fundamentally questioned the strategic justification for afforestation and, to some extent, the import-saving justification for forestry. However, in rediscovering the employment argument, a curious one at a time of almost full employment (unemployment stood at just 244,000 in the UK in the month the report was published: Butler and Butler 1986), the report advocated a continued high rate of planting. The Zuckerman Report was accepted by government and a further round of planting, up to the late 1960s, was announced in July 1958.

Most in the forestry lobby saw the Zuckerman Report as another green light for upland afforestation, and commended the manner in which it justified forestry on land-use grounds. Arguing, correctly, that agriculture over many centuries had effectively denuded the uplands of trees, the report suggested that the balance needed restoring, and that, in this process, due regard should be given to the needs of wildlife and the beauty of the countryside. It is surprising how few in the forestry policy community also saw the increasing potential for conflict inherent in the new justification for forestry. On ecological and landscape grounds, the restoration of a land-use balance in the uplands would surely require broadleaved planting rather than coniferous afforestation.

The coniferisation debate

As we saw in chapter 7, disquiet over the consequences for upland landscapes of larger conifer plantations emerged in the 1930s in the Lake District. As the wartime emergency receded, fresh expressions of concern emerged in the 1950s, accelerating in the 1960s and 1970s. The alliance ranged against the proponents of upland coniferous afforestation was, in some ways, formidable. Nature conservationists were concerned about loss of upland habitats, particularly with respect to bird species associated with heather moorland. But more articulate, initially, were the advocates of stricter controls on forestry because of its implications for landscape and access. The Ramblers' Association emerged as one of the key protagonists in the debate (Ramblers' Association 1972, 1980; Tompkins 1986).

Even the farming lobby was, for many years, out of sympathy with afforestation, arguing that state support meant that the Forestry Commission competed unfairly with local farmers in the land market and that the viability of upland farming enterprises was severely compromised by forestry operations (interestingly research in Scotland has shown upland afforestation to have had a remarkably small impact on the viability and output of agricultural enterprises: Mather and Murray 1988). In Wales, suspicions about the activities of the forestry sector take on a whole new layer of meaning where forestry as represented by the Forestry Commission has been portrayed as a kind of English Stalinism by at least one well-known Welsh literary figure, D.J. Williams (Morgan 1982). The antagonism to afforestation is clearly linked to nostalgia for a vanishing way of life. In the words of an FUW delegate to the 1973 Union AGM, afforestation is the 'arch enemy: I well remember as a child, seeing farms supporting 1,000 sheep being bought up by the Forestry Commission'. An FUW committee recommended in 1972 that planting and replanting should be discontinued within National Parks and AONBs, and that elsewhere strict landscape criteria should apply and every effort made to replace conifers with hardwoods (Cox, Lowe and Winter 1987). However this position was modified later as members began to reap the benefits of selling land at good prices to forestry companies. In time an uneasy alliance with forestry interests was forged and the FUW embarked upon a collision course with conservation interests over afforestation plans for Llanbrynmair Moors (see Lowe et al. 1986).

In an effort to assuage mounting criticism of insensitive planting programmes, Dame Sylvia Crowe was appointed in 1963 as the Commission's first landscape consultant. She revolutionised the approach to upland afforestation by insisting that planting should follow contour lines and that broadleaved trees and larch replace Sitka spruce in more sensitive and edge locations, and that watercourses should also be treated sensitively (Crowe 1978). By the 1980s, she was able to write confidently about the positive potential for good landscape offered by a newly sensitised forestry (Crowe 1986). But the Ramblers' Association were not easily persuaded by such arguments:

The Ramblers argue further that not only should new planting be more stringently controlled, but that there is nothing inherently bad about having a smaller area under trees than in most other countries. The hills are better used, they argue, for food production and landscape conservation than for growing trees.

(Davidson and Wibberley 1977: 37)

In attempting to outmanoeuvre the ramblers' critique, the Commission also countered by stressing the recreation benefits of forestry. This was first outlined in a Commission policy statement in 1963 and it rapidly became much more than an incidental by-product of timber production.

In a 1972 government White Paper (HM Treasury 1972), discussed in greater detail below, recreation emerged, in cost-benefit analysis, as one of the few positive aspects of afforestation. Forest parks and trails, picnic and camping sites, all contributed to a countryside recreation boom in the 1960s and 1970s and many Commission staff became as involved in the theory and practice of recreation management as in growing trees (Grayson *et al.* 1975). To the list of learned aboricultural texts, regularly published by the Commission, were now added equally erudite monographs on recreation (e.g. Mutch 1968). The Forestry Act 1967, in addition to repealing and up-dating the 1919, 1945 and 1951 Acts, gave a new special responsibility to the Commission to cater for public recreation and to enhance the beauty of the countryside. A year later, the Countryside Act 1968 empowered the Commission to plant and manage for amenity reasons and granted the Commission powers to provide facilities such as campsites, picnic places and visitor centres.

Despite its new responsibilities, the Forestry Commission's annual report for 1971 demonstrated the continuing importance of conifers in planting schemes. Of the 77 million trees planted by the Commission that year, 57 per cent were Sitka spruce; 19 per cent pine (mostly Lodgepole); 6 per cent silver fir, western hemlock and red cedar; and 4.5 per cent larch. The remaining 13 per cent were unidentified save to say that the most common broadleaved species used was beech which accounted for just 280,000 trees.

Thus, by the end of the 1960s the Forestry Commission appeared to have survived the initial assault of the ramblers. Traditional justifications for its planting programme – the need to reduce dependence on expensive imports and the provision of rural employment – appeared intact and to these had been added new amenity and recreation arguments. But more change was afoot, nor was the debate on coniferisation so easily abated.

STOP–GO POLICIES IN THE 1970s AND 1980s

The publication, *Forestry in Great Britain: An Interdepartment Cost/Benefit Study* (HM Treasury 1972), provided a devastating critique of forest economics and the strategic arguments for afforestation. It criticised the Dedication Scheme as having departed from its original intention to encourage replanting on land felled during the 1939–1945 war, whereas 80 per cent of dedication schemes now applied to bare land (this history of the scheme was disputed by the Royal Forestry Society which correctly pointed out that the scheme had its origins, not in the war, but in the conditions of the early 1930s: Garthwaite 1972). In dismissing the rationale for afforestation on grounds of import-saving, the report declared that grants should only be available for proven employment or environmental gains. At the time, the report was seen as a far-reaching attack on forestry, and the initial response from the forestry lobby was one of vigorous opposition, even staged panic. In the words of the Timber Growers' Organisation, it plunged forestry into *the greatest crisis for a generation*, and another apologist for forestry complained that the report was based on dubious assumptions, curious conjectures and questionable arithmetic (Garthwaite 1972).

In retrospect, though, the door for further planting was left more than a little ajar. Indeed, the White Paper proposed that the forestry estate should continue to expand, albeit at a slower rate. Moreover, in some ways the paper was remarkably cautious, failing for instance to compare the cost of employment generation by afforestation with alternative methods of rural regeneration such as factory programmes or tourism (Davidson and Wibberley 1977; Wibberley 1974). And in one respect, it provided a significant boost to forestry. It should be remembered that this was a Conservative government initiative, and whilst the politics of reducing public expenditure provided a powerful incentive for the report, so too did that of providing favourable tax arrangements to encourage private forestry, which were given a boost by the 1970–1974 Conservative government. The 1970s and 1980s marked a period when the main incentives for afforestation switched from a combination of state afforestation and grant aid to private landowners, to tax-induced planting by national forestry companies, such as Fountain Forestry or the Economic Forestry Group, at the behest of the accountants of super-wealthy clients. The precise means of tax minimisation provided for forestry are complex, involving low notional rather than actual incomes and favourable arrangements for capital transfer tax. Helpful explanations of the favourable forestry taxation regimes prior to the dramatic changes in 1988 (discussed below) are given by Stewart (1985) and the Forestry Commission (1985).

The forestry industries took the 1972 White Paper as their cue for a rigorous defence of forestry, stressing its role in stemming rural depopulation and in the environmental benefits that it offered. A newly formed umbrella body for forestry interests, the Forestry Committee Great Britain (forerunner of the Forestry Industry Committee Great Britain), co-ordinated the campaign, which included the commissioning of a report by Professor Wolfe, an economist at Edinburgh University, to scrutinise critically the cost/benefit analysis used by the Treasury team (N.D.G. James 1981).

In the event, forestry emerged relatively unscathed, with just a 10 per cent reduction in its planting targets (which had rarely been fully met in the past in any case). However, the government was insistent that environmental elements should become more central. In a statement to the House of Commons by the forestry minister in the Ministry of Agriculture, in October 1973, the following points were made:

- To qualify for grants, owners would be required not only to manage their woodlands in accordance with the principles of good forestry but there would also have to be 'effective integration with agriculture and environmental safeguards together with such opportunities for recreation as may be appropriate'.
- Where a 'significant proportion of hardwoods were planted a higher grant would be paid'.
- In future when dealing with grants to private estates, the Commission would consult with the agricultural departments and the planning authorities, regarding the amenity and land use aspects of the proposals as submitted by the woodland owner.
- The RACs would be reconstituted so that their membership included representatives of agricultural, planning and amenity interests as well as those of the forestry industry.

(quoted in N.D.G. James 1981: 263)

The implications of these new conditions had barely had time to register when the rate of private planting went into free fall as a result of the determination of the Labour government, elected in 1974, to reform capital taxation. The Capital Transfer Tax 1975 had serious repercussions for private forest planting which declined from 24,400 hectares per annum in 1972 to just under 9,000 hectares in 1977. Once again, though, the forestry lobby showed itself equal to the challenge. After intensive lobbying, Labour agreed in 1976 to set up an Interdepartmental Review of Forestry Taxation and Grants, and by 1978 enough changes had been introduced for the Labour Chancellor, Dennis Healey, to be congratulated in the magazine *Forestry and British Timber*, for 'learning the error of his ways' (Tompkins 1989: 93). By the late 1970s, the Forestry Commission (1977) was once again urging massive expansion of afforestation as a *prudent investment*, a view bolstered by a bullish Reading University report (CAS 1980), which reinstated the argument for forestry to stem imports. Despite economic critiques (Bowers n.d., 1982; Miller 1981), the government accepted, in general, the imports argument, although not endorsing either the Commission's or the Reading study's figures.

The projected rise in demand for timber, coupled with likely pressures on world forests, lay at the heart of a new government forestry policy statement in December 1980, which outlined a continuing expansion of forestry at about the same rate (25–30,000 hectares p.a.) as the previous twenty-five years. At the same time the government signalled that the Commission would be expected to dispose of some of its land, this requirement being enshrined in legislation in the Forestry Act 1981. The disposals policy was further reinforced in 1984 but, to date, the Commission as such has not been privatised. Whilst Commission afforestation had virtually ceased, private forestry expanded rapidly during the decade, as shown in Figure 11.1.

By 1980, the earlier concerns over the impact of upland afforestation on landscape and recreation opportunities had largely been replaced by concerns over the ecological impact. A House of Lords report (1980) reinforced this concern and the early and mid-1980s saw a spate of hard-hitting reports critical of forestry policies from a conservation perspective (Grove 1983; RSPB 1985; Tompkins 1986), including a report by one of the Reading study authors, in effect publicly recanting his earlier views (Stewart 1987). A somewhat more measured approach was taken by Denne, Bown and Abel (1986), who urged a continuing programme of forestry expansion but with much stricter environmental safeguards, including a three-tier system of land use designation with, at the apex, heritage sites (e.g. SSSIs) where there would be a strong presumption against forestry.

The forestry lobby was not slow to respond to the new environmental onslaught. In the uplands it countered with scientific arguments of its own (Timber Growers UK 1986). It issued guidelines for its members on *forestry practice in harmony with nature and the community* (Timber Growers UK 1985). Moreover the more politically astute of private forestry proponents welcomed the diversion caused by one of the Forestry Commission's own responses to environmental criticism – the launching in 1985 of a new Broadleaved Woodland Policy with grants to encourage broadleaved plantations in the lowlands as well as the uplands (Watkins 1986). The Broadleaved policy had its immediate origins in a symposium organised by the Commission and the Institute of Chartered Foresters (Malcolm, Evans and Edwards 1982) and in a subsequent consultation paper (Forestry Commission 1984).

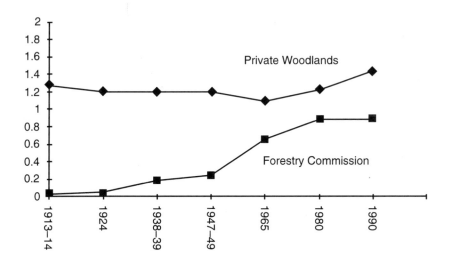

Figure 11.1 Areas of woodland (million hectares) in Great Britain

Source: Grayson 1993: 45

For the first time since the establishment of the Commission, policy attention was directed firmly at the lowlands. At a time of focus on alternative uses for 'surplus' agricultural land (Cox *et al.* 1986) the broadleaved debate and initiative provided a welcome, but short-lived, respite for hard-pressed upland forestry interests.

The impact of the broadleaved policy, which in its original form lasted only to 1988, as with its successors (the Woodland Grant Scheme and the Farm Woodland Scheme), has been modest in terms of the total areas planted. But between 1982/83 and 1987/88 there were impressive increases in the use of broadleaves as a proportion of new planting and restocking in Great Britain as shown in Figure 11.2.

In 1986, it was 'business as usual' in terms of the periodic battles between the essentially upland forestry interest and its critics. This time the critique of the economics of forestry was provided by PIEDA plc (1986) in a report to the National Audit Office, which broadly replicated the 1972 Treasury study. The National Audit Office subsequently published its own critical review of the Commission (NAO 1986). The reports rejected investment in upland forestry on import-saving, employment and environmental grounds. As usual the response from the industry was quick and effective:

> For much of the time the forestry lobby appears to be occupied in a game of musical chairs. When a crisis occurs, and the music stops, there is a remarkably speedy and effective concentration on the immediate problem at hand.... Further proof that it is an attack on its economic credibility that galvanises the forestry world, came after the publication

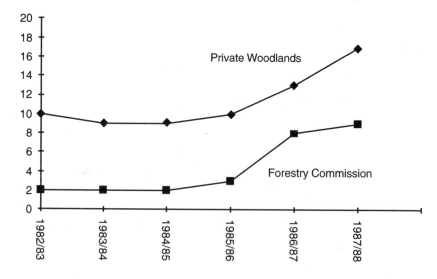

Figure 11.2 Percentage use of broadleaves for new planting and restocking in Great Britain

Source: Forestry Commission 1989a: 4

of the highly critical NAO report of 1986. Once again the lobby regrouped, this time into the Forestry Industry Committee of Great Britain (FICGB), in 'one of the most significant public relations exercises ever undertaken by the industry' (Forestry and British Timber magazine February 1987).

(Tompkins 1989: 92–93)

A rapid response was given by a firm of consultants commissioned by the FICGB (Firn Crichton Roberts 1987) and in the fullness of time a more substantial 'glossy' case for the defence appeared, this time putting much emphasis on the importance of forestry not so much for the provision of direct rural employment but as part of a wider timber complex, with considerable up and down stream economic impact (FICGB 1987). But in contrast to earlier occasions, there was not the immediate positive and half-apologetic response from government that the forestry lobby had come to expect in such circumstances. Instead Chancellor Nigel Lawson, perhaps with half an eye on wooing the increasingly vociferous environmental lobby, delivered a body blow to private forestry by putting an end to tax-led forestry investments, with dramatic consequences for planting rates, not-withstanding the introduction of more generous planting grants the follow-ing year (Johnson 1992). In the same year, a joint statement by the Secretary of State for the Environment and the Minister of Agriculture effectively put a ban on further afforestation in the English uplands, although by that time the demand for land for afforestation purposes had slumped throughout the UK as a result of the taxation changes.

THE 1990s: FORESTRY AND THE ENVIRONMENT

By the 1990s, therefore, forestry had become inextricably linked with environmental issues with much of the debate being on the future manage-ment of existing forests and woodlands rather than on new planting schemes. When the Agriculture Committee of the House of Commons (1990) turned its attention to forestry in the 1989/90 session, these issues were very much to the fore. Over 500 pages of evidence are neatly presented by Pryor et al. (1992) in Table 11.2, which summarises the opposing views of the two groups. The 1990 report provided the trigger, not on this occasion for a lobbying response, but for the Forestry Commission itself to seek to regain the initiative in a debate which was fast slipping away from it. In September 1991, the Commission produced a policy statement which accepted the principle of multiple objectives and asserted the importance of environmen-tally sustainable forestry and the delivery of public benefits (Forestry Commission 1991). In a new twist to the debate it also stressed the role forestry could play in absorbing carbon dioxide and thus combating global warming (see also Cmnd 2429, 1994). In addition, the statement announced an internal reorganisation which led to the separation of the Forestry Authority from its commercial arm, Forest Enterprise, in 1994.

Table 11.2 Forest industry and environmental concerns

Issue	Forest industry concerns	Environmental concerns
New planting	Continuity of future timber supply to maintain and develop the established industry's structure and viability.	Loss of semi-natural habitats and access to them; damage to the landscape and historic sites; acidification of water catchments.
Replanting	Loss of potential timber yield and reduction of profitability from requirements to use broadleaves and native species.	Use of inappropriate species, perpetuating unattractive landscapes and loss of semi-natural species composition.
Management	The increasing complexity of management and the opportunity costs of incorporating environmental requirements.	Loss of conservation value, non-woodland habitats and other features, through insensitive management or neglect.
Felling	Extension of rotations due to limitations on rates of felling and size of felling areas, and design which increase costs.	Insensitive felling design and rates, detracting from attractive landscapes, and causing losses of mature timber habitats.
Public benefits and consultation	Too much uncertainty and expensive time involved in consultation over both planting and felling proposals; low public opinion and understanding of forestry.	Inadequate public involvement and accommodation of objections; insufficient public access and social benefits from publicly subsidised forests; threats to state ownership and loss of access.

Source: Pryor *et al.* 1992: 19

In many respects the Commission's initiative was successful and anticipated many of the key issues highlighted by the Commons Environment Committee in another scrutiny already under way (House of Commons 1993). Amongst the Environment Committee's recommendations were the following:

- The government should set adequate priorities for achieving multiple forestry objectives.
- Forestry policy should be co-ordinated with other rural policies.
- There is a need for national and UK forestry strategies.
- The Countryside Commission and English Nature should be involved at an early stage in all aspects of forestry policy.
- Indicative Forestry Strategies should be developed by local authorities and used to stimulate the conservation and management of existing woodland resource as well as its expansion.
- The protection of ancient and semi-natural woodlands should be strengthened.

- The Forestry Commission's environmental guidelines should be adhered to at all stages of woodland management and the results of monitoring made public.

Thus, despite the voluminous evidence from the FICGB, Timber Growers UK and the Institute of Chartered Foresters, which amounted to an apology for continued upland conifer planting, the report came down firmly on the side of further reforms to improve the environmental record of forestry.

Few of the proposals presented particular difficulties for the Commission in its new guise as an authority (not an enterprise) anxious to display its environmental credentials. For by now the Commission was facing a new threat not from environmentalists but from within government itself. It was widely rumoured in 1993 and early 1994 that Forest Enterprise would be privatised. The Forestry Commission in a new alliance with environmental groups mounted a campaign to persuade government of the need to retain the state forests within the public sector for reasons of maintaining public access and implementing many of the new multiple objectives policies that had now been accepted. In August 1994, the publication of the conclusions of a two-year policy review by the government, which rejected privatisation (Cmnd 2644), was a notable success story for this new alliance. The privatisation threat was transmuted into the much more modest proposal for a Next Steps Agency. The environmental groups immediately swung into action – by November the CPRE, in addition to making its own lengthy submission in response to the policy review statement, had joined with the RSPB in publishing a policy document examining in detail the Next Steps proposals and how they should be implemented (Richards 1994).

Before leaving forestry policy developments, and perhaps having given the impression that the environmentalists had won the day, it is worth noting that among the new policy commitments announced by the forestry minister, Ian Lang, in July 1994 in anticipation of the policy review report, was a firm commitment to increase grants for coniferous planting, especially on the better land. For all the emphasis on broadleaves and sustainability, the timber imperative is still strong and the forestry lobby whilst bruised remains active.

THE FORESTRY LOBBY: A CLOSED POLICY COMMUNITY?

In reviewing the main developments in forestry policy this century, this chapter has demonstrated the strength of the forestry lobby. It is now time to sum up, with a little more detail, the nature of this lobby and the extent to which other interests are able to influence policy. In contrast to agricultural policy making, there has been little in the way of systematic analysis of forestry by political scientists. The nearest to political analysis are the works of anti-forestry polemicists such as Shoard (1987) and Tompkins (1989). Shoard confines her remarks to highlighting the landowning bias within the Commission and the failure to take on board amenity and trade union interests. Tompkins' approach is more thorough and helpful, in that he seeks to identify the boundaries and membership of the community:

> There is a range of overlapping memberships amongst the various groups that make up the forestry lobby, which disguises the fact that it has as its core a mere 150 people. Senior staff are active and number about 15 individuals. Sawmilling and timber processing industries field about ten prominent representatives. The higher echelons of the afforestation companies expend a great deal of effort in lobbying, and there are probably about ten key campaigners. There is also a group of perhaps five academics from the forestry universities. The real heart of the forestry lobby lies with the major landowners. They fill the ranks of the TGUK and the CLA, and sit on the Regional Advisory Committees (RACs). Overall, the forestry lobby can be seen to consist of just 50 most conspicuous activists.
>
> (Tompkins 1989: 80)

At the heart of the lobby have been the Timber Growers' Organisation and Timber Growers Scotland, which merged in 1983 to form Timber Growers UK. TGUK has never been a large body, with just 2,654 members in 1983. Its close organisational links with the NFU, at whose London headquarters it has an office, belie the fact that it has never successfully sought many members from the ranks of farmers and small landowners with small areas of woodland. It has been, and remains, dominated by large landowners, including the forestry companies (Tompkins 1989). Its high subscription rates, in marked contrast to the less influential Royal Forestry Society, suggest that its main purpose is to act as a highly specialist single-interest group committed to large-scale timber growing. The secret of the timber lobby has been that, in normal times, its day-to-day work has been conducted by a small group of timber growers. It is the TGUK that has negotiated with government on highly specialised matters of taxation and grants. Only in times of crisis has a much wider and potentially rather more unruly constituency been assembled to press the forestry case. The composition of the FICGB, at its formation in 1987, is shown in Table 11.3. Despite the number of organisations involved in the FICGB, forestry is a small world. There are numerous dual memberships and little inherent conflict within the Committee. Indeed were it not so, such a grouping would be counter-productive. It would be, for example, unthinkable in agriculture where, despite the closeness of the policy community, there are on the outer fringes many disparate voices.

There are probably three main reasons why, despite its tight-knit nature, the forestry policy community has failed to withstand such major policy changes in the last decade and is now but a shadow of its former self. First, the community was almost too tight for its own good. Convinced of the rightness of its own case, it failed to recognise the strength of growing

Table 11.3 The Forestry Industry Committee of Great Britain 1987

Institutions	*Forestry companies*
Timber Growers UK	Economic Forestry Group
Institute of Chartered Foresters	Fountain Forestry
Association of Professional Foresters	Tilhill
Royal Forestry Society	Scottish Woodlands
Royal Scottish Forestry Society	
Royal Institution of Chartered Surveyors	*Forestry surveyors and land agents*
Forestry Training Council	John Clegg and Co
Agricultural and Allied Workers Group of the	Bidwells
Transport and General Workers Union	Lowther Scott Harden
Department of Forestry and Natural	Savills
Resources, Edinburgh University	Woosnam and Tyler
Department of Forestry, Aberdeen University	
British Timber Merchants' Association	*Timber-using industries*
Home Timber Merchants' Association	Aaronson Bros
Scotland	Caderboard
UK and Ireland Particleboard Association	Highland Forest Products
British Paper and Board Industry Federation	Shotton Paper
Forestry Section of the Horticultural Trades	St Regis Paper (UK)
Association	Thames Board
Scottish Council for Development and	Western Softwoods
Industry	

Source: Tompkins 1989: 94

opposition. Nor did it sufficiently see the need to shift ground in order to accommodate new interests. Second, its complete dependence upon government financial policies, whether through grant aid or taxation, made it, economically, highly vulnerable and dependent. The whole planting industry could be transformed by fairly straightforward changes to tax or grant rates, a point never lost on environmental groups used to dealing with the vastly more complex worlds of agriculture or energy or transport where policy shifts so often fail to result in the desired changes. Third, it was vulnerable in political terms too. The administrative arrangements which so long worked to its advantage were transparent enough to be vulnerable to a minor revolution *if* that was what government decided. There is of course room for continuing debate on the extent to which the Forestry Commission has been revolutionised in the last few years, but certainly it has shifted profoundly and few would now argue that the Commission serves only as a front for private timber interests.

A good case in point is the composition of the RACs. The RACs have been mentioned several times in this chapter, but their full significance has yet to be explained. Their original terms of reference (under the Forestry Act 1967) were to advise and help the commissioners in many ways, including the following:

- to advise and help the commissioners in understanding the social effect of any of the Commission activities;
- to advise the commissioners as to the development of facilities for recreation and other social amenities;
- to advise on the effect of the commissioners' activities on the countryside and other land users, especially in agriculture and forestry, so as to ensure the continued good relationship with all other regional bodies directly interested in the land, rural protection, nature conservation and the welfare of the countryside;
- to advise the commissioners on local authority regional strategies and structure plans and local plans in respect of forestry;
- to assist the commissioners in reconciling differences of view arising in connection with applications for grant aid for private woodlands or for felling licences, when requested to do so by the commissioners' Conservator for the region, by consulting with the parties concerned as necessary and advising the Conservator of the Committee's findings.

The RACs key political role has been this last one, and in 1974 they were given a further conciliatory role – all cases of dispute have to go to the RAC. For many years the RACs were seen as nothing short of window dressing. Their proceedings were conducted in private and membership was predominantly confined to forestry interests. During the 1980s their work came under intense scrutiny and for some the conclusions were short and to the point: 'It is hard to conceive a more authoritarian, secretive or biased procedure, compared to which the procedures of town and country planning are models of democracy, openness and fairness' (MacEwen and MacEwen 1982: 291). Others were more sanguine (e.g. Hetherington 1988), but the more radical environmentalists called for planning controls to be extended to forestry operations and to replace the work of the RACs (Shoard 1980, 1987). The main counter to this has been to strengthen the role of environmental interests in the RACs. In 1985 they were required to balance environmental and forestry interests in their deliberations. And, crucially, in 1991 it was announced that henceforth RAC meetings and their recommendations would be published (but the meetings continued to take place in private), and that environmental representation would be increased on the RACs. The new composition was four members from forestry/timber interests, four from environmental interests, and four from other groups such as MAFF and the trades unions. In some ways this increased representation was a hollow victory, for forestry had changed so much already that in England, for example, no disputed afforestation case had been put to an RAC since 1987. But in many ways the broadening role of the RACs provides a good example of how policy arrangements which once served to bolster one interest can be recast to strengthen another.

CONCLUSIONS

Despite the changes of recent years, forestry remains a contentious area of rural policy. Few would deny that the forestry lobby remains powerful and that the inherent strength of such a tight-knit policy community means that any opportunities for policy changes that would give advantages to private sector forestry will be grasped. New alliances with the farming lobby over alternative land uses, and with those concerned with energy and climate change, offer potential justifications for an increased rate of upland afforestation in the future. Meanwhile, much attention will be focused on local authorities, urged to produce indicative forestry strategies, on the efforts of the Countryside Commission, MAFF and others to encourage planting in the lowlands, and the impact of the changed status of Forest Enterprise. As Fiona Reynolds, the Director of CPRE, and Barbara Young, Chief Executive of the RSPB, explain in their foreword to their joint report on a Next Steps forestry agency:

> The privatisation of the Forestry Commission's woodlands remains a real possibility in the future. This report concludes that structural and management changes in Forest Enterprise need to be driven by a desire for improved efficiency and transparency rather than by a desire to pave the way for future privatisation.
>
> (Richards 1994: 5)

Privatisation, should it take place in the future, would undoubtedly result in a renewed round of debate on planning controls over forestry operations. Increasingly the debate will shift from planting to management of existing woods. Much of the planting inspired by the policies of the Commission during this century is now coming to maturity. Timber production in Great Britain increased by 50 per cent between 1980 and 1994 and is set to double by the year 2010 (Cmnd 2644, 1994). The manner in which harvesting is carried out and the nature of replanting will be one of the major policy issues in the future. It is somewhat ironic that the forestry lobby may well find itself increasingly fighting to be allowed to cut down the very trees which it had to battle so hard to plant in the first place. The forestry policy community is likely to remain beleaguered.

12

CONCLUSIONS: EMERGING THEMES IN RURAL POLICY

INTRODUCTION

The forces determining both the nature and the pace of agricultural policy change that have emerged in the early 1990s are both varied and complex. They are, moreover, in certain respects contradictory and the tensions arising from these contradictions are at the heart of a continuing debate on policy which seems set to continue well into the next century. The re-politicisation of agriculture seems, at present, to be irreversible. The main forces at work are as follows:

- the restructuring and redefinition of the British countryside as a sphere of leisure consumption rather than primary production.
- international and national concerns and imperatives regarding environmental sustainability and biodiversity.
- European policy initiatives to reform the CAP.
- the UK Conservative government's commitment to free market principles, wealth creation, limits on public spending and a repudiation of corporatism.
- the continuing strength of the UK farm and landowning lobby within a policy community that retains at least some degree of closure.

This final chapter examines each of these themes briefly in turn, so as to highlight some of the key emerging issues for students of agricultural and environmental politics.

A RESTRUCTURED COUNTRYSIDE

This has not been a central theme of this volume (although chapter 7 represents a statement about the relevant cultural history of the countryside), but underlying many of the policy changes observed and discussed have been assumptions about the changing demands made upon the countryside. The restructuring thesis (see Marsden *et al.* 1990 and 1993; Murdoch and Marsden 1994) concerns the geographical distribution both of economic activity and of people. Rural areas in the western world have become sites for new forms of economic activity, such as high-tech and service businesses, and new sites for the relocation of manufacturing from the old urban-industrial centres.

This together with long-distance commuting and retirement has meant a dramatic repopulation of all but the remotest countryside during the last two decades. The new populations present both possibilities and challenges to the farming community. They provide a stimulus for new forms of economic activity on farms, stimulating the demand for on-farm recreation facilities, farm-processed foods, etc. But the traditional market in farm property is disrupted, with a growing demand for farmhouses and buildings for residential rather than productive purposes – 'positional goods' (Hirsch 1976) in a highly segmented market place.

The new populations make demands upon the environment too, demands for conservation, controls over pollution, and for access for recreation and leisure. The countryside as a 'social construction' is interpreted in contrasting ways by different groups within an increasingly fragmented and pluralistic society (Clark *et al.* 1994). The future pace and nature of these trends will be an important factor shaping the demands placed upon agriculture and articulated within the environmental movement. Marsden *et al.* 1993 have highlighted the importance of regional and local variation in the nature of restructuring, suggesting that any set of national policies will have different policy consequences, and give rise to different pressures on the policy process, according to the 'developmental trajectories of rural localities'. They identify four main sets of parameters likely to determine these trajectories:

- *economic parameters*, most notably the structure of the local economy – its buoyancy and diversity (rate of growth, level of unemployment etc.), and the role of the state in the local economy (level of dependence upon state agencies and state financial support etc.);
- *social parameters*, including demographic structure, rate of population change, influence of the 'middle' class, level of commuting and proportion of retirees;
- *political parameters*, including ideals of representation (who is a legitimate representative); forms of participation (e.g. level of interest-group activity); type of politics (whether, for example, organised around production interests or the protection of positional goods); and
- *cultural parameters*, including dominant attitudes towards property rights (e.g. land as heritage (stewardship), as productive asset, as fictitious asset) and a sense of locality/community ('belonging').

(Marsden *et al.* 1993: 186–187)

THE POLITICS OF SUSTAINABILITY

Whatever the demands made upon the agricultural environment at the local level, the imperative set by global concerns is also important. The commitments facing the UK under the terms of the Fifth Environmental Action Programme of the EU (1992–2000) and the UN Rio Convention on Biological Diversity of 1992 are likely increasingly to influence the conduct of agricultural politics in the UK. The EU programme, entitled 'Towards sustainability', represents a further development of a framework designed to facilitate environmental planning at all stages of the policy process across all policy sectors and with long-term targets for sustainability and environmental regeneration:

- It focuses on the agents and activities which deplete natural resources and otherwise damage the environment, rather than wait for problems to emerge;
- It endeavours to initiate changes in current trends and practices which are detrimental to the environment, so as to provide optimal conditions for socio-economic well-being and growth for the present and future generations;
- It aims to achieve such changes in society's patterns of behaviour through the optimum involvement of all sectors of society in a spirit of shared responsibility, including public administration, public and private enterprise, and the general public (as both individual citizens and consumers);
- Responsibility will be shared through a significant broadening of the range of instruments to be applied contemporaneously to the resolution of particular issues or problems.

(para. 11 of Executive Summary)

Hitherto the implementation of EU environmental policy has been hindered by the lack of consistent and compatible data and information covering Europe as a whole. The European Environment Agency (EEA) established in Copenhagen in October 1993 should help to remedy this, as its chief aim is to provide the member states of the Union with information at a European level. It will also seek to monitor the success or otherwise of EU environmental policies. It is to be expected that, given time, the scrutiny offered by the EEA will be an important element in policy debate.

Sustainable Development: The UK Strategy (Cmnd 2426) launched by the Prime Minister, John Major, at the Mansion House on 25 January 1994, along with *Biodiversity: The UK Action Plan* (Cmnd 2428) and papers on climate change and forestry (Cmnds 2427 and 2429), provide an attempt to forge a coherent and holistic policy covering commitments under both the EU programme and the Rio Convention. However, a continuing reliance on disparate policy measures within only the most general of overall policy frameworks represents a departure from the principles of biodiversity and sustainability as understood by many environmentalists (Winter 1994). In particular the UK's reliance on special sites appears to be in breach of Article 8 (c) of the Rio Convention on Biological Diversity which states that 'each contracting party shall, as far as possible and appropriate, regulate or manage biological resources important for the conservation of biological diversity whether within *or outside* protected areas, with a view to ensuring their conservation and sustainable use' (Cmnd 2127, 1993) (my emphasis).

Moreover, the reports highlight continuing and disturbing evidence of a loss of habitats and of landscape features. The Countryside Survey 1990, undertaken by the Institute of Terrestrial Ecology and the Institute of Freshwater Ecology, discovered that between 1978 and 1990 there had been a 13 per cent loss of species richness in semi-improved grasslands, 14 per cent in woodlands and an 11 per cent loss in upland grass (Department of the Environment 1993). The study also uncovered a net loss of 23 per cent of hedgerows between 1984 and 1990, either as a result of removal or degradation. The targets set by the government in *Biodiversity: The UK Action Plan* (Cmnd 2428) are primarily directed towards the continuation of well established procedures, the improvement of monitoring and the honouring of commitments. For example, the government commits itself to compliance with the timetable for the designation of SACs under the Habitats Directive by the year 2004, and to designating new Natural Heritage Areas in Scotland under the terms of the Natural Heritage (Scotland) Act 1991. But few, if any, major policy innovations are fore-shadowed. The reaction of many environmentalists was hostile:

> Despite the presence of the Prime Minister (his first environmental outing since the 1992 Earth Summit) the Mansion House event lacked political substance. There was no new money. No new targets. No changes of responsibility. No policy announcements by the departments ... such as Industry, Transport, Agriculture.
>
> (Rose 1994: 68)

However great the disappointment felt by many environmentalists with progress to date, few could deny that the issue of sustainability is here to stay. As a principle for economic and environmental management its influence on future policy is likely to be profound. The precautionary principle in pollution control and the accepted need to find new uses for land that will increase biodiversity and help to absorb CO^2 are just two examples of how fundamentally policies may change. Already the concept of sustainability itself is being subject to a more rigorous conceptual and empirical analysis with notions such as *environmental impact coefficients* being developed to give more precise meaning (Owens and Cowell 1994). Land inevitably figures highly in such discussions and the use of land will become central to future research efforts designed to facilitate the development of more sustainable economic systems.

REFORMING THE CAP

If sustainability is likely to require a relatively high degree of regulation and government intervention, so too does this appear to be the future path for the CAP. The original expectation that the 1992 reform would be extensively reviewed in 1996 now seems to have passed. With the expected GATT imperative to cut commodity support still further, it is hard to avoid the conclusion that there will be a further pressure for a support policy based on environmental support. In 1992–1993, when the reform policies were being introduced, there seemed, for a while, to be strong pressures bringing environmental and agricultural interests together. Indeed, the reform was widely heralded as an opportunity for a new rapprochement between the two sets of interests. In the event, the agri-environmental elements were a pale shadow of what was hoped for by environmentalists. The main feature of the new agriculture is the highly bureaucratic nature of the new support system. Farmers are confronted by a bewilderingly complicated system of controls. The cost to the public purse remains high and few of the expected environmental benefits have materialised. It can only be a matter of time before a further package of reform is launched.

CONSERVATIVE POLICIES

The UK government elected in 1983 continues a commitment, albeit somewhat weaker than in the Thatcher years, to free market principles, and to removing the last vestiges of corporatism. Therefore within the context of a European policy which promotes high spending and intervention, it is seeking ways of shedding domestic policies that lie outside the CAP framework. A number of the shibboleths of UK agrarian corporatism appear to have been destroyed. The demise of the Milk Marketing Board in 1994 represented a fundamental break with the market control policies and state-sanctioned monopoly based on corporatist policy making of the past. It remains to be seen how far the new arrangements correspond with a free market place, but, whatever the outcome, the symbolic importance of dismantling a structure supported by successive governments for sixty years, is profound.

Less dramatic has been the abolition of the annual review of agriculture. Stripped of its early price-fixing tasks, the annual review remained an opportunity for the NFU to engage with government in a routine consulta-tion exercise. The reform of agricultural holdings legislation in 1995 brought to an end nearly 100 years of increasing tenant security (although none of the legislative provisions are retrospective). The Agricultural Wages Board has come under threat, but so far has survived. Agricultural research has been pushed further into the market place, although, somewhat paradoxically, those research priorities that remain in the gift of the state are increasingly centrally directed.

Together these changes amount, domestically, to a continuing dismantling of the corporatist and interventionist regime. And the continuing de-regulationist philosophy is important in terms of the context of CAP and environmental reforms which seem to point the way to greater bureaucracy and control. It would almost seem as though the UK is attempting to strip away its own traditional policy engagements in the agricultural sector leaving only those which are necessitated by CAP reform and the environmental policies of the Community and international agreement. So far the UK government has resisted the notion of renationalising agricultural policy (Grant 1995), at least openly, but it remains an option that should not be entirely lost sight of. Nor should it be forgotten that the UK has consistently argued for market-led reform of the CAP and even now is attempting to contain and reduce national expenditure at a time when CAP expenditure is rising.

THE FARM LOBBY

It is fitting that the final paragraph of the book should refer to the farm lobby, which has been such a major concern throughout the text. Notwithstanding the enormous changes that have shaken the NFU in recent years, it remains very much at the forefront of policy development. It is a central player in the policy community. On environmental issues, when the government has been forced to adopt interventionist policies, the Union has applied pressure to ensure its commitment to a non-directive policy style through its assertion of the rights of individuals, especially their property rights. The Union has sought the maximum freedom for owners and managers of land to interpret regulations, a philosophy of voluntarism. In many instances of policy it has been successful. Whilst its influence has declined and there are more and more examples of these principles being breached, the legacy of corporatism and self-regulation is strong. The NFU is still a force to be reckoned with in the formulation and implementation of policies for agriculture and the environment. On a range of issues its position is no longer unassailable, but nor have other interest groups emerged capable of challenging the NFU with regard to the depth of its involvement in agricultural policy across such a wide range of issues. In the closing paragraph of *The State and the Farmer*, penned more than thirty years ago, Self and Storing wrote that 'the ending of the close partnership between Government and Union could only be welcomed'. The partnership, as it operated in the 1950s, has ended. There are now far too many issues that divide for it to be otherwise, but the legacy of partnership is deeply felt on both sides and continues to influence greatly the conduct of agricultural policy. It is likely to do so for a considerable time to come.

BIBLIOGRAPHY

Abercrombie, P. (1926) *The Preservation of Rural England*, London: Hodder and Stoughton.
Abrams, M.H. (1971) *Natural Supernaturalism: Tradition and Revolution in Romantic Literature*, London: Oxford University Press.
Adams, W.M. (1984) *Implementing the Act: A Study of Habitat Protection under Section II of the Wildlife and Countryside Act 1981*, Oxford: BANC and WWF.
—— (1986) *Nature's Place: Conservation Sites and Countryside Change*, London: Allen & Unwin.
—— (1991) 'SSSIs: who cares?' *Ecos*, 12 (1), 59–64.
Adams, W.M., Bourn, N.A.D. and Hodge, I. (1992) 'Conservation in the wider countryside: SSSIs and wildlife habitat in eastern England', *Land Use Policy*, 9 (4), 235–248.
AGES (1993a) *Register of Agro-Environmental R&D 1992/93*, London: Report compiled for the Priorities Board by the Advisory Group on the Environment Sector.
—— (1993b) *Second Report of the Advisory Group on the Environment Sector*, to the Priorities Board for Agriculture and Food R&D.
Allen, D.E. (1969) *The Victorian Fern Craze*, London: Hutchinson.
—— (1976) *The Naturalist in Britain: A Social History*, London: Allen Lane.
Andersen, S.S. and Eliassen, K.A. (eds) (1993) *Making Policy in Europe: The Europeification of National Policy-making*, London: Sage.
Anderson, P. (1987) 'The figures of descent', *New Left Review*, 161, 20–77.
Ansell, D.J. and Tranter, R.B. (1992) 'The five-year set-aside scheme in England and Wales: an initial assessment', *Farm Management*, 8 (1), 19–31.
Armstrong, Sir R. (1986) 'Keepers of the cabinet', *Daily Telegraph*, 8 December.
Ashby, A.A., Enfield, R.R. and Lloyd, E.M.H. (1925) 'Stabilization of agricultural prices', *Ministry of Agriculture Economic Series*, 2.
Association of Agriculture (1988) *Studying an Issue: Nitrates in the Water Supply*, London: Association of Agriculture.
Astor, Viscount and Murray, K.A.H. (1932) *Land and Life: The Economic National Policy for Agriculture*, London: Victor Gollancz.
—— (1933) *The Planning of Agriculture*, London: Oxford University Press.
Astor, Viscount and Rowntree, S.B. (1939) *British Agriculture*, London: Penguin.
Averyt, W.F. (1977) *Agropolitics in the European Community*, New York: Praeger.
Bachrach, P. and Baratz, M.S. (1963) 'Decisions and nondecisions: an analytical framework', *American Political Science Review*, 57 (3), 632–641.
Baker, S. (1973) *Milk to Market*, London: Heinemann.
Baldock, D. (1988) *The Nitrates Issue: A Case Study of the Anglian Water Authority, UK*, London: Institute for European Environmental Policy.
Baldock, D. and Beaufoy, G. (1992) *Green or Mean? Assessing the Environmental Value of the CAP Reform 'Accompanying Measures'*, London: CPRE and IEEP.
Baldock, D., Cox, G., Lowe, P. and Winter, M. (1990) 'Environmentally Sensitive Areas: incrementalism or reform?', *Journal of Rural Studies*, 6 (2), 143–162.
Baldwin, N. (1985) 'The House of Lords', unpublished PhD thesis, University of Exeter.
Ball, S. (1991) 'Law and the countryside: the Environmental Protection Act 1990 and nature conservation', *Land Management and Environmental Law Report*, 3 (3), 81–84.
Ball, S. and Bell, S. (1991) *Environmental Law*, London: Blackstone Press.
Barnes, D.G. (1930) *A History of the English Corn Laws 1660–1846*, London: Routledge.
Barnett, L.M. (1985) *British Food Policy in the First World War*, London: Allen & Unwin.
Barrett, S. and Fudge, C. (1981) 'Examining the policy–action relationship', 3–32 in S. Barrett

and C. Fudge (eds), *Policy and Action: Essays on the Implementation of Public Policy*, London: Methuen.

Barrett, S. and Hill, M. (1984) 'Policy, bargaining and structure in implementation theory: towards an integrated perspective', *Policy and Politics*, 12 (3), 219–240.

Bate, R. (1991) *Romantic Ecology: Wordsworth and the Environmental Tradition*, London: Routledge.

Bateman, D.I. (1989) 'Heroes for present purposes? – a look at the changing idea of communal land ownership in Britain', *Journal of Agricultural Economics*, 40 (3), 269–289.

Bebbington, D.W. (1989) *Evangelicalism in Modern Britain: A History from the 1730s to the 1980s*, London: Unwin Hyman.

Beckett, J.V. (1986) *The Aristocracy in England 1660–1914*, Oxford: Basil Blackwell.

Beer, S. (1956) 'Pressure groups and parties in Britain', *American Political Science Review*, 50 (1), 1–23.

—— (1965) *Modern British Politics*, London: Faber & Faber.

Bentley, A.F. (1949) *The Process of Government: A Study of Social Pressures*, Evanston: Principia Press of Illinois (first published 1908).

Beresford, T. (1975) *We Plough the Fields*, Harmondsworth: Penguin.

Berghoff, H. (1990) 'Public schools and the decline of the British economy 1870–1914', *Past and Present*, 129, 148–167.

Birch, A.H. (1964) *Representative and Responsible Government*, London: Allen & Unwin.

Blacksell, M. and Gilg, A. (1981) *The Countryside: Planning and Change*, London: Allen & Unwin.

Blake, R. (1955) *The Unknown Prime Minister: The Life and Times of Andrew Bonar Law, 1858–1923*, London: Eyre & Spottiswoode.

Blewett, N. (1965) 'The franchise in the United Kingdom, 1885–1918', *Past and Present*, 32, 27–56.

—— (1968) 'Free Fooders, Balfourites, Whole Hoggers. Factionalism within the Unionist Party, 1906–1910', *Historical Journal*, 11 (1), 95–124.

Blunden, J. and Curry, N. (eds) (1990) *A People's Charter? Forty Years of the 1949 National Parks and Access to the Countryside Act*, London: HMSO.

Body, R. (1982) *Agriculture: The Triumph and the Shame*, London: Temple Smith.

—— (1984) *Farming in the Clouds*, London: Temple Smith.

—— (1987) *Red or Green for Farmers*, Saffron Walden: Broad Leys Publishing.

Booth, A. (1987) 'Britain in the 1930s: a managed economy', *Economic History Review*, 40 (4), 499–522.

Bottomore, T. (1966) *Elites and Society*, Harmondsworth: Penguin.

Bouquet, M. (1985) *Family, Servants and Visitors: The Farm Household in Nineteenth and Twentieth Century Devon*, Norwich: Geo Books.

—— (1987) 'Bed, breakfast and an evening meal: commensality in the nineteenth and twentieth century farm household in Hartland', 93–104 in M. Bouquet and M. Winter (eds), *Who from Their Labours Rest: Conflict and Practice in Rural Tourism*, Aldershot: Avebury.

Bowers, J.K. (1982) 'Is afforestation economic?' *Ecos*, 3 (1), 4–7.

—— (1985) 'British agricultural policy since the Second World War', *Agricultural History Review*, 33 (1), 66–76.

—— (n.d.) 'Do we need more forests?' University of Leeds, School of Economic Studies Discussion Paper No.137.

Bowers, J.K. and Cheshire, P. (1983) *Agriculture, the Countryside and Land Use*, London: Methuen.

Bowers, J.K., O'Donnell, K. and Whatmore, S. (1988) *Liquid Assets: The Likely Effects of Privatisation of the Water Authorities on Wildlife Habitats and Landscape*, London: CPRE, RSPB and WWF.

Bowler, I.R. (1979) *Government and Agriculture*, London: Longman.

—— (1985) *Agriculture under the Common Agricultural Policy: A Geography*, Manchester: Manchester University Press.

—— (1987) '"Non-market failure" in agricultural policy – a review of the literature for the European Community', *Agricultural Administration and Extension*, 26 (1), 1–15.

Bowman, J.C. (1992) 'Improving the quality of our water: the role of regulation by the National Rivers Authority', *Public Administration*, 70 (4), 565–575.

Boyce, R.W.D. (1988) *British Capitalism at the Crossroads 1919–1932*, Cambridge: Cambridge University Press.

Brand, J. (1992) *British Parliamentary Parties*, Oxford: Clarendon Press.

Brenner, R. (1982) 'The agrarian roots of European capitalism', *Past and Present*, 97, 16–113.

Brooking, T.W.H. (1977) 'Agrarian businessmen organise: a comparative study of the origins and early phase of development of the National Farmers' Union of England and Wales and the New Zealand Farmers' Union, ca 1880–1929', unpublished PhD thesis, University of Otago, New Zealand.

Brown, B.H. (1943) *The Tariff Reform Movement in Great Britain 1881–1895*, New York: AMS Press.

Bryden, J. (1991) 'Rural policy changes in the European Community', Paper presented to Research Programme on Farm Structures and Pluriactivity, Calabria Review Meeting, Nethy Bridge: Arkleton Trust.

Bryden, J., Hawkins, E., Gilliatt, J., MacKinnon, N. and Bell, C. (1992) *Farm Household Adjustment in Western Europe 1987–1991*, Final Report on the Research Programme on Farm Structures and Pluriactivity, Vol. 1, Nethy Bridge: Arkleton Trust.

Buckwell, A.E., Harvey, D.R., Thompson, K.T. and Parton, K.A. (1982) *The Costs of the Common Agricultural Policy*, London: Croom Helm.

Budge, I. (1983) *The New British Political System: Government and Society in the 1980s*, London: Longman.

Burch, M. (1979) 'Policy making in central government', in B. Jones and D. Kavanagh (eds), *British Politics Today*, Manchester: Manchester University Press.

Burrel, M. (1942) 'War-time food production: the work of War Agricultural Executive Committees, West Sussex', *Journal of the Royal Agricultural Society of England*, 103, 70–75.

Burrow, J.W. (1968) 'Editor's introduction', 11–48, in C. Darwin, *The Origin of Species*, Harmondsworth: Penguin.

Butler, D. and Butler, G. (1986) *British Political Facts 1900–1985*, London: Macmillan.

Butterwick, M. and Rolfe, E.N. (1968) *Food, Farming and the Common Market*, Oxford: Oxford University Press.

Cain, P.J. and Hopkins, A.G. (1993) *British Imperialism: Innovation and Expansion 1688–1914*, London: Longman.

Campbell, J. (1993) *Edward Heath*, London: Jonathan Cape.

Cannadine, D. (1990) *The Decline and Fall of the British Aristocracy*, New York and London: Yale University Press.

Carpenter, L.P. (1976) 'Corporatism in Britain, 1930–45', *Journal of Contemporary History*, 11 (1), 3–25.

Carson, R. (1963) *Silent Spring*, London: Hamish Hamilton.

Carter, N. and Lowe, P. (1994) 'Environmental politics and administrative reform', *Political Quarterly*, 65 (3), 263–274.

Carter, N., Klein, R. and Day, P. (1992) *How Organizations Measure Success: The Use of Performance Indicators in Government*, London: Routledge.

CAS (1980) *Strategy for the UK Forestry Industry*, University of Reading, Centre for Agricultural Strategy Report No.6.

Cawson, A. (1986) *Corporatism and Political Theory*, Oxford: Basil Blackwell.

Cawson, A. and Saunders, P. (1983) 'Corporatism, competitive politics and class struggle', 8–27 in R. King (ed.), *Capital and Politics*, London: Routledge & Kegan Paul.

Centre for Rural Studies (1990) *A Review of the Countryside Commission's Countryside Adviser Policy 1988*, Royal Agricultural College, Cirencester: CRS Occasional Paper No. 9.

Chambers, J.D. and Mingay, G.E. (1966) *The Agricultural Revolution 1750–1880*, London: Batsford.

Chapman, D. (1944) *Agricultural Information and the Farmer*, MAF New Series No.38.

Chapman, J. and Seeliger, S. (1991) 'The influence of the agricultural executive committees in the first world war: some evidence from West Sussex', *Southern History*, 13 (1), 105–122.

Cherry, G. (1975) *Environmental Planning 1939–1969, Volume II: National Parks and Recreation in the Countryside*, London: HMSO.

Chester, D.N. (1954) 'The Crichel Down Case', *Public Administration*, 32 (Winter), 389–401.

Clark, G., Darrall, J., Grove-White, R., MacNaughten, P. and Urry, J. (1994) *Leisure Landscapes. Leisure, Culture and the English Countryside: Challenges and Conflicts*, London: CPRE.

Clark, K. (1969) *Civilisation*, London: BBC Books and John Murray.

Clarke, R. (1973) *Ellen Swallow: The Woman Who Founded Ecology*, Chicago: Chicago University Press.

Cloke, P. and MacLaughlin, B. (1989) 'Politics of the Alternative Land Use and Rural Economy (ALURE) proposals in the UK: crossroads or blind alley?' *Land Use Policy*, 6 (3), 235–248.

Clutterbuck, C. and Lang, T. (1982) *More Than We Can Chew: The Crazy World of Food and Farming*, London: Pluto Press.

Cmnd 1319 (1902) *Report of the Departmental Committee Appointed by the Board of*

Agriculture to Enquire into and Report upon British Forestry.

Cmnd 4460 (1909) *Second Report (on Afforestation) of the Royal Commission Appointed to Inquire into and to Report on Certain Questions Affecting Coast Erosion, the Reclamation of Tidal Lands and Afforestation in the United Kingdom.*

Cmnd 6713 (1913) *Report of the Committee Appointed by the Board of Agriculture and Fisheries to Advise on Matters Relating to the Development of Forestry.*

Cmnd 9079 (1918) *Final Report of the Agricultural Policy Sub-Committee of the Reconstruction Committee* (Selborne Committee).

Cmnd 8881 (1918) *Final Report of the Reconstruction Committee Forestry Sub-Committee* (Acland Committee).

Cmnd 1582 (1922) *Second Interim Report of the Committee on National Expenditure* (Geddes Report).

Cmnd 6378 (1942) *Report of the Committee on Land Utilisation in Rural Areas* (Scott Committee), London: HMSO.

Cmnd 6433 (1943) *Report of the Committee on Post-War Agricultural Education in England and Wales* (Luxmoore Committee), London: Ministry of Agriculture and Fisheries.

Cmnd 6447 (1943) *Post-War Forestry Policy, Report by HM Forestry Commissioners*, London: Forestry Commission.

Cmnd 6500 (1944) *Post-War Forestry Policy, Private Woodlands, Supplementary Report by HM Forestry Commissioners*, London: Forestry Commission.

Cmnd 6628 (1945) *National Parks in England and Wales* (Dower Report), London: HMSO.

Cmnd 6631 (1945) *National Parks: A Scottish Survey* (Ramsay Committee), Edinburgh: HMSO.

Cmnd 7121 (1947) *Report of the National Parks Committee (England and Wales)* (Hobhouse Committee), London: HMSO.

Cmnd 7122 (1947) *Report of the Wildlife Conservation Special Committee* (Huxley Committee), London: HMSO.

Cmnd 7207 (1947) *Report of the Special Committee on Footpaths and Access to the Countryside* (Hobhouse Committee), London: HMSO.

Cmnd 7235 (1947) *National Parks and the Conservation of Nature in Scotland* (Ramsay Committee), Edinburgh: HMSO.

Cmnd 9732 (1956) *Report of the Committee Appointed to Review the Provincial and Local Organization and Procedures of the Ministry of Agriculture, Fisheries and Food* (Arton Wilson Committee), London: MAFF.

Cmnd 1249 (1960) *Agriculture – Report on Talks between the Agriculture Departments and the Farmers' Unions*, London: MAFF.

Cmnd 6020 (1975) *Food from Our Own Resources*, London: MAFF.

Cmnd 6371 (1976) *Air Pollution Control: An Integrated Approach*, Royal Commission on Environmental Pollution, London: HMSO.

Cmnd 7458 (1979) *Farming and the Nation*, London: Ministry of Agriculture, Fisheries and Food.

Cmnd 7644 (1979) *Agriculture and Pollution*, Seventh Report of the Royal Commission on Environmental Pollution, London: HMSO.

Cmnd 9149 (1984) *Tackling Pollution: Experience and Prospects*, Royal Commission on Environmental Pollution, London: HMSO.

Cmnd 1200 (1990) *This Common Inheritance*, London: HMSO.

Cmnd 1966 (1992) *Freshwater Quality*, Royal Commission on Environmental Pollution, London: HMSO.

Cmnd 2127 (1993) *Convention on Biological Diversity*, Rio Declaration, London: HMSO.

Cmnd 2203 (1993) *The Government's Expenditure Plans 1993/94 to 1995/96. Departmental Report by the Ministry of Agriculture, Fisheries and Food and Intervention Board*, London: HMSO.

Cmnd 2426 (1994) *Sustainable Development: The UK Strategy*, London: HMSO.

Cmnd 2427 (1994) *Climate Change: The UK Programme*, London: HMSO.

Cmnd 2428 (1994) *Biodiversity: The UK Action Plan*, London: HMSO.

Cmnd 2429 (1994) *Sustainable Forestry: The UK Programme*, London: HMSO.

Cmnd 2644 (1994) *Our Forests – The Way Ahead: Enterprise, Environment and Access*, Conclusions from the Forestry Review, Edinburgh: HMSO.

Cmnd 2803 (1995) *The Government's Expenditure Plans 1995–96 to 1997–98: Departmental Report by the Ministry of Agriculture, Fisheries and Food and the Intervention Board*, London: HMSO.

Colman, D. (1994) 'Comparative evaluation of environmental policies: ESAs in a policy context', 219–246 in M. Whitby (ed.), *Incentives for Countryside Management: The Case of Environmentally Sensitive Areas*, Wallingford: CAB International.

Colman, D. and Lee, N. (1988) *Evaluation of the Broads Grazing Marshes Conservation Scheme*, Department of Agricultural Economics, University of Manchester.

Commission of the European Communities (1979) *The Agricultural Policy of the European Community*, Luxembourg: European Documentation Series.

—— (1980) *Reflections on the Common Agricultural Policy*, Luxembourg: European Documentation Series.

—— (1981) *Report on the Mandate*, Luxembourg: Bulletin EC, Supplement 1/81.

—— (1982) 'Guidelines for European agriculture', 63–107 in ECC, *A New Impetus for the Common Policies: Follow-up to the Mandate of 30 May 1980*, Bulletin of the European Communities, Supplement 4/81, Luxembourg: Office for Official Publications of the EC.

—— (1985) *Perspectives for the Common Agricultural Policy*, Luxembourg: Office for Official Publications of the EC.

—— (1988) *The Future of Rural Society*, Luxembourg: Office for Official Publications of the EC.

—— (1991) *The Development and Future of the Common Agricultural Policy*, Bulletin of the European Communities, Supplement 4/88, Luxembourg: Office for Official Publications of the EC.

—— (1993) *The Agricultural Situation in the Community 1992 Report*, Luxembourg: Office for Official Publications of the EC.

—— (1994) *Twenty-Third Financial Report on the European Agricultural Guidance and Guarantee Fund*, Luxembourg: Office for Official Publications of the EC.

Cooke, G.W. (ed) (1981) *Agricultural Research 1931–1981*, London: Agricultural Research Council.

Cooper, A.F. (1989) *British Agricultural Policy 1912–36: A Study in Conservative Politics*, Manchester: Manchester University Press.

Courthope, G.L. (1944) 'Post-war agricultural policy', *Journal of the Royal Agricultural Society of England*, 105, 70–71.

Cox, G. and Lowe, P. (1983a) 'A battle not the war: the politics of the Wildlife and Countryside Act', *Countryside Planning Yearbook*, 4, 48–77.

—— and —— (1983b) 'Countryside politics: goodbye to goodwill?', *Political Quarterly*, 54 (3), 268–282.

Cox, G., Lowe, P. and Winter, M. (1985) 'Changing directions in agricultural policy: corporatist arrangements in production and conservation policies', *Sociologia Ruralis*, 25 (2), 130–154.

——, —— and —— (1986a) 'From state direction to self regulation: the historical development of corporatism in British agriculture', *Policy and Politics*, 14 (4), 475–490.

——, —— and —— (1986b) 'Agriculture and conservation in Britain: a policy community under siege', 181–215 in G. Cox, P. Lowe and M. Winter (eds), *Agriculture: People and Policies*, London: Allen & Unwin.

——, —— and —— (1987) 'Political regionalism and the land: the case of the Farmers' Union of Wales', unpublished working paper.

——, —— and —— (1988) 'Private rights and public responsibilities: the prospects for agricultural and environmental controls', *Journal of Rural Studies*, 4 (4), 323–337.

——, —— and —— (1989) 'The farm crisis in Britain', 113–134 in D. Goodman and M. Redclift (eds), *The International Farm Crisis*, London: Macmillan.

——, —— and —— (1990a) 'Agricultural regulation and the politics of milk production', 169–198 in C. Crouch and R. Dore (eds), *Corporatism and Accountability: Organized Interests in British Public Life*, Oxford: Clarendon Press.

——, —— and —— (1990b) 'The political management of the dairy sector in England and Wales', 82–111 in T. Marsden and J. Little (eds), *Political, Social and Economic Perspectives on the International Food System*, Aldershot: Avebury.

——, —— and —— (1990c) *The Voluntary Principle in Conservation: A Study of the Farming and Wildlife Advisory Group*, Chichester: Packard.

—— and —— (1991) 'The origins and early development of the National Farmers' Union', *Agricultural History Review*, 39 (1), 30–47.

Cox, G., Flynn, A., Lowe, P. and Winter, M. (1986) *Alternative Uses of Agricultural Land in England and Wales*, Berlin: International Institute for Environment and Society.

Cripps, Sir J. (1979) *The Countryside Commission: Government Agency or Pressure Group?* Town Planning Discussion Paper No.21, University College London.

Crowe, Dame S. (1978) *The Landscape of Forests and Woods*, Forestry Commission Booklet No. 44, London: HMSO.

———— (1986) 'The forest's potential for recreation and landscape', 144–146 in D. Jenkins (ed.), *Symposium on Trees and Wildlife in the Scottish Uplands*, Abbots Ripton: Institute of Terrestrial Ecology.

Cunningham, G. (1963) 'Policy and practice', *Public Administration*, 41 (3), 229–237.

Curry, N. (1985) 'Countryside recreation sites policy: a review', *Town Planning Review*, 56 (1), 70–89.

———— (1994) *Countryside Recreation, Access and Land Use Planning*, London: E. & F. N. Spon.

Dahl, R.A. (1961) *Who Governs?* New Haven: Yale University Press.

Danziger, R. (1988) *Political Powerlessness: Agricultural Workers in Post-War England*, Manchester: Manchester University Press.

Daunton, M.J. (1989) '"Gentlemanly capitalism" and British industry 1820–1914', *Past and Present*, 122, 119–158.

Davey, B., Josling, T.E. and McFarquhar, A. (eds) (1976) *Agriculture and the State*, London: Macmillan.

Davidson, J. and Wibberley, G. (1977) *Planning and the Rural Environment*, Oxford: Pergamon Press.

Day, W. (1963) *Fair Prospect for Farming*, private publication.

———— (1968) 'We need trade union leaders', *Farmer and Stockbreeder*, 19 March, 23.

Dearlove, J. and Saunders, P. (1984) *Introduction to British Politics*, Cambridge: Polity Press.

Dellheim, C. (1982) *The Face of the Past: The Preservation of the Medieval Inheritance in Victorian England*, Cambridge: Cambridge University Press.

Denman, D.R. (1958) *Origins of Ownership*, London: Allen & Unwin.

Denman, R. and Denman, J. (1990) *A Study of Farm Tourism in the West Country*, Exeter: West Country Tourist Board.

Denne, T., Bown, M.J.D. and Abel, J.A. (1986) *Forestry: Britain's Growing Resource*, London: CEED.

Densham, H.A.C. (1989) *Scammell and Densham's Law of Agricultural Holdings*, London: Butterworth.

Department of the Environment (1974) *Second Report of Steering Committee on Water Quality.*

———— (1983) *Agriculture and Pollution*, The Government's Response to the Seventh Report of the Royal Commission on Environmental Pollution, London: HMSO.

———— (1986) *Nitrate in Water*, Report of the Nitrate Co-ordination Group, London: HMSO.

———— (1987) *Planning Policy Guidance on Nature Conservation*, London: DoE.

———— (1988) *Assessment of Groundwater Quality in England and Wales*, prepared for the DoE by Sir William Halcrow and Partners in association with Lawrence Gould Consultants.

———— (1991) *Policy Appraisal and the Environment: A Guide for Government Departments*, London: HMSO.

———— (1992) *The Countryside and the Rural Economy*, PPG7, London: HMSO.

———— (1993) *Countryside Survey 1990 Main Report*, London: Department of the Environment.

Dewey, C.J. (1974) 'The rehabilitation of the peasant proprietor in nineteenth century economic thought', *History of Political Economy*, 6 (1), 17–47.

Dewey, P.E. (1989) *British Agriculture in the First World War*, London: Routledge.

Diamond, J.M. (1975) 'The island dilemma: lessons of modern biogeographic studies for the design of natural reserves', *Biological Conservation*, 7 (2), 129–145.

Dilnot, A.W. and Morris, C.N. (1982) 'The distributional effects of the Common Agricultural Policy', *Fiscal Studies*, 3 (2), 92–100.

Dintenfass, M. (1992) *The Decline of Industrial Britain 1870–1980*, London: Routledge.

Dobson, A. (1990) *Green Political Thought*, London: Harper Collins.

Doig, A. (1989) 'The resignation of Edwina Currie: a word too far', *Parliamentary Affairs*, 42 (3), 317–329.

Donoughue, B. (1987) *Prime Minister: The Conduct of Policy under Harold Wilson and James Callaghan*, London: Jonathan Cape.

Douglas, R. (1976) *Land, People and Politics: A History of the Land Question in the United Kingdom 1878–1952*, London: Allison & Busby.

Drewry, G. (ed.) (1989) *The New Select Committees*, Oxford: Clarendon Press.

Drummond, I.M. (1974) *Imperial Economic Policy 1917–1939*, London: George Allen & Unwin.

Dunbabin, J.P.D. (1963a) 'The politics of the establishment of county councils', *Historical Journal*, 6 (2), 238–250.

———— (1963b) 'The "Revolt of the Field": the agricultural labourers' movement in the 1870s', *Past and Present*, 26, 68–97.

—— (1968) 'The incidence and organisation of agricultural trade unionism in the 1870s', *Agricultural History Review*, 16, 114–141.

Dunleavy, P. (1990) 'Government at the centre', 97–125 in P. Dunleavy, A. Gamble and G. Peel (eds), *Developments in British Politics 3*, London: Macmillan.

Dunleavy, P. and O'Leary, B. (1987) *Theories of the State: The Politics of Liberal Democracy*, London: Macmillan.

Dunleavy, P. and Rhodes, R.A.W. (1990) 'Core executive studies in Britain', *Public Administration*, 68 (1), 3–28.

Easton, D. (1953) *The Political System*, New York: Knopf.

Edwards, A.M., Gasson, R.M., Haynes, J.E. and Hill, G.P. (1994) *Socio-Economic Evaluation of the Capital Grant Element of the Farm Diversification Grant Scheme*, Wye College, Farm Business Unit Occasional Paper No. 22.

Edwards, R. (1991) *Fit for the Future: Report of the National Parks Review Panel*, Cheltenham: Countryside Commission.

Ekins, P. (ed.) (1986) *The Living Economy*, London: Routledge & Kegan Paul.

Eldon, J. (1988) 'Agricultural change, conservation and the role of advisors', *Ecos*, 9 (4), 14–20.

Emy, H.V. (1972) 'The impact of financial policy on English party politics before 1914', *Historical Journal*, 15 (1), 103–121.

Ensor, Sir R. (1936) *England 1870–1914*, Oxford: Clarendon Press.

Ernle, Lord (1912, 6th edn 1961) *English Farming Past and Present*, London: Heinemann.

Evans, D. (1992) *A History of Nature Conservation in Britain*, London: Routledge.

Evans, J.L. (1936) 'The financing of agricultural education', *Agricultural Progress*, 13 (1), 28–35.

Eversley, Lord (1910) *Commons, Forests and Footpaths*, London: Cassell.

Eyerman, R. and Jamison, A. (1989) 'Environmental knowledge as an organizational weapon: the case of Greenpeace', *Social Science Information*, 28 (1), 99–119.

Fairbrother, N. (1972) *New Lives New Landscapes*, Harmondsworth: Penguin.

Farrer, A. (1994, 1st edn 1964) *Saving Belief*, London: Mowbray.

Fedden, H.R. (1968) *The Continuing Purpose*, London: Longman.

Fennell, R. (1979) *The Common Agricultural Policy of the European Community*, London: Granada.

—— (1985) 'A reconsideration of the objectives of the Common Agricultural Policy', *Journal of Common Market Studies*, 2 (3), 257–276.

—— (1987) 'Reform of the CAP: shadow or substance?', *Journal of Common Market Studies*, 26 (1), 61–77.

FICGB (1987) *Beyond 2000: The Forest Industry of Great Britain*, London: Forestry Industry Committee of Great Britain.

—— (1989) *Options for British Forestry*, London: Forestry Industry Committee of Great Britain.

Firn Crichton Roberts Ltd (1987) *The Forestry Industry Response to the National Audit Office Report*, prepared for the FICGB, Edinburgh.

Fisher, J.R. (1978) 'The Farmers' Alliance: an agricultural protest movement of the 1880s', *Agricultural History Review*, 26 (1), 15–25.

Fleming, D. (1961) 'Charles Darwin, the anaesthetic man', *Victorian Studies*, 4 (3), 219–236.

Fletcher, T.W. (1973) 'The great depression of English agriculture, 1873–1896', 30–55 in P.J. Perry (ed.), *British Agriculture 1875–1914*, London: Methuen.

Flynn, A. (1986) 'Agricultural policy and party politics in post-war Britain', 216–236 in G. Cox, P. Lowe and M. Winter (eds), *Agriculture: People and Policies*, London: Allen & Unwin.

—— (1989) 'Rural working class interests in party policy-making in post-war England', unpublished PhD thesis, University College London.

Forestry Commission (1977) *The Wood Production Outlook in Great Britain*, London: Forestry Commission.

—— (1984) *Broadleaves in Britain: A Consultation Paper*, Edinburgh: Forestry Commission.

—— (1985) *Taxation of Woodlands*, Leaflet No. 12, London: HMSO.

—— (1989a) *Broadleaves Policy: Progress 1985–88*, Edinburgh: Forestry Commission.

—— (1989b) *Taxation of Woodlands*, Bulletin 64, London: HMSO.

—— (1991) *Forestry Policy for Great Britain*, Edinburgh: Forestry Commission.

Fox, W. (1984) 'Deep ecology: a new philosophy of our time?', *The Ecologist*, 14 (5/6), 194–200.

From, J. and Stava, P. (1993) 'Implementation of community law: the last stronghold of national control?' 55–67 in S.S Andersen and K.A. Eliassen (eds), *Making Policy in Europe: The Europeification of National Policy-making*, London: Sage.

Fry, G.K. (1986) 'Inside Whitehall', 88–106 in H. Drucker, P. Dunleavy, A. Gamble and G. Peele (eds), *Developments in British Politics 2*, London: Macmillan.

Gamble, A. (1994) 'The British *ancien régime*', in H. Kastendiek and R. Stinshoff (eds), *Changing Conceptions of Constitutional Government, Developments in British Politics and the Constitutional Debate Since the 1960s*, Bochum: Universitäts Verlag Dr. N. Brockheyer.

Gardner, B. (1987) 'The Common Agricultural Policy: the political obstacle to reform', *Political Quarterly*, 58 (2), 167–179.

Garner, R. (1993) 'Political animals: a survey of the animal protection movement in Britain', *Parliamentary Affairs*, 46 (3), 333–352.

Garthwaite, P.F. (1972) 'Forestry policy – 1972: statement by the Royal Forestry Society of England and Wales', *Quarterly Journal of Forestry*, 66 (4), 289–295.

General Secretariat of the Council of the European Union (1993) *Fortieth Review of the Council's Work, 1st January – 31st December 1992*, Luxembourg: Office for Official Publications of the EC.

Giddens, A. (1973) *The Class Structure of the Advanced Societies*, London: Hutchinson.

Giddings, P.J. (1974) *Marketing Boards and Ministers: A Study of Agricultural Marketing Boards as Political and Administrative Instruments*, Farnborough: Saxon House.

—— (1985) 'The agriculture committee', 57–71 in G. Drewry (ed.), *The New Select Committees: A Study of the 1979 Reforms*, Oxford: Clarendon Press.

Glasier, J.B. (1921) *William Morris and the Early Days of the Socialist Movement*, London: Longman, Green & Co.

Goldsmith, E. (ed.) (1971) *Can Britain Survive?*, London: Tom Stacey.

—— (1972) *A Blueprint for Survival*, London: Tom Stacey.

Goldsmith, M.J.F. (1963) 'The genesis of the Milk Marketing Board and its work, 1933–1939', unpublished MA thesis, University of Manchester.

Gollin, A. (1965) *Balfour's Burden: Arthur Balfour and Imperial Preference*, London: Anthony Blond.

Gosse, E. (1949) *Father and Son* (first published 1907), Harmondsworth: Penguin.

Gough, I. (1979) *The Political Economy of the Welfare State*, London: Macmillan.

Grant, W. (1978) 'Industrialists and farmers: British interests and the European Community', *West European Politics*, 1 (1), 89–106.

—— (1981) 'The politics of the green pound 1974–79', *Journal of Common Market Studies*, 19 (4), 313–329.

—— (1983) 'The National Farmers' Union: the classic case of incorporation?' 129–143 in D. Marsh (ed.), *Pressure Politics*, London: Junction Books.

—— (1985a) 'Introduction', 1–31 in W. Grant (ed.), *The Political Economy of Corporatism*, London: Macmillan.

—— (ed.) (1985b) *The Political Economy of Corporatism*, London: Macmillan.

—— (1989) *Pressure Groups, Politics and Democracy in Britain*, London: Philip Allan.

—— (1995) 'The limits of Common Agricultural Policy reform and the option of renationalization', *Journal of European Public Policy*, 2 (1), 1–18.

Gray, A., Jenkins, B., Flynn, A. and Rutherford, B. (1991) 'The management of change in Whitehall: the experience of the FMI', *Public Administration*, 69 (1), 41–59.

Grayson, A.J. (1993) *Private Forestry Policy in Western Europe*, Wallingford: CAB International.

Grayson, A.J., Sidaway, R.M. and Thompson, F.P. (1975) 'Some aspects of recreation planning in the Forestry Commission', 89–107 in G. Searle (ed.), *Recreational Economics and Analysis*, London: Longman.

Green, E.H.H. (1985) 'Radical conservatism: the electoral genesis of tariff reform', *Historical Journal*, 28 (3), 667–692.

Grove, R. (1983) *The Future for Forestry*, Cambridge: British Association of Nature Conservationists.

Hall, A.D. (1942) *Reconstruction and the Land*, London: Macmillan.

Hallet, G. (1959) 'The economic position of British agriculture', *Economic Journal*, 69 (3), 522–540.

Halliday, J. (1987) *The Effect of Milk Quotas on Milk Producing Farms: A Study of Registered Milk Producers in the Honiton and Torrington Areas of Devon*, Exeter: Devon County Council and University of Exeter.

—— (1988) 'Dairyfarmers take stock: a study of milk producers' reactions to quota in Devon', *Journal of Rural Studies*, 4 (3), 193–202.

Ham, C. and Hill, M. (1993) *The Policy Process in the Modern Capitalist State*, Hemel Hempstead: Harvester Wheatsheaf.

Hammond, R.J. (1962) *Food*, London: HMSO and Longman, Green & Co.

Hanson, A.H. and Walles, M. (1981, 4th edn) *Governing Britain*, London: Fontana.

Hardy, D. (1991) *From Garden Cities to New Towns: Campaigning for Town and Country Planning, 1899–1946*, London: E. & F.N. Spon.

Harris, J. (1972) *Unemployment and Politics: A Study in English Social Policy 1886–1914*, Oxford: Clarendon Press.

—— (1982) 'Bureaucrats and businessmen in British food control, 1916–19', 135–156 in K. Burk (ed.), *War and State: The Transformation of British Government, 1914–1919*, London: Allen & Unwin.

Harrison, R.J. (1980) *Pluralism and Corporatism: The Political Evolution of Modern Democracies*, London: George Allen & Unwin.

Harvey, D. (1982) 'National interests and the CAP', *Food Policy* (August), 174–190.

Harvey, D.R. and Thomson, K.J. (1985) 'Costs, benefits and the future of the Common Agricultural Policy', *Journal of Common Market Studies*, 24 (1), 2–20.

Haudricourt, A. (1973) 'Botanical nomenclature and its translation', 265–273 in M. Teich and R. Young (eds), *Changing Perspectives in the History of Science*, London: Heinemann.

Hawes, D. (1992) 'Parliamentary select committees: some case studies in contingent influence', *Policy and Politics*, 20 (3), 227–236.

Heclo, H. (1972) 'Review article: policy analysis', *British Journal of Political Science*, 2 (1), 83–108.

Henderson, Sir W. (1981) 'British agricultural research and the Agricultural Research Council', 3–113, in G.W. Cooke (ed.), *Agricultural Research 1931–1981*, London: ARC.

Hennessy, P. (1990) *Whitehall*, London: Fontana.

Hetherington, M.J. (1988) 'Afforestation consultations in Northern Scotland', *Scottish Forestry*, 42 (3), 185–191.

Hill, B. (1989) *Farm Incomes, Wealth and Agricultural Policy*, Aldershot: Gower.

—— (1990) 'Incomes and wealth', 135–160 in D. Britton (ed.), *Agriculture in Britain: Changing Pressures and Policies*, Wallingford: CAB International.

Hill, B. and Ray, D. (1987) *Economics for Agriculture*, London: Macmillan.

Hill, B.E. (1984) *The Common Agricultural Policy: Past, Present and Future*, London: Methuen.

Hill, G. (1993) 'Environmental legislation in relation to pollution control and the achievement of water quality objectives – an overview of recent developments', *Environmental Policy and Practice*, 3 (1), 37–50.

Hill, M., Aaronovitch, S. and Baldock, D. (1989) 'Non-decision making in pollution control in Britain: nitrate pollution, the EEC drinking water directive and agriculture', *Policy and Politics*, 17 (3), 227–240.

Hirsch, F. (1976) *The Social Limits to Growth*, Cambridge, Mass.: Harvard University Press.

Hirst, P. (1977) 'Economic classes and politics', 125–154 in A. Hunt (ed.), *Class and Class Structure*, London: Lawrence & Wishart.

HM Treasury (1972) *Forestry in Great Britain: An Interdepartment Cost/Benefit Study*, Norwich: HMSO.

—— (1988) *Policy Evaluation: A Guide for Managers*, London: HMSO.

Hobhouse, L.T. (1911) *Liberalism*, London: Butterworth.

Hodge, I.D., Adams, W.M. and Bourn, N.A.D. (1994) 'Conservation policy in the wider countryside: agency competition and innovation', *Journal of Environmental Planning and Management*, 37 (2), 199–213.

Hogwood, B. and Gunn, L.A. (1984) *Policy Analysis for the Real World*, Oxford: Oxford University Press.

Holmes, C.J. (1988) 'Science and the farmer: the development of the agricultural advisory service in England and Wales, 1900–1939', *Agricultural History Review*, 36 (1), 77–86.

Hood, C. (1982) 'Government bodies and government growth', 44–68 in A. Barker (ed.), *Quangos in Britain: Government and the Networks of Public Policy Making*, London: Macmillan.

Horn, P. (1984) *The Changing Countryside in Victorian and Edwardian England and Wales*, London: Athlone Press.

Hoskins, W.G. (1938) 'The ownership and occupation of the land in Devonshire, 1650–1800', unpublished PhD thesis, University of London.

Hoskins, W.G. and Stamp, L.D. (1963) *The Common Lands of England and Wales*, London: Collins.

Houghton, W. (1957) *The Victorian Frame of Mind*, New Haven: Yale University Press.

House of Commons (1983) *Organisation and Financing of Agricultural Research & Development*, First Report of the Select Committee on Agriculture, Session 1982–83, HC Paper 38-1.

—— (1985a) *Operation and Effectiveness of Part 2 of the Wildlife and Countryside Act*, First

Report of the Environment Committee Session 1984–85, HC 6.

—— (1985b) *The UK Government Agricultural Development and Advisory Services, Including Lower Input Farming*, 6th Report of the Agriculture Committee Session 1984–85, HC 502.

—— (1987) *Pollution of Rivers and Estuaries*, 3rd Report of the Environment Committee Session 1986–87, HC 183.

—— (1988) *Pollution of Rivers and Estuaries*, Government's Response to 3rd Report of the Environment Committee Session 1987–88, HC 543.

—— (1990) *Land Use and Forestry*, 2nd Report of the Agriculture Committee Session 1989–90, HC 16.

—— (1993) *Forestry and the Environment*, 1st Report of the Environment Committee Session 1992–93, HC 257.

House of Lords (1980) *Scientific Aspects of Forestry*, Second Report of the Select Committee on Science and Technology (Sherfield Report) Session 1979–80, HL 381.

—— (1984a) *Agriculture and the Environment*, 20th Report of the Select Committee on the European Communities Session 1983–84, HL 247.

—— (1984b) *Agricultural and Environmental Research*, 4th Report of the Select Committee on Science and Technology Session 1983–84, HL 272.

—— (1987) *Financing the Community*, 4th Report of the Select Committee on the European Communities Session 1987–88, HL 14.

—— (1989) *Nitrates in Water*, 16th Report of the Select Committee on the European Communities, Session 1988–89, HL 73.

Howarth, R.W. (1985) *Farming for Farmers*, London: Institute of Economic Affairs.

Howarth, W. (1989) 'Water pollution: improving the legal controls', *Journal of Environmental Law*, 1 (1), 25–37.

—— (1992) 'Agricultural pollution and the aquatic environment', 51–72 in W. Howarth and C.P. Rodgers (eds), *Agriculture, Conservation and Land Use: Law and Policy Issues for Rural Areas*, Cardiff: University of Wales Press.

Howkins, A. (1991) *Reshaping Rural England: A Social History 1850–1925*, London: Harper Collins.

Hughes, G. (1994) 'ESAs in the context of a "culturally sensitive area": the case of the Cambrian mountains, 135–152 in M. Whitby (ed.), *Incentives for Countryside Management: The Case of Environmentally Sensitive Areas*, Wallingford: CAB International.

ICI (1986) *Nitrates and our Environment*, Billingham: ICI Fertilisers.

Ilbery, B.W. (1990) 'Adoption of the arable set-aside scheme in England', *Geography*, 76 (1), 69–73.

—— (1992) 'Agricultural policy and land diversion in the European Community', 153–166 in A. Gilg (ed.), *Progress in Rural Policy and Planning Volume 2*, London: Belhaven.

Ilbery, B.W. and Bowler, I.R. (1993) 'The farm diversification grant scheme: adoption and nonadoption in England and Wales', *Environment and Planning C: Government and Policy*, 11 (2), 161–170.

Ionescu, G. and Gellner, E. (eds) (1969) *Populism: Its Meaning and National Characteristics*, London: Weidenfeld & Nicolson.

IUCN (1975) *World Directory of National Parks and Other Protected Areas*, Morges, Switzerland: IUCN.

James, N.D.G. (1981) *A History of English Forestry*, Oxford: Basil Blackwell.

James, S. (1992) *British Cabinet Government*, London: Routledge.

Jenkins, T.N. (1990) *Future Harvests. The Economics of Farming and the Environment: Proposals for Action*, London: CPRE and WWF.

Jenkins, W.I. (1978) *Policy Analysis*, London: Martin Robertson.

Jessop, B. (1982) *The Capitalist State*, Oxford: Martin Robertson.

Jewell, T.K. (1991) 'Water law: agricultural water pollution issues and NRA enforcement policy', *Land Management and Environmental Law Report*, 3 (4), 110–114.

Jogerst, M.A. (1991) 'Backbenchers and select committees in the British House of Commons: can Parliament offer useful roles for the frustrated?', *European Journal of Political Research*, 20 (1), 21–38.

Johnson, J.A. (1992) 'A harvest of discontent: some perceptions of the impacts of lowland forest management of the fiscal changes of 1988', *Quarterly Journal of Forestry*, 86 (3), 150–162.

Jones, A. (1991) 'The impact of the EC's set-aside programme: the response of farm businesses in Rendsburg-Eckenforde, Germany', *Land Use Policy*, 8 (2), 108–124.

Jones, B. (1991) 'The policy making process', 501–520 in B. Jones (ed.), *Politics UK*, London: Philip Allan.

Jones, B. and Moran, M. (1991) 'Introduction: explaining politics', 1–22 in B. Jones (ed.), *Politics UK*, London: Philip Allan.

Jones, G.E. (1963) 'Sources of information and advice available to United Kingdom farmers: description and appraisal', *Sociologia Ruralis*, 3 (1), 52–68.

Jones, G.W. (1989) 'A revolution in Whitehall? Changes in British central government since 1979', *West European Politics*, 12 (3), 238–261.

Jones, J.B. (1990) 'Party committees and all-party groups', 117–136, in M. Rush (ed.), *Parliament and Pressure Politics*, Oxford: Oxford University Press.

Jordan, A. (1993) 'Integrated pollution control and the evolving style and structure of environmental regulation in the UK', *Environmental Politics*, 2 (3), 405–427.

Jordan, A.G. (1990) 'The pluralism of pluralism: an anti-theory?' *Political Studies*, 38 (2), 286–301.

Jordan, A.G. and Richardson, J. (1987) *British Politics and the Policy Process*, London: Allen & Unwin.

Jordan, A.G. and Schubert, K. (1992) 'A preliminary ordering of policy network labels', *European Journal of Political Research*, 21 (1), 7–27.

Jordan, A.G., Maloney, W.A. and McLaughlin, A.M. (1994) 'Characterizing agricultural policy-making', *Public Administration*, 72 (4), 505–526.

Josling, T.E. (1970) 'Agriculture and import saving: a cautionary note', in *Agriculture and Import Saving*, Hill Samuel & Co. Ltd, Occasional Paper No.5.

Judge, D. (1990) *Parliament and Industry*, Aldershot: Dartmouth.

—— (1992) 'The effectiveness of the post-1979 select committee system: the verdict of the 1990 procedure committee', *Political Quarterly*, 63 (1), 91–100.

—— (1993) *The Parliamentary State*, London: Sage.

Kerridge, E. (1969) *Agrarian Problems in the Sixteenth Century and After*, London: Allen & Unwin.

King, A. and Clifford, S. (1985) *Holding Your Ground*, London: Maurice Temple Smith.

Kjeldahl, R. and Tracy, M. (eds) (1994) *Renationalisation of the Common Agricultural Policy?* Copenhagen: Institute of Agricultural Economics.

Koester, U. and Tangermann, S. (1990) 'The why, how and consequences of agricultural policies: the European Community', in F. Sanderson (ed.), *Agricultural Protectionism in the Industrial World*, Baltimore: Johns Hopkins University Press.

Koning, N. (1994) *The Failure of Agrarian Capitalism: Agrarian Politics in the United Kingdom, Germany, the Netherlands and the USA 1846–1919*, London: Routledge.

Kropotkin, P. (1898) *Fields, Factories and Workshops*, London: Thomas Nelson (1912 edn).

Lambi, I.N. (1963) *Free Trade and Protection in Germany, 1868–1879*, Wiesbaden: Frank Steiner.

Latham, E. (1952) 'The group basis of politics', *American Political Science Review*, 46 (2), 376–397.

Lawrence Gould Consultants Ltd (1989) *Conserving the Countryside: Costing it out*, London: CPRE.

Lee, J.M. (1963) *Social Leaders and Public Persons: A Study of County Government in Cheshire since 1888*, Oxford: Clarendon Press.

Lehmbruch, G. and Schmitter, P.C. (eds) (1982) *Patterns of Corporatist Policy-Making*, London: Sage.

Leneman, L. (1989) 'Land settlement in Scotland after World War I', *Agricultural History Review*, 37 (1), 52–64.

Liefferink, J.D., Lowe, P. and Mol, A.P.J. (1993) 'The environment and the European Community: the analysis of European integration', 1–13 in J.D. Liefferink, P. Lowe and A.P.J. Mol (eds), *European Integration and Environmental Policy*, London: Belhaven Press.

Lindblom, C. (1959) 'The science of "muddling through"', *Public Administration Review*, 19 (2), 79–88.

—— (1965) *The Intelligence of Democracy*, New York: Free Press.

—— (1979) 'Still muddling, not yet through', *Public Administration Review*, 39 (6), 517–525.

Lloyd, T., Watkins, C. and Williams, D. (1995) 'Turning farmers into foresters via market liberalisation', *Journal of Agricultural Economics* , 46 (3), 361–370.

Lowe, P. (1976) 'Amateurs and professionals: the institutional emergence of British plant ecology', *Journal of the Society for the Bibliography of Natural History*, 7, 517–535.

—— (1983) 'Values and institutions in the history of British nature conservation', 329–352 in A. Warren and F.B. Goldsmith (eds), *Conservation in Perspective*, Chichester: John Wiley & Sons.

Lowe, P. and Goyder, J. (1983) *Environmental Groups in Politics*, London: Allen & Unwin.

Lowe, P. and Ward, N. (1993) 'Risk, morality and the social construction of farm pollution: engaging with the risk society', paper presented to Annual Conference of the Rural Economy and Society Study Group.

Lowe, P. and Winter, M. (1987) 'Alternative perspectives on the alternative land-use debate', in N.R. Jenkins and M. Bell (eds), *Farm Extensification: Implications of EC Regulation 1760/87*, Merlewood Research and Development Paper No.112.

—— and —— (1988) 'Conservation – more than appeasement', 164–188 in J. Blunden and N. Curry (eds), *A Future for Our Countryside*, Oxford: Blackwell.

Lowe, P., Clark, J., Seymour, S. and Ward, N. (1992) *Pollution Control on Dairy Farms: An Evaluation of Current Policy and Practice*, London: SAFE Alliance.

Lowe, P., Cox, G., MacEwen, M., O'Riordan, T. and Winter, M. (1986) *Countryside Conflicts: The Politics of Farming, Forestry and Conservation*, London: Gower/Maurice Temple Smith.

Lowe, R. (1978) 'The erosion of state intervention in Britain 1917–24', *Economic History Review*, 31 (2), 270–286.

Lowenthal, D. (1985) *The Past is a Foreign Country*, Cambridge: Cambridge University Press.

—— (1991) 'British national identity and the English landscape', *Rural History*, 2 (2), 205–230.

Lowerson, J. (1980) 'Battles for the countryside', 258–280 in F. Gloversmith (ed.), *Class, Culture and Social Change: A New View of the 1930s*, Brighton: Harvester Press.

Luckin, B. (1990) *Questions of Power: Electricity and Environment in Inter-war Britain*, Manchester: Manchester University Press.

Mabey, R. (1980) *The Common Ground*, London: Arrow Books.

MacCarthy, F. (1994) *William Morris: A Life for Our Time*, London: Faber & Faber.

McCrone, G. (1962) *The Economics of Subsidising Agriculture*, London: Allen & Unwin.

MacEwen, A. and MacEwen, M. (1982) *National Parks: Conservation or Cosmetics?*, London: Allen & Unwin.

McInerney, J. and Turner, M. (1991) *Patterns, Performance and Prospects in Farm Diversification*, University of Exeter, Agricultural Economics Unit Report No. 236.

McIntosh, R.P. (1985) *The Background of Ecology: Concept and Theory*, Cambridge: Cambridge University Press.

Mackenzie, O. (1980, 1st edn 1921) *A Hundred Years in the Highlands*, Edinburgh: National Trust for Scotland.

McKenzie, R.T. (1958) 'Parties, pressure groups and the British political process', *Political Quarterly*, 29 (1), 5–16.

MacKenzie, W.J.M. (1955) 'Pressure groups in British government', *British Journal of Sociology*, 6 (2), 133–148.

McQuiston, J.R. (1973) 'Tenant right: farmer against landlord in Victorian England 1847–1883', *Agricultural History*, 47 (1), 95–113.

Madden, M. (1957) 'The National Union of Agricultural Workers, 1906–1956', unpublished B.Lit. thesis, University of Oxford.

Madgwick, P. (1991) *British Government: The Central Executive Territory*, London: Philip Allan.

MAFF (1985) *Code of Good Agricultural Practice*, London: HMSO.

—— (1991) *Code of Good Agricultural Practice for the Protection of Water*, London: MAFF.

—— (1992) *Code of Good Agricultural Practice for the Protection of Air*, London: MAFF.

—— (1993a) *At the Farmer's Service*, London: HMSO.

—— (1993b) *Code of Good Agricultural Practice for the Protection of Soil*, London: MAFF.

—— (1993c) *Solving the Nitrate Problem: Progress in Research and Development*, London: MAFF.

—— (1993d) *Pilot Nitrate Sensitive Areas Scheme: Report on the First Three Years*, London: MAFF.

—— (1994) *Environment Research Strategy and Requirements Document 1994–96*, London: MAFF.

MAFF and DoE (1995) *Environmental Land Management Schemes in England*, Consultation Document, London: MAFF and DoE.

Maguire, J. (1978) *Marx's Theory of Politics*, Cambridge: Cambridge University Press.

Malcolm, D.C., Evans, J. and Edwards, P.N. (eds) (1982) *Broadleaves in Britain: Future Management and Research*, Farnham: Institute of Chartered Foresters.

Maloney, W.A. and Richardson, J. (1994) 'Water policy-making in England and Wales: policy communities under pressure', *Environmental Politics*, 3 (4), 110–137.

Maloney, W.A., Jordan, G.H. and McLaughlin, A.M. (1994) 'Interest groups and public policy:

the insider/outsider model revisited', *Journal of Public Policy*, 14 (1) 17–38.

Marren, P. (1990) *Woodland Heritage*, Newton Abbot: David & Charles.

Marrison, A.J. (1977) 'The development of a tariff reform policy during Joseph Chamberlain's first campaign', 214–241 in W.H. Chaloner and B.M. Ratcliffe (eds), *Trade and Transport*, Manchester: Manchester University Press.

—— (1986) 'The Tariff Commission, agricultural protection and food taxes, 1903–13', *Agricultural History Review*, 34 (2), 171–187.

Marsden, T., Lowe, P. and Whatmore, S. (eds) (1990) *Rural Restructuring: Global Processes and their Responses*, London: David Fulton.

Marsden, T., Murdoch, J., Lowe, P., Munton, R. and Flynn, A. (1993) *Constructing the Countryside*, London: UCL Press.

Marsh, D. and Rhodes, R.A.W. (1992a) 'The implementation gap', 170–187 in D. Marsh and R.A.W. Rhodes (eds), *Implementing Thatcherite Policies*, Buckingham: Open University Press.

—— and —— (eds) (1992b) *Policy Networks in British Government*, Oxford: Oxford University Press.

Marsh, J. (1982) *Back to the Land: The Pastoral Impulse in England, from 1880 to 1914*, London: Quartet Books.

Marsh, J.S. and Swanney, P.J. (1980) *Agriculture and the European Community*, London: Allen & Unwin.

Martin, D. (1981) *John Stuart Mill and the Land Question*, Hull: University of Hull Publications.

Martin, R.M. (1983) 'Pluralism and the new corporatism', *Political Studies*, 31 (1), 86–102.

Marx, K. and Engels, F. (1952) *Manifesto of the Communist Party*, Moscow: Progress Publishers (first published in 1848).

Mather, A.S. (1985) 'The rise and fall of government-assisted land settlement in Scotland', *Land Use Policy*, 217–224.

Mather, A.S. and Murray, N.C. (1988) 'The dynamics of rural land use change: the case of private sector afforestation in Scotland', *Land Use Policy*, 5 (1), 103–120.

Matless, D. (1990a) 'Definitions of England, 1928–89: preservation, modernism and the nature of the nation', *Built Environment*, 16 (3), 179–191.

—— (1990b) 'Ages of English design: preservation, modernism and tales of their history, 1926–1939', *Journal of Design History*, 3 (4), 203–212.

—— (1991) 'Ordering the land: the "preservation" of the English countryside, 1918–39', unpublished PhD thesis, University of Nottingham.

—— (1993a) 'Appropriate geography: Patrick Abercrombie and the energy of the world', *Journal of Design History*, 6 (3), 167–178.

—— (1993b) 'One man's England: W.G. Hoskins and the English culture of landscape', *Rural History*, 4 (2), 187–207.

Matthews, A.H.H. (1915) *Fifty Years of Agricultural Politics, Being a History of the Central Chamber of Agriculture, 1865–1915*, London: King & Son.

May, T. and Nugent, N. (1982) 'Insiders, outsiders and thresholders: corporatism and pressure group strategies in Britain', paper presented to the Annual Conference of the Political Studies Association, University of Kent.

Meadows, D.H., Meadows, D.L., Randers, J. and Behrens, W. III (1983) *The Limits to Growth*, London: Pan.

Meester, G. and Van der Zee, F.A. (1993) 'EC decision-making, institutions and the Common Agricultural Policy', *European Review of Agricultural Economics*, 20 (2), 131–150.

Mellanby, K. (1967) *Pesticides and Pollution*, London: Collins.

Mertz, S.W. (1989) 'The European Economic Community Directive on Environmental Assessments: how will it affect United Kingdom developers?' *Journal of Planning and Environmental Law*, 483–498.

Middlemas, K. (1979) *Politics in Industrial Society: The Experience of the British System since 1911*, London: André Deutsch.

Miliband, R. (1973) *The State in Capitalist Society*, London: Quartet Books (first published in 1969).

—— (1977) *Marxism and Politics*, Oxford: Oxford University Press.

Miller, R. (1981) *State Forestry for the Axe*, London: Institute of Economic Affairs.

Mills, C.W. (1956) *The Power Elite*, New York: Oxford University Press.

Minay, C. (1990) 'The Development Commission and English rural development', 211–225 in H. Buller and S. Wright (eds), *Rural Development: Problems and Practices*, Aldershot: Avebury.

MMB (1956) *The Work of the Milk Marketing Board of England and Wales*, Thames Ditton: MMB.

———— (1989) *Permanent Producer Sample 1988–89*, Thames Ditton: MMB.

Mollett, J.A. (1960) 'The Wheat Act of 1932: a forerunner of modern farm price support programmes', *Agricultural History Review*, 8 (1), 20–35.

Moore, N.W. (1987) *The Bird of Time: The Science and Politics of Nature Conservation*, Cambridge: Cambridge University Press.

Moore, S. (1991) 'The agrarian conservative party in parliament', *Parliamentary History*, 10 (2), 342–362.

Morgan, K.O. (1979) *Consensus and Disunity: The Lloyd George Coalition Government 1918–1922*, Oxford: Clarendon Press.

———— (1981) *Rebirth of a Nation: Wales, 1880–1980*, Oxford: Oxford University Press.

———— (1982) *Wales 1880–1980*, Oxford: Oxford University Press, and Cardiff: University of Wales Press.

Morris, C.N. (1980) *The Common Agricultural Policy*, London: Institute of Fiscal Studies.

Mosca, G. (1939) *The Ruling Class*, London: McGraw Hill.

Mowat, C.L. (1955) *Britain between the Wars*, London: Methuen.

Moyer, H.W. and Josling, T.E. (1990) *Agricultural Policy Reform: Politics and Process in the EC and the USA*, Hemel Hempstead: Harvester Wheatsheaf.

Murdoch, J. (1992) 'Representing the region: Welsh farmers and the British state', 160–181 in T. Marsden, P. Lowe and S. Whatmore (eds), *Labour and Locality: Uneven Development in the Labour Process*, London: David Fulton.

Murdoch, J. and Marsden, T. (1994) *Reconstituting Rurality*, London: UCL Press.

Murray, K.A.H. (1955) *Agriculture*, London: HMSO.

Mutch, A. (1983) 'Farmers' organizations and agricultural depression in Lancashire 1890–1900', *Agricultural History Review*, 31 (1), 26–36.

Mutch, W.E.S. (1968) *Public Recreation in National Forests*, Forestry Commission Booklet No. 21, HMSO.

Naess, A. (1973) 'The shallow and the deep, long-range ecology movement. A summary', *Inquiry*, 16, 95–100.

———— (1989) *Ecology, Community and Lifestyle*, Cambridge: Cambridge University Press.

NAO (1986) *Review of Forestry Commission Objectives and Achievements*, Report by the Comptroller and Auditor General of the National Audit Office, London: HMSO.

———— (1994) *Protecting and Managing Sites of Special Scientific Interest in England*, Report by the Comptroller and Auditor General of the National Audit Office, London: HMSO.

Nash, E.F. (1965) *Agricultural Policy in Britain*, Cardiff: University of Wales Press.

Neave, G. (1988) 'On the cultivation of quality, efficiency and enterprise: an overview of recent trends in higher education in Western Europe', *European Journal of Education*, 23 (1/2), 7–23.

Neville-Rolfe, E. (1984) *The Politics of Agriculture in the European Community*, London: Policy Studies Institute.

———— (1990) 'British agricultural policy and the EC', 173–199 in D. Britton (ed.), *Agriculture in Britain: Changing Pressures and Policies*, Wallingford: CAB International.

Newby, H. (1977) *The Deferential Worker*, London: Allen Lane.

———— (1987) *Country Life: A Social History of Rural England*, London: Weidenfeld & Nicolson.

NFU (1919) *NFU General Council*, Institute of Agricultural History, University of Reading.

Nicoll, W. and Salmon, T.C. (1990) *Understanding the European Communities*, London: Philip Allan.

Noël, E. (1988) *Europe*, Washington: Office of the EC.

Norton, P. (1991a) 'The constitution in flux', 277–292 in B. Jones (ed.), *Politics UK*, London: Philip Allan.

———— (1991b) 'Parliament 1 – The House of Commons', 314–355 in B. Jones (ed.), *Politics UK*, London: Philip Allan.

———— (1993) *Does Parliament Matter?* Hemel Hempstead: Harvester Wheatsheaf.

Norton, P. and Aughey, A. (1981) *Conservatives and Conservatism*, London: Maurice Temple Smith.

NRA (1992) *The Influence of Agriculture on the Quality of Natural Waters in England and Wales*, Bristol: National Rivers Authority.

NRA/MAFF (1990) *Water Pollution from Farm Waste 1989, England and Wales*, London: MAFF.

Ockenden, J. and Franklin, M. (1995) *European Agriculture: Making the CAP Fit the Future*, London: Pinter.

O'Connor, J. (1973) *The Fiscal Crisis of the State*, New York: St Martin's.

Offe, C. (1972) 'Political authority and the class structure – an analysis of late capitalist societies', *International Journal of Sociology*, 2 (1), 73–108.

—— (1975) 'The theory of the capitalist state and the problem of policy formation', 125–144 in L. Lindberg, R. Alford, C. Crouch and C. Offe (eds), *Stress and Contradiction in Modern Capitalism*, Lexington: D.H. Heath.

Offer, A. (1981) *Property and Politics, 1871–1914: Landownership, Law, Ideology and Urban Development in England*, Cambridge: Cambridge University Press.

O'Riordan, T. (1976) *Environmentalism*, London: Pion.

O'Riordan, T. and Weale, A. (1989) 'Administrative reorganization and policy change: the case of Her Majesty's Inspectorate of Pollution', *Public Administration*, 67 (3), 277–294.

Orwin, C.S. and Darke, W.F. (1935) *Back to the Land*, London: King & Son.

Orwin, C.S. and Whetham, E.H. (1964) *A History of British Agriculture 1846–1914*, London: Longman.

Owens, S. (1990) 'The unified pollution inspectorate and best practicable environmental option in the United Kingdom', 169–198 in N. Haigh and F. Irwin (eds), *Integrated Pollution Control in Europe and North America*, Washington DC: The Conservation Foundation.

Owens, S. and Cowell, R. (1994) 'Lost land and limits to growth: conceptual problems for sustainable land use change', *Land Use Policy*, 11 (3), 168–180.

Packenham, R. (1970) 'Legislatures and political development', 521–582 in A. Kornberg and L.D. Musolf (eds), *Legislatures in Developmental Perspective*, Durham NC: Duke University Press.

Panitch, L. (1980) 'Recent theorizations of corporatism: reflections on a growth industry', *British Journal of Sociology*, 31 (2), 159–187.

—— (1981) 'Trade unions and the capitalist state', *New Left Review*, 125, 21–43.

Parker, D.J. and Penning-Rowsell, E.C. (1980) *Water Planning in Britain*, London: Allen & Unwin.

Pasmore, A. (1976) *Verderers of the New Forest: A History of the New Forest 1877–1977*, Hampshire: Pioneer Publications.

Pearce, J. (1981) *The Common Agricultural Policy*, London: Routledge & Kegan Paul.

Pelling, H. (1967) *Social Geography of British Elections 1885–1910*, London: Macmillan.

PEP (1949) 'Agricultural executives', *Planning*, 294, 31 January.

Pepper, D. (1986) *The Roots of Modern Environmentalism*, London: Routledge.

Pepperall, R.A. (1950) *A Biography of Sir Thomas Baxter*, Wells, Somerset: Clare, Son & Co.

Perkin, H. (1969) *The Origins of Modern English Society, 1780–1880*, London: Routledge & Kegan Paul.

—— (1989) *The Rise of Professional Society: England since 1880*, London: Routledge.

Perry, P.J. (ed.) (1973) *British Agriculture 1875–1914*, London: Methuen.

Peterken, G. (1981) *Woodland Conservation and Management*, London: Chapman and Hall.

Petit, M., de Benedictis, M., Britton, D., de Groot, M., Henrichsmeyer, W. and Leshi, F. (1987) *Agricultural Policy Formation in the European Community: The Birth of Milk Quotas and CAP Reform*, Amsterdam: Elsevier.

Phillips, A. (1993) 'The national parks – will they be fit for the future?' *Ecos*, 14 (3/4), 30–36.

Phillips, H. and Hatton, C. (1994) 'Implementing the Habitats Directive in the UK', *Ecos*, 15 (1), 17–22.

Phillips, P.W.B. (1990) *Wheat, Europe and the GATT*, London: Pinter Publishers.

PIEDA plc. (1986) *Forestry in Great Britain*, Report to the National Audit Office.

—— (1993) *Assessment of Conservation Advice to Farmers*, London: Department of the Environment.

Pimlott, B. (1992) *Harold Wilson*, London: HarperCollins.

Pollitt, C. (1993) 'Occasional excursions: a brief history of policy evaluation in the UK', *Parliamentary Affairs*, 46 (3), 353–362.

Ponting, C. (1986) *Whitehall: Tragedy and Farce*, London: Sphere Books.

Porchester, Lord (1977) *A Study of Exmoor*, London: HMSO.

Porritt, J. (1984) *Seeing Green*, Oxford: Basil Blackwell.

Porritt, J. and Winner, D. (1988) *The Coming of the Greens*, London: Fontana.

Potter, C. (1983) *Investing in Rural Harmony*, Godalming: WWF.

—— (1986) 'Processes of countryside change in lowland England', *Journal of Rural Studies*, 2 (3), 187–195.

—— (1988) 'Environmentally sensitive areas in England and Wales: an experiment in countryside management', *Land Use Policy*, 5 (3), 301–313.

Poulantzas, N. (1973) *Political Power and Social Classes*, London: New Left Books.

—— (1975) *Classes in Contemporary Capitalism*, London: New Left Books.

Powell, L.B. 'Full harvest: the story of the National Farmers' Union of England and Wales', unpublished MS held at the NFU Library in Knightsbridge.

Pressman, J. and Wildavsky, A. (1973) *Implementation*, Berkeley: University of California Press.

Presthus, R. (1964) *Men at the Top*, Oxford: Oxford University Press.

Pringle, D. (1994) *The First 75 Years: A Brief Account of the History of the Forestry Commission 1919–1994*, Edinburgh: Forestry Commission.

Pryor, R. with Land Use Consultants and Forestry Investment Management (1992) *Future Forestry: A New Direction for Forest Policy*, London: Wildlife Link.

Pye-Smith, C. and North, R. (1984) *Working the Land*, London: Maurice Temple Smith.

Pye-Smith, C. and Rose, C. (1984) *Crisis and Conservation: Conflict in the British Countryside*, Harmondsworth: Penguin.

Rackham, O. (1980) *Ancient Woodland*, London: Edward Arnold.

Ramblers' Association (1972) *The Future of Forestry*, London.

—— (1980) *Afforestation: The Case against Expansion*, London.

Redclift, M. and Benton, T. (eds) (1994) *Social Theory and the Global Environment*, London: Routledge.

Reynolds, F. and Sheate, W.R. (1992) 'Reorganization of the conservation authorities', 73–89 in W. Howarth and C.P. Rodgers (eds), *Agriculture, Conservation and Land Use: Law and Policy Issues for Rural Areas*, Cardiff: University of Wales Press.

Rhodes, R.A.W. (1988) *Beyond Westminster and Whitehall: The Sub-Central Governments of Britain*, London: Unwin Hyman.

—— (1991) 'Theory and methods in British public administration: the view from political science', *Political Studies*, 34 (3), 533–554.

Richards, S. (1994) *A Step Forward for Our Forests?* London: CPRE and RSPB.

Richardson, J.J. and Jordan, A.G. (1979) *Governing under Pressure: The Policy Process in a Post-Parliamentary Democracy*, Oxford: Martin Robertson.

Richardson, J.J., Jordan, A.G. and Kimber, R.H. (1978) 'Lobbying, administrative reform and policy styles', *Political Studies*, 26 (1), 47–64.

Richardson, J.J., Maloney, W.A. and Rudig, W. (1992) 'The dynamics of policy change: lobbying and water privatization', *Public Administration*, 70 (2), 157–175.

Ritschel, D. (1991) 'A corporatist economy in Britain? Capitalist planning for industrial self-government in the 1930s', *English Historical Review*, 418, 41–65.

Roberts, I. and Tie, G. (1982) 'The emergence of the EEC as a net exporter of grain', *Quarterly Review of the Rural Economy*, 4 (4), 295–304.

Rodgers, C.P. (1992) 'Land management agreements and agricultural practice: towards an integrated legal framework for conservation law', 139–164 in W. Howarth and C.P. Rodgers (eds), *Agriculture, Conservation and Land Use: Law and Policy Issues for Rural Areas*, Cardiff: University of Wales Press.

Rose, C. (1994) 'Thank you Prime Minister', *Parliamentary Brief*, 2 (6), 68–69.

RSNC (1989) *Losing Ground*, Lincoln: Royal Society for Nature Conservation.

RSPB (1985) *Forestry in the Flow Country: The Threat to Birds*, Sandy: Royal Society for the Protection of Birds.

Rubinstein, W.D. (1981a) *Men of Property*, London: Croom Helm.

—— (1981b) 'New men of wealth and the purchase of land in nineteenth-century Britain', *Past and Present*, 92, 125–147.

—— (1992) 'Cutting up rich: a reply to F.M.L. Thompson', *Economic History Review*, 45 (2), 350–361.

—— (1993) *Capitalism, Culture and Decline in Britain 1750–1990*, London: Routledge.

Rudig, W. (ed.) (1990) *Green Politics*, Edinburgh: Edinburgh University Press.

Runciman, W.G. (1993) 'Has British capitalism changed since the First World War?', *British Journal of Sociology*, 44 (1), 53–67.

Russell, A.K. (1962) 'The General Election of 1906', unpublished D.Phil. thesis, University of Oxford.

—— (1973) *Liberal Landslide*, Newton Abbot: David & Charles.

Ryan, P. (1991) 'The European Community's environment policy: meeting the challenge of the 90s', *European Environment*, 1 (6), 1–13.

Ryle, G.B. (1969) *Forest Service*, Newton Abbot: David & Charles.

Sandbach, F.R. (1978) 'The early campaign for a National Park in the Lake District', *Transactions of the Institute of British Geographers*, 3 (4), 498–514.

Sargent, J.A. (1985) 'Corporatism and the European Community', 229–253 in W. Grant (ed.),

The Political Economy of Corporatism, London: Macmillan.

Saul, S.B. (1969) *The Myth of the Great Depression*, London: Macmillan.

Saunders, P. (1984) *We Can't Afford Democracy Too Much: Findings from a Study of Regional State Institutions in South-East England*, University of Sussex, Urban and Regional Studies Working Paper 43.

Saward, M. (1990) 'Cooption and power: who gets what from formal incorporation', *Political Studies*, 38 (4), 588–602.

Schmitter, P.C. (1974) 'Still the century of corporatism?', *Review of Politics*, 36 (1), 85–131.

Schmitter, P.C. and Lehmbruch, G. (eds) (1979) *Trends Toward Corporatist Intermediation*, London: Sage.

Schumacher, E. (1976) *Small is Beautiful*, London: Sphere.

Scott, J. (1982) *The Upper Classes: Property and Privilege in Britain*, London: Macmillan.

Self, P. and Storing, H. (1962) *The State and the Farmer*, London: George Allen & Unwin.

Selman, P. (1992) *Environmental Planning*, London: Paul Chapman.

Sewell, W. and Barr, L. (1978) 'Water administration in England and Wales', *Water Resources Bulletin*, 14, 337–348.

Seymour, S., Cox, G. and Lowe, P. (1992) 'Nitrates in water: the politics of the polluter pays principle', *Sociologia Ruralis*, 32 (1), 82–103.

Shaw, M. (1991) 'National Parks – towards 2000', *Ecos*, 12 (2), 55–59.

Sheail, J. (1974) 'The role of the War Agricultural and Executive Committees in the food production campaign of 1915–1918 in England and Wales', *Agricultural Administration*, 1, 141–154.

—— (1975) 'The concept of national parks in Great Britain 1900–1950', *Transactions of the Institute of British Geographers*, 52 (1), 41–56.

—— (1976a) 'Land improvement and reclamation: the experiences of the First World War in England and Wales', *Agricultural History Review*, 24 (2), 110–125.

—— (1976b) *Nature in Trust: the History of Nature Conservation in Britain*, Glasgow: Blackie.

—— (1981) *Rural Conservation in Inter-War Britain*, Oxford: Clarendon Press.

—— (1985) *Pesticides and Nature Conservation: The British Experience 1950–1975*, Oxford: Clarendon Press.

—— (1987) *Seventy-Five Years in Ecology: The British Ecological Society*, Oxford: Basil Blackwell.

—— (1988) 'The great divide: an historical perspective', *Landscape Research*, 13 (1), 2–5.

—— (1993) 'The agricultural pollution of watercourses: the precedents set by the beet-sugar and milk industries', *Agricultural History Review*, 41 (1), 31–43.

Sheate, W.R. and Macrory, R.B. (1989) 'Agriculture and the EC Environmental Assessment Directive: lessons for community policy-making', *Journal of Common Market Studies*, 28 (1), 68–81.

Shell, D. (1992) *The House of Lords*, Hemel Hempstead: Harvester Wheatsheaf.

Shoard, M. (1980) *The Theft of the Countryside*, London: Temple Smith.

—— (1987) *This Land is Our Land*, London: Paladin.

Shucksmith, M. (1990) *Housebuilding in Britain's Countryside*, London: Routledge.

Shucksmith, M. and Winter, M. (1990) 'The politics of pluriactivity in Britain', *Journal of Rural Studies*, 6 (4), 429–435.

Smith, M.J. (1989a) 'Land nationalisation and the agricultural policy community', *Public Policy and Administration*, 4 (3), 9–21.

—— (1989b) 'The Annual Review: the emergence of a corporatist institution', *Political Studies*, 37 (1), 81–96.

—— (1990a) 'Pluralism, reformed pluralism and neopluralism: the role of pressure groups in policy-making', *Political Studies*, 38 (2), 302–322.

—— (1990b) *The Politics of Agricultural Support in Britain*, Aldershot: Dartmouth.

—— (1991) 'From policy community to issue network: salmonella in eggs and the new politics of food', *Public Administration*, 69 (2), 235–255.

—— (1992) 'CAP and agricultural policy', 138–151 in D. Marsh and R.A.W. Rhodes (eds), *Implementing Thatcherite Policies*, Buckingham: Open University Press.

—— (1993) *Pressure, Power and Policy: State Autonomy and Policy Networks in Britain and the United States*, London: Harvester Wheatsheaf.

Smith, M.S. (1980) *Tariff Reform in France 1860–1900*, Ithaca: Cornell University Press.

Snyder, F. (1985) *Law of the Common Agricultural Policy*, London: Sweet & Maxwell.

South West Water Authority (1986) *Environmental Investigation of the River Torridge*, SWWA: Department of Environmental Services.

Southey, R. (1829) *Sir Thomas More: or Colloquies on the Progress and Prospects of Society*, London: John Murray.

Stanyer, J. (1967) *County Government in England and Wales*, London: Routledge & Kegan Paul.

Stapledon, R.G. (1935) *The Land Now and Tomorrow*, London: Faber & Faber.

Stewart, P. (1985) 'British forestry policy: time for a change?', *Land Use Policy*, 2 (1), 6–29.

—— (1987) *Growing against the Grain: United Kingdom Forestry Policy*, London: CPRE.

Street, A.G. (1937) *Farming England*, London: Batsford.

Sturmey, S.G. (1955) 'Owner-farming in England and Wales, 1900 to 1950', *Manchester School*, 23, 246–268.

SWA (1985) *Report on Thanet Nitrate Investigation*, Southern Water Authority.

Swinbank, A. (1978) *The British Interest and the Green Pound*, Reading: Centre for Agricultural Strategy Paper No. 6.

Sykes, A. (1979) *Tariff Reform in British Politics 1903–1913*, Oxford: Clarendon Press.

Tansley, A.G. (ed.) (1911) *Types of British Vegetation*, Cambridge: Cambridge University Press.

—— (1935) 'The use and abuse of vegetational concepts and terms', *Ecology*, 16 (3), 284–307.

—— (1939a) *The British Islands and their Vegetation*, Cambridge: Cambridge University Press.

—— (1939b) 'British ecology during the past quarter-century', *Journal of Ecology*, 27 (2), 513–530.

Thomas, K. (1983) *Man and the Natural World*, London: Allen Lane.

Thomas, R.H. (1983) *The Politics of Hunting*, Aldershot: Gower.

Thompson, E.P. (1993) *Customs in Common*, Harmondsworth: Penguin.

Thompson, F.M.L. (1963) *English Landed Society in the Nineteenth Century*, London: Routledge.

—— (1990) 'Life after death: how successful nineteenth-century businessmen disposed of their fortunes', *Economic History Review*, 43 (1), 40–61.

—— (1992) 'Stitching it together again', *Economic History Review*, 45 (2), 362–375.

Thomson, K. J. (1994) 'Chapter A: EC agriculture past and present', 42–75 in *European Economy, Reports and Studies No. 4 EC Agricultural Policy for the 21st Century*, European Commission Directorate-General for Economic and Financial Affairs.

Timber Growers UK (1985) *The Forestry and Woodland Code*, London: TGUK.

—— (1986) *Afforestation and Nature Conservation Interactions*, London: TGUK.

Tomlinson, J. (1990) *Public Policy and the Economy since 1900*, Oxford: Clarendon Press.

Tompkins, S.C. (1986) *The Theft of the Hills: Afforestation in Scotland*, London: Ramblers Association.

—— (1989) *Forestry in Crisis: The Battle for the Hills*, London: Christopher Helm.

Tracy, M. (1982) *Agriculture in Western Europe: Challenge and Response 1880–1980*, London: Granada.

—— (1989) *Government and Agriculture in Western Europe 1880–1988*, Hemel Hempstead: Harvester Wheatsheaf.

Traill, B. (1982) 'Taxes, investment incentives and the cost of agricultural inputs', *Journal of Agricultural Economics*, 33 (1), 1–12.

Trollope, A. (1980) *Doctor Thorne*, Oxford: Oxford University Press (first published 1858).

Truman, D. (1951) *The Governmental Process*, New York: Alfred A. Knopf.

Tsinisizelis, M. (1985) 'The politics of the Common Agricultural Policy: a study of interest group politics', unpublished PhD thesis, University of Manchester.

Tuckwell, S.B. and Knight, M.S. (1988) 'Guidelines for the agricultural use of water supply catchments to minimise leaching of nitrate', *Water Supply*, 6, 295–302.

Turner, E.S. (1964) *All Heaven in a Rage*, London: Martin Joseph.

Turner, M. (1992) 'Output and prices in UK agriculture, 1867–1914, and the Great Agricultural Depression reconsidered', *Agricultural History Review*, 40 (1), 38–51.

Tyldesley, D. (1986) *Gaining Momentum: An Analysis of the Role and Performance of Local Authorities in Nature Conservation*, Oxford: Pisces Publications.

Vacher's Parliamentary Companion, published quarterly, Berkhamsted: Vachers Publications.

Vickery, A.V. (1958) 'Agricultural policies and programmes in England and Wales, 1900–1921', unpublished D.Phil. thesis, University of Oxford.

Wagstaff, H. (1983) 'Capital investment, unemployment, and the meaning of efficiency: implications for agricultural policy', paper presented at the Annual Conference of the Agricultural Economics Society, University of Exeter.

Wallace, H. (1983) 'Negotiation, conflict, and compromise: the elusive pursuit of common policies', 43–80 in H. Wallace, W. Wallace and C. Webb (eds) (2nd edn), *Policy Making in the European Community*, Chichester: John Wiley & Sons.

—— (1990) 'Britain and Europe', 150–172 in P. Dunleavy, A. Gamble and G. Peel (eds), *Developments in British Politics 3*, London: Macmillan.

Ward, N., Clark, J., Lowe, P. and Seymour, S. (1993) *Water Pollution from Agricultural Pesticides*, University of Newcastle upon Tyne: Centre for Rural Economy Research Report.

Ward, S.B. (1976) 'Land reform in England 1880–1914', unpublished PhD thesis, University of Reading.

Watkins, C. (1981) 'The development of tree preservation and felling control legislation in England and Wales', *Quarterly Journal of Forestry*, 75 (4), 220–226.

—— (1983) 'The public control of woodland management', *Town Planning Review*, 54 (4), 437–459.

—— (1986) 'Recent changes in government policy towards broadleaved woodland', *Area*, 18 (2), 117–122.

—— (1990) *Woodland Management and Conservation*, Newton Abbot: David & Charles.

Watkins, C., Williams, D. and Lloyd, T. (in press) 'Constraints on farm woodland planting in England', *Forestry*, 69.

Watson Committee (1956) *Report of the Committee on the Marketing of Woodland Produce*, Forestry Commission.

Watson, J.S. (1960) *The Reign of George III 1760–1815*, Oxford: Clarendon Press.

Weale, A. (1992) *The New Politics of Pollution*, Manchester: Manchester University Press.

Weale, A., O'Riordan, T. and Kramme, L. (1991) *Controlling Pollution in the Round*, London: Anglo-German Foundation.

Webster, S. and Felton, M. (1993) 'Targeting for nature conservation in agricultural policy', *Land Use Policy*, 10 (1), 67–82.

Wheelock, V. (1986) *The Food Revolution*, Marlow: Chalcombe Publications.

Whetham, E. (1972) 'The Agriculture Act 1920 and its repeal – the "Great Betrayal"', *Agricultural History Review*, 22 (1), 36–49.

—— (1978) *The Agrarian History of England and Wales, Vol. VIII 1914–1939*, Cambridge: Cambridge University Press.

Whitby, M. (1994) 'What future for ESAs?' 253–271 in M. Whitby (ed.), *Incentives for Countryside Management: The Case of Environmentally Sensitive Areas*, Wallingford: CAB International.

Whitby, M. and Lowe, P. (1994) 'The political and economic roots of environmental policy in agriculture', 1–24 in M. Whitby (ed.), *Incentives for Countryside Management: The Case of Environmentally Sensitive Areas*, Wallingford: CAB International.

Wibberley, G.P. (1974) 'Land use and rural planning', 213–218 in *Proceedings of the Royal Society Meeting on Forests and Forestry in Britain*, London.

Wiener, M.J. (1981) *English Culture and the Decline of the Industrial Spirit 1850–1980*, Cambridge: Cambridge University Press.

Wilkinson, W.B. (1976) 'The nitrate problem and groundwater', *Water*, November.

Williams, D., Lloyd, T. and Watkins, C. (1994) *Farmers Not Foresters: Constraints on the Felling and the Planting of New Farm Woodland*, University of Nottingham, Department of Geography Working Paper 27.

Williams, H.T. (1960) *Principles for British Agricultural Policy*, London: Oxford University Press.

Williams, M.V. (1985) 'National park policy 1942–1984', *Journal of Planning and Environment Law*, 359–377.

Williams, R. (1960) *Culture and Society 1780–1950*, London: Chatto & Windus.

—— (1973) *The Country and the City*, London: Chatto & Windus.

Williams, W.H. (1965) *The Commons, Footpaths and Open Spaces Preservation Society: A Short History of the Society and its Work*, Henley: The Commons Society.

Williamson, P.J. (1985) *Varieties of Corporatism*, Cambridge: Cambridge University Press.

—— (1989) *Corporatism in Perspective: An Introductory Guide to Corporatist Theory*, London: Sage.

Wilmot, S. (1993) 'Agriculture and pollution in Victorian England', paper delivered to the British Agricultural History Society, London.

Wilson, G. (1977) *Special Interests and Policy Making: Agricultural Policies and Politics in Britain and the USA 1956–70*, London: John Wiley.

Winter, M. (1984) 'Corporatism and agriculture in the UK: the case of the Milk Marketing Board', *Sociologia Ruralis*, 24 (2), 106–119.

—— (1985a) 'Administering land-use policies for agriculture: a possible role for County

Agriculture and Conservation Committees', *Agricultural Administration*, 18 (4), 235–249.
—— (1985b) 'County agricultural committees: a good idea for conservation?', *Journal of Rural Studies*, 1 (3), 205–209.
—— (1986) 'The survival and re-emergence of family farming: a study of the Holsworthy area of West Devon', unpublished PhD thesis, Open University.
—— (1987) 'Private tourism in the English and Welsh Uplands: farming, visitors, and property', 22–34 in M. Bouquet and M. Winter (eds), *Who from Their Labours Rest? Conflict and Practice in Rural Tourism*, Aldershot: Avebury.
—— (1990a) 'Conservation and environmental pressures on the countryside', 335–347 in *Faith in the Countryside*, Report of the Archbishops' Commission on Rural Areas.
—— (1990b) 'Land use policy in the UK: the politics of control', *Land Development Studies*, 7, 3–14.
—— (1994) 'Grey areas in greening Britain', *Parliamentary Brief*, 2 (6), 73–74.
—— (1995) *Networks of Knowledge*, Godalming: Worldwide Fund for Nature.
Winter, M., Richardson, C., Short, C. and Watkins, C. (1990) *Agricultural Land Tenure in England and Wales*, London: Royal Institution of Chartered Surveyors.
Withrington, D. and Jones, W. (1992) 'The enforcement of conservation legislation: protecting Sites of Special Scientific Interest', 90–107 in W. Howarth and C.P. Rodgers (eds), *Agriculture, Conservation and Land Use: Law and Policy Issues for Rural Areas*, Cardiff: University of Wales Press.
Wolff, R.P., Barrington, Moore, Jr and Marcus, H. (1965) *A Critique of Pure Tolerance*, Boston: Beacon Press.
Woodward, Sir L. (1962, 2nd edn) *The Age of Reform 1815–1870*, Oxford: Clarendon Press.
Wormell, P. (1978) *Anatomy of Agriculture*, London: Harrap.
Wright, P. (1985) *On Living in an Old Country*, London: Verso.
—— (1991) *A Journey Through Ruins: The Last Days of London*, London: Radius.
Yearley, S. (1991) *The Green Case: A Sociology of Environmental Issues, Arguments and Politics*, London: HarperCollins.
—— (1992) 'Green ambivalence about science: legal-rational authority and the scientific legitimation of a social movement', *British Journal of Sociology*, 43 (4), 511–532.
Young, S.C. (1992) 'The different dimensions of green politics', *Environmental Politics*, 1 (1), 9–44.
Zebel, S.H. (1967) 'Joseph Chamberlain and the genesis of tariff reform', *Journal of British Studies*, 7 (1), 131–157.
Zuckerman, Sir S. (1957) *Forestry, Agriculture and Marginal Land*, Zuckerman Committee Report, Norwich: HMSO.

INDEX